AF616112

MAIN LIBRARY WHITEKNIGHTS
90 0571861 3

ELEPEN

Protocols for Gene Analysis

Methods in Molecular Biology

John M. Walker, SERIES EDITOR

31. **Protocols for Gene Analysis,** edited by ***Adrian J. Harwood, 1994***
30. **DNA–Protein Interactions,** edited by ***G. Geoff Kneale, 1994***
29. **Chromosome Analysis Protocols,** edited by ***John R. Gosden, 1994***
28. **Protocols for Nucleic Acid Analysis by Nonradioactive Probes,** edited by ***Peter G. Isaac, 1994***
27. **Biomembrane Protocols:** *II. Architecture and Function,* edited by ***John M. Graham and Joan A. Higgins, 1994***
26. **Protocols for Oligonucleotide Conjugates,** edited by ***Sudhir Agrawal, 1994***
25. **Computer Analysis of Sequence Data:** *Part II,* edited by ***Annette M. Griffin and Hugh G. Griffin, 1994***
24. **Computer Analysis of Sequence Data:** *Part I,* edited by ***Annette M. Griffin and Hugh G. Griffin, 1994***
23. **DNA Sequencing Protocols,** edited by ***Hugh G. Griffin and Annette M. Griffin, 1993***
22. **Optical Spectroscopy, Microscopy, and Macroscopic Techniques,** edited by ***Christopher Jones, Barbara Mulloy, and Adrian H. Thomas, 1994***
21. **Protocols in Molecular Parasitology,** edited by ***John E. Hyde, 1993***
20. **Protocols for Oligonucleotides and Analogs,** edited by ***Sudhir Agrawal, 1993***
19. **Biomembrane Protocols:** *I. Isolation and Analysis,* edited by ***John M. Graham and Joan A. Higgins, 1993***
18. **Transgenesis Techniques,** edited by ***David Murphy and David A. Carter, 1993***
17. **Spectroscopic Methods and Analyses,** edited by ***Christopher Jones, Barbara Mulloy, and Adrian H. Thomas, 1993***
16. **Enzymes of Molecular Biology,** edited by ***Michael M. Burrell, 1993***
15. **PCR Protocols,** edited by ***Bruce A. White, 1993***
14. **Glycoprotein Analysis in Biomedicine,** edited by ***Elizabeth F. Hounsell, 1993***
13. **Protocols in Molecular Neurobiology,** edited by ***Alan Longstaff and Patricia Revest, 1992***
12. **Pulsed-Field Gel Electrophoresis,** edited by ***Margit Burmeister and Levy Ulanovsky, 1992***
11. **Practical Protein Chromatography,** edited by ***Andrew Kenney and Susan Fowell, 1992***
10. **Immunochemical Protocols,** edited by ***Margaret M. Manson, 1992***
9. **Protocols in Human Molecular Genetics,** edited by ***Christopher G. Mathew, 1991***
8. **Practical Molecular Virology,** edited by ***Mary K. L. Collins, 1991***
7. **Gene Transfer and Expression Protocols,** edited by ***Edward J. Murray, 1991***
6. **Plant Cell and Tissue Culture,** edited by ***Jeffrey W. Pollard and John M. Walker, 1990***
5. **Animal Cell Culture,** edited by ***Jeffrey W. Pollard and John M. Walker, 1990***

Methods in Molecular Biology • 31

Protocols for Gene Analysis

Edited by

Adrian J. Harwood

Medical Research Council, Cambridge, UK

Humana Press  **Totowa, New Jersey**

999 Riverview Drive, Suite 208
Totowa, New Jersey 07512

Printed in the United States of America. 10 9 8 7 6 5 4 3 2 1

Library of Congress Cataloging in Publication Data

Main entry under title:

Methods in molecular biology.

Protocols for gene analysis / edited by Adrian J. Harwood.
p. cm.—(Methods in molecular biology; 31)
Includes bibliographical references and index.
ISBN 0-89603-258-2
1. Gene mapping. 2. Nucleotide sequence. I. Harwood, Adrian J.
II. Series: Methods in molecular biology (Totowa, N.J.); 31
QH445.2.P76 1994
574.87'322—dc20 94-2365
CIP

Preface

It is now twenty years since Cohen and Boyer's first steps into DNA cloning. In the time since then, there has been an ever increasing acceleration in the development and application of the cloning methodology. With the recent development of the polymerase chain reaction, a second generation of the technology has been born, enabling the isolation of DNA (and in particular, genes) with little more information than the partial knowledge of the sequence. In fact, DNA sequencing is now so advanced that it can almost be carried out on the industrial scale. As a consequence of these advances, it now appears feasible to sequence whole genomes, including one the size of the human. What are we going to do with this information? The future of basic molecular biology must lie in the ability to analyze DNA (and especially the genes within it) starting at the DNA level. It is for these problems that *Protocols for Gene Analysis* attempts to offer solutions.

So you have a piece of DNA, possibly a gene—what do you do next? The first section of this book contains a number of "basic" techniques that are required for further manipulation of the DNA. This section is not intended to be a comprehensive collection of methods, but merely to serve as an up-to-date set of techniques. I refer you to other volumes in the *Methods in Molecular Biology* series for further recombinant DNA techniques.

The rest of the volume is broken up into discrete sections that should be perused according to one's individual research interests. Part 2 details a number of methods for in vitro mutagenesis—a direct means of investigating function, whether it is the control of gene expression or the function of the gene product. Part 3 describes a number of methods and electrophoretic techniques for the elucidation of genomic structure. Part 4 offers an excursion into an area that is becoming more important as the number of those genes involved in inherited diseases increases. To study the polymorphic nature of the genetic pool within a population, it is necessary to be able rapidly to detect the sequence variation within it. A number of technical innovations are described that make this possible. Part 5 returns to the more general area of gene expression, and includes a number of techniques vital to the understanding of gene transcription.

The final two sections take the reader far onto another molecular biology frontier. Both ultimately concern the interaction between different genetic elements, and how it is now possible to step from one to another. Part 6 describes the identification of elements within DNA sequences that bind to proteins, and how this information can be used to isolate the genes encoded by these proteins. In a similar manner, Part 7 describes methods for protein production from cloned DNA and the identification of novel proteins through their association with that DNA. It also takes the process one step farther and describes how ligand binding alone can be used to isolate proteins of interest. At this point, the analysis has returned to square one.

Protocols for Gene Analysis has been prepared in the same manner as the previous volumes of the *Methods in Molecular Biology* series. Each chapter is written in a recipe-like format and designed for direct practical use within the laboratory. Each chapter therefore stresses the practical details, is readily replicable, describes variant versions of the technique, and discusses the problems (and solutions) encountered by their expert authors.

Finally, I would like to thank my colleagues for helpful suggestions and the diligent contributors for all their work and patience through the editorial process.

Adrian J. Harwood

Contents

Contributors

PETER J. AINSWORTH • *Child Health Research Institute and Department of Pediatrics, Children's Hospital of Western Ontario and University of Western Ontario, London, Ontario, Canada*
JACQUELINE K. BARTON • *Division of Chemistry and Chemical Engineering, California Institute of Technology, Pasadena, CA*
DOMINIQUE BELIN • *Department of Pathology, University of Geneva Medical School CMU, Geneva, Switzerland*
PAMELA A. BENFIELD • *The DuPont Merck Pharmaceutical Company, Experimental Station, Wilmington, DE*
ANGELIKA BODENTEICH • *Laboratory of Biochemical Genetics, NIMH Neuroscience Center at St. Elizabeth's Hospital, Washington, DC*
JACQUELINE BOULTWOOD • *Molecular and Cytogenetic Haematology Unit, Department of Haematology, John Radcliffe Hospital, Headington, Oxford, UK*
LAMBERT BUSQUE • *Division of Hematology and Oncology, Harvard Medical School, Brigham and Women's Hospital, Boston, MA*
ING-MING CHIU • *Departments of Molecular Genetics and Internal Medicine, The Ohio State University, Davis Medical Research Center, Columbus, OH*
ANDREW R. CLARK • *Department of Medicine, University of Birmingham, Queen Elizabeth Hospital, Birmingham, UK*
DENISE CLARK • *Basic Sciences Division, Fred Hutchinson Cancer Research Center, Seattle, WA*
IAN G. COWELL • *Imperial Cancer Research Fund, Gene Transcription Laboratory, Hammersmith Hospital, Du Cane, London, UK*
JOANNE CROWE • *QIAGEN Inc., Chatsworth, CA*
HEINZ DÖBELI • *F. Hoffman-La Roche Ltd., Pharmaceutical Research–New Technologies, Basel, Switzerland*

KEVIN DOCHERTY • *Department of Medicine, University of Birmingham, Queen Elizabeth Hospital, Birmingham, UK*
LUCY DRURY • *Imperial Cancer Research Fund, Clare Hall Laboratories, South Mimms, Herts, UK*
IAN DURRANT • *Research and Development, Amersham International, Amersham, Bucks, UK*
GREG S. ELGAR • *Molecular Genetics Unit, MRC Centre, Cambridge, UK*
SUE FOWLER • *Research and Development, Amersham International, Amersham, Bucks, UK*
REINER GENTZ • *F. Hoffmann-La Roche Ltd., Pharmaceutical Research—New Technologies, Basel, Switzerland*
D. GARY GILLILAND • *Division of Hematology and Oncology, Harvard Medical School, Brigham and Women's Hospital, Boston, MA*
PATRICIA P. HARLOW • *The DuPont Merck Pharmaceutical Company, Experimental Station, Wilmington, DE*
JUTTA HARDERS • *DIAGEN GmbH, Hilden, Germany*
DANIEL L. HARTL • *Department of Genetics, Washington University School of Medicine, St. Louis, MO*
JANET HARWOOD • *Imperial Cancer Research Fund, Clare Hall Laboratories, South Mimms, Herts, UK*
CHRISTOPH HELLER • *Laboratoire de Physico Chimie Théorique, Ecole Superieur de Physique et de Chimie Industrielles, Paris, France*
KARSTEN HENCO • *DIAGEN GmbH, Hilden, Germany*
STEVEN HENIKOFF • *Howard Hughes Medical Institute, Fred Hutchinson Cancer Research Center, Seattle, WA*
ERICH HOCHULI • *F. Hoffman-La Roche Ltd., Pharmaceutical Research–New Technologies, Basel, Switzerland*
HELEN C. HURST • *Imperial Cancer Research Fund, Gene Transcription Laboratory, Hammersmith Hospital, Du Cane, London, UK*
GRACE M. HOBSON • *Alfred I. DuPont Institute, Wilmington, DE*
MASAYORI INOUYE • *Department of Biochemistry, University of Medicine and Dentistry of New Jersey, Robert Wood Johnson Medical School at Rutgers, Piscataway, NJ*
JEAN-PIERRE JOST • *FMI, Basel, Switzerland*

JIE KANG • *DIAGEN GmbH, Hilden, Germany*
SCOTT L. KLAKAMP • *Division of Chemistry and Chemical Engineering, California Institute of Technology, Pasadena, CA*
EDWARD L. KUFF • *Laboratory of Biochemistry, National Cancer Institute, National Institutes of Health, Bethesda, MD*
CLAUDE G. LERNER • *Abott Laboratories, Anti-Infective Research Division of Pharmaceutical Discovery, Abott Park, IL*
MICHAEL A. LYDAN • *Department of Zoology, Erindale College, University of Toronto, Mississauga, Ontario, Canada*
ROBERTO MANTOVANI • *DOI/LGME, Faculté de Médecine, Strasbourg Cedex, France*
FRANK MERANTE • *Department of Genetics, Research Institute, The Hospital for Sick Children, Toronto, Ontario, Canada*
CARL R. MERRIL • *Laboratory of Biochemical Genetics, NIMH Neuroscience Center at St. Elizabeth's Hospital, Washington, DC*
JUDY A. MIETZ • *Laboratory of Biochemistry, National Cancer Institute, National Institutes of Health, Bethesda, MD*
LLOYD G. MITCHELL • *Laboratory of Biochemical Genetics, NIMH Neuroscience Center at St. Elizabeth's Hospital, Washington, DC*
RUPERT MUTZEL • *Fakultät für Biologie, Universität Konstanz, Konstanz, Germany*
RENÉ L. MYERS • *Departments of Molecular Genetics and Internal Medicine, The Ohio State University, Davis Medical Research Center, Columbus, OH*
DANTON H. O'DAY • *Department of Zoology, Erindale College, University of Toronto, Mississauga, Ontario, Canada*
HOWARD OCHMAN • *Department of Genetics, Washington University School of Medicine, St. Louis, MO*
GERALDINE A. PHEAR • *Department of Radiology, Health Sciences Center, University of Utah, Salt Lake City, UT*
JACQUES PHILIPPE • *Department of Genetics and Microbiology, Centre Medical Universitaire, Geneva, Switzerland*
MATTHEW L. POULIN • *Departments of Molecular Genetics and Internal Medicine, The Ohio State University, Davis Medical Research Center, Columbus, OH*

GERALD PROTEAU • *Department of Zoology, Erindale College, University of Toronto, Mississauga, Ontario, Canada*
SANDEEP RAHA • *Department of Biochemistry, Erindale College, University of Toronto, Mississauga, Ontario, Canada*
JUTA K. REED • *Departments of Biochemistry and Chemistry, Erindale College, University of Toronto, Mississauga, Ontario, Canada*
DETLEV RIESNER • *Institut für Physikalische Biologie, Heinrich-Heine-Universität, Düsseldorf, Germany*
DAVID I. RODENHISER • *Child Health Research Institute and Department of Pediatrics, Children's Hospital of Western Ontario and University of Western Ontario, London, Ontario, Canada*
HANS PETER SALUZ • *HKI, Jena, Germany*
GUSTAV SCHONFELD • *Department of Internal Medicine, Washington University School of Medicine, St. Louis, MO*
VALERIE SCOTT • *Department of Medicine, University of Birmingham, Queen Elizabeth Hospital, Birmingham, UK*
BARTON E. SLATKO • *New England Biolabs, Inc., Beverly, MA*
RAI AJIT K. SRIVASTAVA • *Department of Internal Medicine, Washington University School of Medicine, St. Louis, MO*
DIETRICH STÜBER • *F. Hoffman-La Roche Ltd., Pharmaceutical Research–New Technologies, Basel, Switzerland*
MICHAEL K. TROWER • *Molecular Genetics Unit, MRC Centre, Cambridge, UK. Present Address: Biochemical Targets Department, Microbiology Division, Glaxo Group Research Ltd., Greenford, Middlesex, UK*
STÉPHANE VIVILLE • *Wellcome Trust Institute of Cancer and Developmental Biology, Cambridge, UK*
ULRICH WIESE • *Institut für Physikalische Biologie, Heinrich-Heine-Universität, Düsseldorf, Germany*

PART I

Basic Recombinant DNA Techniques

CHAPTER 1

Transformation of Bacteria by Electroporation

Lucy Drury

1. Introduction

The use of an electrical field to reversibly permeabilize cells (electroporation) has become a valuable technique for transfer of DNA into both eukaryotic and prokaryotic cells. Many species of bacteria have been successfully electroporated *(1)* and many strains of *E. coli* are routinely electrotransformed to efficiencies of 10^9 and 10^{10} transformants/µg DNA. Frequencies of transformation can be as high as 80% of the surviving cells and DNA capacities of nearly 10 µg of transforming DNA/mL are possible *(2)*.

The benefit of attaining such high efficiency of transformation is apparent, for example, in the case of plasmid libraries. It is often preferable to construct a library in a plasmid owing to its small size and flexibility. In addition, it is invaluable where the use of a shuttle vector is required for the subsequent transfection of eukaryotic cells. Chemical methods of making cells transformation competent are unable to produce high enough efficiencies to make this kind of library possible.

Several commercial machines are available that deliver either a square wave pulse or an exponential pulse. Since most of the published data has been obtained using an exponential waveform, this discussion will be confined to that pulse shape. An exponential pulse is generated by the discharge of a capacitor. The voltage decays over time as a function of the time constant τ.

$$\tau = RC$$

From: *Methods in Molecular Biology, Vol. 31: Protocols for Gene Analysis*
Edited by: A. J. Harwood Copyright ©1994 Humana Press Inc., Totowa, NJ

R is the resistance in ohms (Ω), C is the capacitance in Farads, and τ is the time constant in seconds.

The potential applied across a cell suspension will be experienced by any cell as a function of field strength ($E = V/d$, where d is the distance between the electrodes) and the length of the cell. A voltage potential develops across the cell membrane; when this exceeds a threshold level the membrane breaks down in localized areas and the cell becomes permeable to exogenous molecules. The permeability produced is reversible provided the magnitude and the duration of the electrical field does not exceed some critical limit, otherwise the cell is irreversibly damaged. Since there is an inverse relationship between field strength and cell size, prokaryotes require a higher field strength for permeabilization than do eukaryotic cells. If the voltage and therefore the field strength is reduced, a longer pulse time is required to obtain the maximum efficiency of transformation, however, this range of compensation is limited *(2)*. Increasing the field strength causes a decrease in cell viability and maximum transformation efficiencies are usually attained when about 30–40% of the cells survive.

In this chapter I will describe and discuss the methodology of bacterial electroporation with particular reference to *E. coli.*

2. Materials

2.1. Making Electrocompetent Bacteria

1. A suitable strain of *E. coli:* I find MC1061, or its *rec*$^-$ derivative, WM1100, transform with the highest efficiency. *See* ref. *1* for other strains.
2. L-Broth: 1% Bacto tryptone, 0.5% Bacto yeast extract, 0.5% NaCl.
3. HEPES: 0.1 m*M* HEPES, pH 7.0. This may be replaced by distilled H_2O.
4. Distilled H_2O: Sterilized by autoclaving.
5. 10% Glycerol (v/v): In sterile distilled H_2O.

2.2. Electroporation of Competent Bacteria

6. Electroporator: Transformation requires a high voltage electroporation device, such as the Bio-Rad (Richmond, CA) gene pulser apparatus used with the pulse controller, and cuvets with 0.2 cm electrode gap.
7. TE: 10 m*M* Tris-HCl, pH 8.0, 1 m*M* EDTA.
8. SOC: 2% Bacto tryptone, 0.5% Bacto yeast extract, 10 m*M* NaCl, 2.5 m*M* KCl, 10 m*M* $MgSO_4$, 20 m*M* glucose.

3. Methods

3.1. Making Electrocompetent Bacteria

1. Grow an overnight culture of the chosen strain in L-broth or any other suitable rich medium.
2. The next day, inoculate 1 L of L-broth with 10 mL of the overnight culture and grow at 37°C with good aeration; the best results are obtained with rapidly growing cells (*see* Note 1).
3. When culture reaches an OD_{600} of 0.5–1.0, place on ice. The optimum cell density may vary for each different strain but I have found that usually about 0.5 is the best.
4. Leave on ice for 15–30 min.
5. Centrifuge the bacteria for 10 min at $4000g_{max}$ keeping them at 4°C. Remove the supernatant and discard.
6. Resuspend the cells in an equal volume of either H_2O or 0.1 m*M* HEPES, previously chilled on ice (*see* Note 2).
7. Spin down cells at 4°C and resuspend in half the volume of ice-cold H_2O or HEPES. Care must be taken since the cells form a very loose pellet in these low ionic solutions.
8. Harvest at 4°C once again and resuspend in 20 mL of ice-cold 10% glycerol.
9. Harvest for the last time and resuspend in 2–3 mL of 10% glycerol. The final cell concentration should be about 3×10^{10} cells/mL.

The cells may be used fresh or frozen on dry ice and stored at –70°C where they will remain competent for about 6 mo. Cells may be frozen and thawed several times with little loss of activity (*see* Note 3).

3.2. Electroporation of Competent Bacteria

1. Chill the cuvets and the cuvet carriage on ice (*see* Note 4).
2. Set the apparatus to the 25 μF capacitor, 2.5 kV, and set the pulse controller unit to 200 Ω.
3. Thaw an aliquot of cells on ice, or use freshly made cells.
4. To a cold, 1.5-mL polypropylene tube, add 40 μL of the cell suspension and 1–5 μL of DNA in H_2O or a low ionic strength buffer such as TE. Mix well and leave on ice for about 1 min (*see* Notes 5–8). There is no advantage in a longer incubation time (*see* Note 9).
5. Transfer the mixture of cells and DNA to a cold electroporation cuvet and tap the suspension down to the bottom.

6. Apply one pulse at the above settings. This should result in a pulse of 2.5 kV/cm with a time constant of 4.8 ms (the field strength will be 12.5 kV/cm).
7. Immediately add 1 mL of SOC medium to the cuvet. Resuspend the cells and remove to a 17 × 100-mm polypropylene tube; incubate the cell suspension at 37°C for 1 h (*see* Note 10). Shaking the tubes at 225 rpm during this incubation may improve recovery of transformants.
8. Plate out appropriate dilutions on selective agar.

There are a number of other problems that may be encountered (*see* Notes 11–15).

4. Notes

1. To achieve highly electrocompetent *E. coli,* the cells must be fast growing and harvested at early to mid-log phase.
2. Washing and resuspending the bacteria in solutions of low ionic concentration is important to avoid arcing in the cuvet owing to conduction at the high voltages required for electroporation.
3. A 10% glycerol solution provides ideal cryoprotection for *E. coli* cells at –70°C. The cell suspension is frozen by aliquoting and placing in dry ice. Quick freezing in liquid nitrogen may be deleterious *(3)*. Several rounds of careful freeze thawing on ice does not seem to affect the level of the cells competence to a great extent.
4. Because of the high field strength necessary, it is best to perform the electroporation at 0–4°C for most species of bacteria. Electroporation of *E. coli* performed at room temperature results in a 100-fold drop in efficiency. This may be related to the state of the cell membranes, or may be a result of the additional joule heating that occurs during the pulse *(4)*.
5. Transformation efficiency may be adjusted by changing the cell concentration. Raising cell concentration from 0.8 to 8 × 10^9/mL increases transformation efficiency by 10- to 20-fold *(4)*. A steady increase in the number of transformants obtained has been found at cell concentrations of up to 2.8 × 10^{10}/mL using a fixed concentration of DNA *(3)*.
6. Transforming DNA must be presented to the cells as a solution of low ionic strength. As mentioned in Note 2, high ionic strength solutions cause arcing in the cuvet or a very short pulse time with resulting cell death and loss of sample. Salts, such as CsCl and ammonium acetate, must be kept to 10 m*M* or less. It is advisable to have the DNA dissolved in TE or H_2O. This is particularly relevant after a ligation because the ligation buffer has an ionic concentration too high for use directly in an electroporation. The DNA must be precipitated in ethanol/sodium

acetate (carrier tRNA can be used in the precipitation without affecting the transformation frequency); alternatively the ligation can be diluted 1/100 and 5 µL used for electroporation *(5).*

7. The concentration of transforming DNA present during an electroporation is directly related to the proportion of cells that are transformed. With *E. coli* this relationship holds over several orders of magnitude, and at high DNA concentrations (up to 7.5 µg/mL) nearly 80% of the surviving cells are transformed *(2).* This is in contrast to chemically treated competent cells where saturation occurs at DNA concentration 100-fold lower, and where a much smaller fraction of the cells are competent to become transformed *(6).* For purposes where a high efficiency but a low frequency of transformation is required (for library construction where cotransformants are undesirable) a DNA concentration of less than 10 ng/mL and a cell concentration of less than 3×10^{10} is appropriate. Alternatively, when a high frequency of transformation is required, use 1–10 mg/mL, which transforms most of the surviving cells *(2).*
8. The size and topology of the DNA molecules may affect transformation efficiency. It is reported that plasmids of up to 20 kb transform with the same molar efficiency as plasmids of 3 kb and converting these plasmids to a relaxed form does not affect their transforming activity *(2).* Larger molecules can be taken up but at much lower efficiencies, for example linear λ DNA (48 kb) has a molar transformation efficiency of 0.1% that of small plasmids *(2).* No direct comparison between *E. coli* plasmids containing the same origin of replication, promoters, and markers but differing only in size has been published, and in my hands different plasmid constructs transform with different molar efficiencies. Powell et al. *(7)* have compared the uptake of related plasmids in *Streptococcus lactis* and observed no clear relationship between size and molar transformation efficiency.
9. There is no evidence for binding of the DNA to the cell surface during the transformation process, and thus increasing the preshock incubation time up to 30 min makes very little difference to the number of resulting transformants *(2).* In support of this observation, experiments by Calvin et al. *(8)* show that when cells are mixed with radioactively labeled plasmid, only a small percentage of the label remains bound after two washes. In addition, certain species of bacteria, such as *Lactobacillus casei*, secrete nucleases, so increasing the preshock incubation time may be detrimental *(9).*
10. Immediately after the pulse *E. coli* cells are quite fragile and rapid addition of the outgrowth medium greatly enhances their viability and transformation efficiency. Even after 1 min delay the efficiency drops

by 3–5-fold and this increases to 20-fold after 10 min *(2)*. Outgrowth is necessary for the cells to express any resistance marker introduced by the transforming plasmid and is usually for an hour or approximately two cell divisions.

11. There are a number of causes of arcing in the cuvet. One reason could be that the ionic strength of the DNA solution or the cell suspension is too high. It is important that the DNA is resuspended in TE or H_2O. If it is a ligation mixture it must be precipitated with 0.3*M* Na acetate and 2–3 vol of ethanol or diluted 100-fold in TE or H_2O. The same problem can be caused by failure to tap the cell/DNA mixture to the bottom of the cuvet.

    Another likely cause may be that the cuvets and the chamber were chilled on ice and residual H_2O on the surfaces induced an arc. If you are electroporating many samples it is not necessary to chill the carriage between every pulse but it is a good idea to dry the carriage between every few samples since condensation can accumulate and cause arcing.

12. Failure to obtain colonies after transformation will be caused either by problems with the cells or the DNA. It is advisable to make a large quantity of an accurate dilution of a supercoiled plasmid, such as pUC18, to use as a positive control in all experiments. Use this routinely to check the cells you make. (5 pg supercoiled DNA will give about 10,000 transformants if your cells are at efficiencies of 10^9 transformants/µg DNA.)

    If no colonies are obtained from the positive control, ensure that the growth conditions and harvesting of the cells were correct. The most competent cells are made from fast growing cells harvested at early to mid-log phase. Keep all the wash solutions at 4°C and keep the cells cold while harvesting. When making a new strain competent it is best to harvest the cells at a range of densities at an OD_{600} of between 0.4–1.0. We have found the best density usually to be around 0.5. If the electrocompetent cells were previously stored at –70°C, ensure that they are still viable. To do this, plate out an appropriate dilution of the cells on a nonselective plate.

    Should the cells only transform with the control, first check the concentration of your DNA. It may also be possible that the DNA contains toxic contaminants such as phenol or SDS. The viability of the cells after electroporation can be checked by plating a sample on a nonselective plate. A survival of 30–40% would be expected using the parameters set out in the methods, but check against an equivalent aliquot of the cells transformed with the control DNA. If the DNA is contaminated, reprecipitate and wash with 70% ethanol, or use GeneClean™ to remove unwanted chemicals.

13. If a recently prepared batch of cells, already tested for electrocompetence, gives a reduced transformation efficiency, it is likely to be because of problems with the electroporation. It is important that the cuvets and the carriage are chilled so that the starting temperature of the cells is 0–4°C. It is crucial to add the outgrowth medium (kept at room temperature) as quickly as possible to the cells after electroporation.
14. An unexpectedly high apparent transformation, efficiency may have a number of explanations. The simplest explanation is that the selective plates have exceeded their shelf life. DNA contamination can also be a problem owing to the high competency of the cells. It is important to maintain good sterile techniques and careful use of micropipets to avoid crosscontamination with DNA used in previous experiments. Since electroporation can release plasmid from cells, the effects of contamination with previously transformed bacteria will be greatly heightened, especially if the plasmid is present at a high copy number in the cell.
15. The particular problems outlined above apply to *E. coli;* problems encountered with other bacterial species could be owing to the characteristics of that strain. For example, if the bacterium is encapsulated the entry of the DNA may be impeded, and some species secrete nucleases that could destroy the DNA. Certain types of bacteria may require a longer recovery time or a longer time to express the selective marker. If the size of the cell is unusual, it may require a different field strength. To establish electroporation conditions for a novel species, it is best to consult references concerning similar bacterial types (*see* ref. *1* for a list of references), for general parameters from which to further optimize.

References

1. Bacterial species that have been transformed by electroporation. Bio-Rad Laboratories, 1414 Harbor Way South, Richmond, CA 94804. Bulletin 1631, 1990.
2. Dower, W. J., Miller, J. F., and Ragsdale, C. W. (1988) High efficiency transformation of *E. coli* by high voltage electroporation. *Nucleic Acids Res.* **16,** 6127–6145.
3. Dower, W. J. (1990) Electroporation of bacteria: a general approach to genetic transformation, in *Genetic Engineering,* vol. 12 (Setlow, J. K., ed.), Plenum, NY, pp. 275–295.
4. Shigekawa, K. and Dower, W. J. (1988) Electroporation of eukaryotes and prokaryotes: A general approach to the introduction of macromolecules into cells. *Biotechniques* **6,** 742–751.
5. Willson, T. A. and Gough N. M. (1988) High voltage *E. coli* electrotransformation with DNA following ligation. *Nucleic Acids Res.* **16,** 11820.
6. Hanahan, D. (1985) Techniques for transformation of *E. coli,* in *DNA Cloning,* vol. 1 (Rickwood, D. and Hames, B. D., eds.) IRL, Oxford, pp. 109–135.

7. Powell, I. B., Achen, M. G., Hillier, A. J., and Davidson, B. E. (1988) A simple and rapid method for genetic transformation of Lactic streptococci by electroporation. *Appl. Environm. Microbiol.* **54,** 655–660.
8. Calvin, N. M. and Hanawalt P. C. (1988) High efficiency transformation of bacterial cells by electroporation. *J. Bacteriol.* **170,** 2796–2801.
9. Chassy, B. M. and Flickinger, J. L. (1987) Transformation of *Lactobacillus casei* by electroporation. *FEMS Microbiol. Letts.* **44,** 173–177.

CHAPTER 2

Direct Cloning of λgt11 cDNA Inserts Into a Plasmid Vector

Matthew L. Poulin and Ing-Ming Chiu

1. Introduction

Cloning vectors derived from bacteriophage λ are used frequently in the construction of both cDNA and genomic DNA libraries *(1)*. The screening of positive plaques from λ libraries is relatively easy with the plaque lifting technique of Benton and Davis *(2)*. However, isolating and subcloning recombinant inserts from the phage clones of interest can be a tedious task. Additionally, if the insert comprises more than one restriction fragment, the smaller fragments may be missed during the subcloning steps. Both polymerase chain reaction (PCR) *(3,4)* and plasmid rescue using the f1 origin of replication, as in the λZAP systems *(5)*, were developed to circumvent this problem. However, these sophisticated procedures may not exist in every molecular cloning laboratory and most of the existing cDNA libraries are constructed in λgt vectors. Here we describe a direct method of cloning inserts from λgt phage into a pBR322 cloning vector.

The *E. coli* strains Y1088, Y1089, and Y1090, commonly used for λgt phage infections, contain an endogenous plasmid known as pMC9 *(6)*. This 6.1-kb plasmid is a pBR322 derivative with the 1.7-kb *Eco*RI fragment containing the *lac*I and *lac*Z genes cloned into the unique *Eco*RI site *(7)*.

We showed that the endogenous pMC9 DNA is released during the lysis of infected bacteria and can be copurified with the phage DNA *(8,9)* (Fig. 1). Thus, following digestion with *Eco*RI to release the phage's

From: *Methods in Molecular Biology, Vol. 31: Protocols for Gene Analysis*
Edited by: A. J. Harwood Copyright ©1994 Humana Press Inc., Totowa, NJ

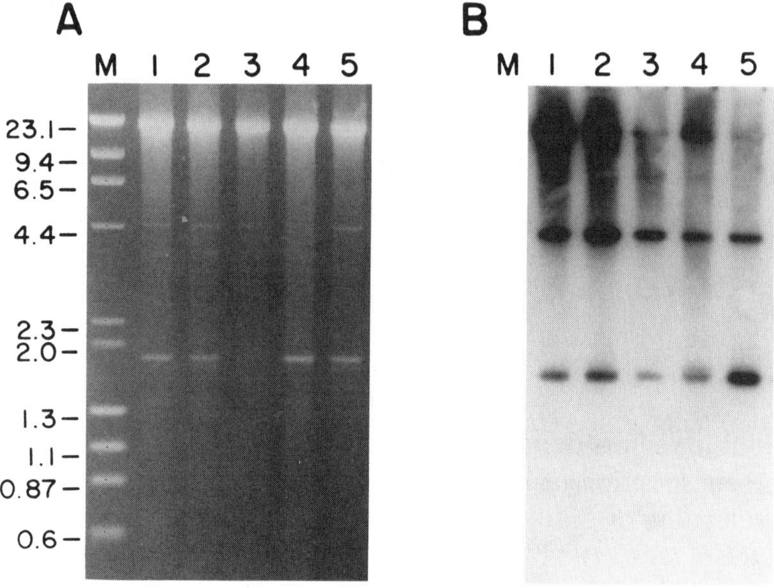

Fig. 1. Purification of recombinant phage DNA from bacterial host Y1088 and digestion with *Eco*RI. **(A)** Phage DNA isolated from five plaques (lanes 1–5) containing the 1.8 kb newt *bek* cDNA insert (arrow head) was digested with *Eco*RI and electrophoresed on a 1% agarose gel. The 4.4-kb and the 1.7-kb fragments from pMC9 can be seen (arrows). **(B)** Southern blot of the gel in A using the pMC9 plasmid as a probe. Arrows indicate the hybridization of the 4.4- and 1.7-kb fragments. M, λ DNA digested with *Hind*III and ϕX DNA digested with *Hae*III. Sizes of the markers are in kb.

cDNA insert, pMC9 is also digested, releasing its 1.7-kb insert. A ligation can be initiated allowing the direct cloning of the λgt11 insert into pBR322. This ligation can result in three different products.

1. The pMC9 itself by either nondigestion or religation of its 1.7-kb *Eco*RI fragment,
2. The self ligation of the 4.4-kb pBR322, or
3. The λgt11 insert ligated into the *Eco*RI site of the pBR322 vector.

By transforming in the presence of isopropyl-β-D-thiogalactoside (IPTG) and 5-bromo-4-chloro-3-idolyl β-D-galactoside (X-gal) the first result can be distinguished from the second and third by the *lac*Z gene product cleaving the X-gal resulting in a blue colony. The last two choices cannot be selected on the basis of color because both will result in white colonies under blue/white selection. One method to distinguish between white colonies that contain an insert and white colonies that do not is microwave colony hybridization *(10)*. Although this is a rapid hybridization technique it requires several days until a result is obtained owing to the hybridization, washing, and exposing steps. A second, more rapid technique is to use the PCR to screen the white colonies by a procedure known as colony PCR *(11)*.

Individual white colonies are boiled in distilled water and are subsequently subjected to the PCR using primers that flank the *Eco*RI site in pBR322. If no insert is present, a fragment of 41 bp will be expected. If the λ phage insert was ligated into the plasmid, then a band the size of the insert will be expected (Fig. 2). A good control for this procedure is to utilize colony PCR on pMC9. This should result in the amplification of a 1.7-kb band (Fig. 2, Lane 2). Once a positive PCR result is obtained, the colony can be grown overnight and a mini plasmid prep can be performed with subsequent sequencing using the PCR primers if so desired. This technique allows one to sample a large number of colonies and obtain results in a relatively short period.

2. Materials

2.1. Phage DNA Extraction

1. NZY: 1% NZ amine A (casein hydrolysate enzymatic; ICN Biochemicals), 0.5% yeast extract, 85 m*M* NaCl, 10 m*M* $MgCl_2$, autoclaved.
2. YT: 0.8% bactotryptone, 0.5% yeast extract, 0.17*M* NaCl, autoclaved.
3. + Amp: Any solution followed by "+ Amp" has 50 μg/mL ampicillin.
4. PSB: 0.1*M* NaCl, 10 m*M* Tris-HCl, pH 7.4, 10 m*M* $MgCl_2$, 0.05% gelatin, autoclaved.
5. CaMg: 10 m*M* $CaCl_2$, 10 m*M* $MgCl_2$, filter sterilized.
6. RNase A: A stock solution of 10 mg/mL was boiled for 15 min and allowed to cool slowly to room temperature. Store at –20°C.
7. $T_{10}E_1$: 10 m*M* Tris-HC1, pH 8.0, 1 m*M* EDTA.
8. Phenol: Tris-HCl buffered phenol at pH 8.0.
9. CIA: Chloroform:isoamyl alcohol (24:1; v/v).
10. 3*M* Sodium acetate, pH 5.2.
11. Absolute ethanol.
12. $T_1E_{0.2}$: 1 m*M* Tris-HCl, pH 8.0, 1 m*M* EDTA.

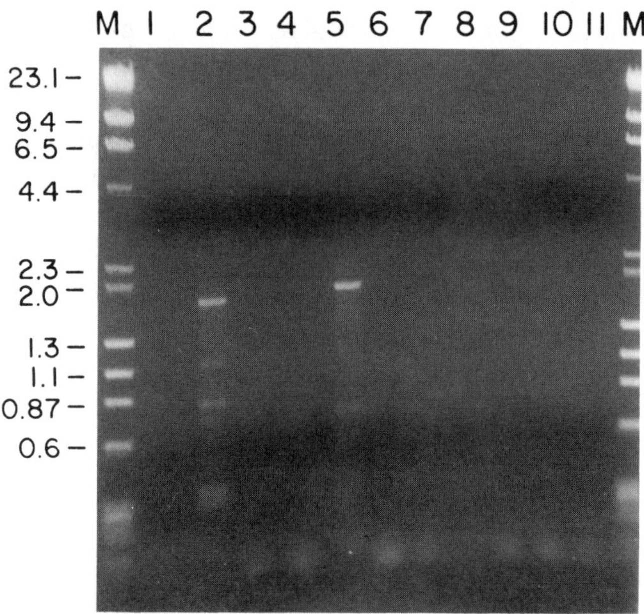

Fig 2. Use of colony PCR to identify positive clones. Ten microliters of the 20 µL PCR reactions were analyzed by agarose gel electrophoresis. Lane 1 represents the H_2O negative control and lane 2 represents the colony PCR of a blue pMC9 colony. Lanes 3–11 are colony PCR on white colonies. Lane 5 represents the 1.8-kb *bek* cDNA insert present in the pBR322 vector. M, λ DNA digested with *Hind*III and ϕX DNA digested with *Hae*III. Sizes of markers are in kb.

2.2. Ligation and Transformation

13. *Eco*RI and 10X buffer H: *Eco*RI at 10 U/µL and 10X buffer at 50 m*M* Tris-HC1, pH 7.5, 10 m*M* $MgCl_2$, 100 m*M* NaCl, 1 m*M* dithioerythritol; store at –20°C.
14. T4 DNA ligase and 10X ligase buffer: T4 DNA ligase at 1 U/µL and 10X ligase buffer at 0.66*M* Tris-HCl, pH 7.5, 5 m*M* $MgCl_2$, 10 m*M* ATP, 10 m*M* dithiothreitol; store at –20°C.
15. SOC: 2% bactotryptone, 0.5% yeast extract, 10 m*M* NaCl, 2.5 m*M* KCl, autoclaved. 10 m*M* $MgCl_2$, 10 m*M* $MgSO_4$, 20 m*M* glucose, filter sterilized.
16. IPTG: 100 m*M* isopropyl-β-D-thiogalactoside in water, stored at –20°C.
17. X-gal: 2% 5-bromo-4-chloro-3-idolyl-β-D-galactoside in dimethylformamide, stored at –20°C.

18. YT + Amp plates: YT media with 1.1% agar, autoclaved. Bring to 55°C and add ampicillin to a final concentration of 50 µg/mL. Pour plates and store at 4°C.

2.3. DNA Characterization by Colony PCR

19. 10X PCR buffer: 0.1*M* Tris-HCl, pH 8.3, 0.5*M* KCl, 15 m*M* $MgCl_2$, 1 mg/mL gelatin.
20. dNTP mix: 1.25 m*M* of each dNTP.
21. PCR primers: Available from New England Biolabs (Beverly, MA). pBR322 *Eco*RI site clockwise primer 5'-GTATCACGAGGCCCT-3' at 100 pmol/µL. pBR322 *Eco*RI site counterclockwise primer: 5'-GATAAGCTGTCAAAC-3' at 100 pmol/µL.
22. *Taq* DNA polymerase: 5 U/µL; store at –20°C.

3. Methods

3.1. Phage DNA Extraction

1. Pick a single positive λgt plaque with a pipet and transfer it to a 1-dram vial containing 1 mL of PSB. Rock overnight. Also, start a fresh overnight culture of Y1088 bacteria in 20 mL of NZY + Amp.
2. In a sterile 1.5-mL Eppendorf tube, combine 100 µL of CaMg, 100 µL of Y1088 bacteria, and 100 µL of λgt phage from the overnight plaque and incubate at 37°C for 30 min.
3. Add the infection to 50 mL of NZY + Amp media in a 250-mL Erlenmeyer flask. Shake at 300 rpm, 37°C overnight (*see* Note 1).
4. Centrifuge the overnight infection at 2000*g*, 4°C for 15 min to clear the lysate of bacterial debris.
5. Transfer 25 mL of the supernatant to two 50-mL polypropylene tubes. Add a few drops of chloroform to one tube and store at 4°C.
6. To the remaining tube containing the 25 mL of phage lysate add 50 µg/mL of RNase A. Incubate at 37°C for 1 h.
7. Centrifuge at 85,000–90,000*g*, 15°C for 1.5 h in a swinging bucket rotor (*see* Note 2). Decant the supernatant and cut the tube 2 cm from the bottom with a scalpel.
8. Add 200 µL of $T_{10}E_1$ to the bottom of the tube containing the pelleted phage and let stand for 15 min. Gently pipet up and down, avoiding bubbles, until the phage pellet is in suspension. Transfer the 200 µL of phage suspension to an Eppendorf tube. Rinse the bottom of the centrifuge tube with an additional 200 µL of $T_{10}E_1$ and transfer to the Eppendorf tube.
9. Extract DNA from the phage suspension by adding 300 µL of phenol and shaking thoroughly for 5 min. Then add 100 µL of CIA and shake

for 1 min. Microfuge at 16,000*g* for 2 min and collect aqueous layer, leaving the protein interface behind. Repeat this step until little or no interface is observed.

10. Extract once with an equal volume of CIA, microfuge as above, and collect the aqueous layer.
11. Precipitate with 1/10 vol of 3*M* sodium acetate, pH 5.2, and 2.5 vol of 100% EtOH, –20°C, for at least 30 min. Microfuge at 16,000*g*, 4°C for 15 min. Dry pellet in a SpeedVac concentrator and resuspend in 100 µL of $T_1E_{0.2}$.

3.2. Ligation and Transformation

1. In an Eppendorf tube add 20 µL of phage DNA, 5 µL of 10X buffer H, 5 µL of *Eco*RI (10 U/µL), 2 µL of RNase A (10 mg/mL), and 18 µL of ddH_2O. Mix well and incubate at 37°C for 1 h. After digestion remove half (25 µL) of the digest to run on a 1% agarose gel (*see* Note 3).
2. Bring the volume of the other half up to 200 µL with $T_{10}E_1$ and extract once with 200 µL of phenol/CIA. Collect the aqueous layer. Reextract once with an equal volume of CIA. Collect the aqueous layer.
3. Precipitate with 1/10 vol of 3*M* sodium acetate, pH 5.2, and 2.5 vol of 100% EtOH, –20°C, for at least 30 min. Microfuge at 16,000*g*, 4°C for 15 min. Decant the supernatant, wash the pellet in 70% EtOH, and remicrofuge for 5 min. Decant the supernatant, dry the pellet, and redissolve in 11 µL of ddH_2O.
4. In two Eppendorf tubes mix the following two reactions:

digested phage	1 µL	10 µL
10X ligase buffer	2 µL	2 µL
T4 DNA ligase	1 µL	1 µL
ddH_2O	16 µL	7 µL

   Incubate at 14°C overnight.
5. Add 3 µL of the ligated DNA to 50 µL of competent *E. coli* DH5α cells. Incubate on ice for 30 min. Heat shock the cells at 37°C for 45 s, then incubate on ice for 2 min. Add 0.45 mL of SOC to the cells and shake at 225 rpm, 37°C for 1 h.
6. Plate 100 µL of cells with 20 µL of 100 m*M* IPTG and 60 µL of 2% X-Gal onto YT + Amp plates. Incubate at 37°C, inverted, overnight.

3.3. DNA Characterization by Colony PCR

1. Number and pick individual colonies from the overnight plates and resuspend each colony in 50 µL of ddH_2O in an Eppendorf tube. Place plates back in the incubator for 3–5 h to allow the colonies to regrow (*see* Note 4).

2. Boil the colonies for 5 min. Microfuge at 16,000*g* for 2 min. Use 5 µL of the supernatant as template for the PCR.
3. For each PCR reaction of 20 µL mix 8.6 µL dH_2O, 2 µL of 10X PCR buffer, 4 µL of dNTP mix, 0.1 µL of each pBR322 *Eco*RI specific oligonucleotides, 0.2 µL of *Taq* polymerase, and 5 µL of template DNA. Overlay with 20 µL of light mineral oil (*see* Note 5).
4. Subject to 30 rounds of temperature cycling as follows:
   94°C for 1 min
   45°C for 1 min
   72°C for 2 min
   Following the 30 cycles a final extension step of 72°C for 7 min is utilized.
5. Analyze 10 µL of each product by agarose gel electrophoresis.
6. Positive colonies should be repicked and grown overnight in 20 mL of YT + Amp media for mini-prep plasmid isolation and subsequent analysis (*see* Notes 6 and 7).

4. Notes

1. The overnight infection of *E. coli* strain Y1088 should be slightly turbid with bacterial debris found strewn throughout.
2. Steps 7 and 8 of Section 3.1. result in a light brown phage pellet on the bottom of the tube. If you are working with more than one sample, make sure you label the bottom of the tube before cutting it. The phage pellet is sticky. Try not to allow the phage to stick to the pipet tip.
3. The agarose gel of the digested phage should reveal the size of the cDNA insert, the concentration of the phage DNA, and the relative amount of the 4.4-kb pBR322 *Eco*RI fragment present in the phage DNA prep (*see* Fig. 1). If the pBR322 fragment is faint or not visible but a prominent cDNA insert is visible, then a low colony number following transformation can be expected.
4. Low colony number is a potential problem. One way to obtain more colonies is to plate the entire transformation, instead of only 100 µL. Do this by microfuging the remaining bacteria for 10 s before plating. Remove the supernatant and resuspend the bacteria in 100 µL of SOC and plate all 100 µL.
5. When preparing 20 or more PCR reactions at a time it is most efficient to utilize a "master mix." This is when all the necessary components of the reaction, except the template, are mixed together and then aliquoted. This reduces the chance for contamination and error when dealing with so many samples. An example of a master mix for 30 samples is as follows:

60 μL of 10X PCR buffer
120 μL of dNTP mix
3 μL of primer 1 at 100 pmol/μL
3 μL of primer 2 at 100 pmol/μL
258 μL of dH_2O
6 μL *Taq* polymerase at 5U/μL

---

450 μL Total in master mix

Fifteen-microliter aliquots are added to each of the 30 PCR tubes. Five microliters of each boiled colony sample is added to each of the first 29 tubes. The last tube is reserved for 5 μL of the H_2O negative control. Lastly, each sample is overlaid with 20 μL of mineral oil.

6. Positive and negative controls are critical in PCR owing to the sensitivity of the reaction. As a positive control a blue colony can be picked and assayed with the expected result of a 1.7-kb band representing the *lac*I and *lac*Z genes contained in the *Eco*RI site of the pBR322 plasmid. The negative control is the last aliquot of the master mix with 5 μL of ddH_2O added. If this results in the amplification of a fragment, then the assay must be disregarded since any sample lane could result from the same contamination. If no amplification occurs in the positive control then the PCR assay must be reevaluated.
7. We have found that 10–30 white bacterial colonies must be screened to find one that contains the cDNA insert. If plating all of the bacteria from the transformation fails to give enough colonies for the PCR assay, and controls indicate that the ligation and transformation are optimal, then the problem is most likely to be with the DNA preparation. If this occurs an alternate method of phage DNA isolation is as follows: Add 20 mL of phenol to the 25 mL of phage supernatant from Section 3.1., step 5 and shake for 5 min. Then add 5 mL of CIA and shake for 1 min and centrifuge at 2000*g* for 10 min. Repeat this extraction until little or no interface is observed. Extract once with an equal volume of CIA, centrifuge as above, and collect the aqueous layer. Proceed to step 11 of Section 3.1. When picking colonies be sure to number them so they can be reisolated when the PCR identifies a positive clone. After colony picking, place the plates back into the incubator for 3–5 h to allow the colony to regrow.

References

1. Sambrook, J., Fritsch, E. R., and Maniatis, T. (1989) *Molecular Cloning: A Laboratory Manual.* Cold Spring Harbor Laboratory, Cold Spring Habor, NY.
2. Benton, W. D. and Davis, R. W. (1977) Screening λgt recombinant clones by hybridization to single plaque *in situ*. *Science* **196,** 180–182.

3. Nishikawa, B. K., Fowlkes, D. M., and Kay, B. K. (1989) Convenient uses of polymerase chain reaction in analyzing recombinant cDNA clones. *BioTechniques* **7,** 730–735.
4. Poulin, M. L. unpublished results.
5. Short, J. M., Fernandez, J. M., Sorge, J. A., and Huse, W. D. (1988) λ ZAP: a bacteriophage λ expression vector with in vivo excision properties. *Nucleic Acids Res.* **16,** 7583–7600.
6. Huynh, T. V., Young, R. A., and Davis, R. W. (1985) Constructing and screening cDNA libraries in λgt10 and λgt11, in *DNA Cloning Techniques: A Practical Approach* (Glover, D., ed.), IRL, Oxford, p. 49.
7. Calos, M. P., Lebkowski, J. S., and Botchan, M. R. (1983) High mutation frequency in DNA transfected into mammalian cells. *Proc. Natl. Acad. Sci. USA* **80,** 3015–3019.
8. Chiu, I.-M. and Lehtoma K. (1990) Direct cloning of cDNA inserts from λgt11 phage DNA into a plasmid vector by a novel and simple method. *Genet. Anal. Techn. Appl.* **7,** 18–23.
9. Chiu, I.-M., Lehtoma, K., and Poulin, M. L. (1992) Cloning of cDNA inserts from phage DNA directly into a plasmid vector. *Meth. Enzymol.* **216,** 508.
10. Buluwela, L., Boehm, A. F., and Rabbits, T. H. (1989) A rapid procedure for colony screening using nylon filters. *Nucleic Acids Res.* **17,** 452.
11. Gussow, D. and Clackson, T. (1989) Direct clone characterization from plaques and colonies by the polymerase chain reaction. *Nucleic Acids Res.* **17,** 4000.

CHAPTER 3

PCR Cloning Using T-Vectors

Michael K. Trower and Greg S. Elgar

1. Introduction

The polymerase chain reaction (PCR) has revolutionized the way that molecular biologists approach the manipulation of nucleic acids through its ability to amplify specific DNA sequences *(1–3)*. This is achieved by repeated rounds of three different steps: heat denaturation of template DNA, annealing of two convergent oligonucleotide primers to the opposite strands of the DNA template, and then 5' to 3' extension from each of the annealed primers using a thermostable DNA polymerase, usually that of *Thermus aquaticus* (*Taq* polymerase). Since the product from one round acts as a template for the next, each cycle results in the doubling of target DNA so that the desired sequence accumulates exponentially.

The applications to which this technique can be applied are numerous *(4)*, but perhaps one of the most important is the capability to directly clone amplified DNA products into vectors for further analysis. This is commonly achieved by so-called "sticky end" cloning in which restriction endonuclease recognition sites are incorporated into the 5' ends of the PCR primers *(5)*. Following amplification the DNA fragment is purified, digested with the appropriate enzyme(s), and then ligated into an identically restricted vector. Obtaining efficient cleavage at the extreme ends of linear PCR products can be difficult *(6)* and moreover their use can result in the restriction of sites that lie within the amplified DNA fragment. The one particular advantage to this method is that the PCR product can be force cloned using designed restriction sites, so, for example, a DNA fragment such as a leader

From: *Methods in Molecular Biology, Vol. 31: Protocols for Gene Analysis*
Edited by: A. J. Harwood

signal sequence can be fused in-frame to a structural gene for expression studies.

Blunt-end cloning is an alternative procedure for cloning PCR products when precise orientation is not required. The protocol is complicated by the inherent terminal transferase activity of *Taq* polymerase, which tends to add a template independent single deoxyadenosine (A) residue to the 3' ends of the PCR product *(7)*. The PCR product, after purification, must therefore be "flush-ended" by treatment with a DNA polymerase having 3' to 5' exonuclease activity, such as Klenow or T4 DNA polymerase, prior to ligation into a blunt-ended vector *(8)*. Unfortunately, it is common to find on screening that a large proportion of the isolated plasmids lack an insert even when blue/white β-galactosidase color selection is available.

Recently there have been reports of alternative procedures for cloning PCR products in which the terminal transferase activity of *Taq* polymerase is exploited *(9,10)*. These methods are up to 50 times more efficient than blunt ended cloning. The protocols rely on the creation of cloning vectors (T-vectors) that, once linearized, have a single 3' deoxythymidine (T) at each end of their arms. This allows direct "sticky end" ligation of PCR products, containing *Taq* polymerase catalyzed A extensions, without further enzymatic processing.

In the method of Smith et al. *(9)* the vector extensions are generated by endonuclease digestion of a specialized plasmid. By contrast, the procedure described by Marchuk et al. *(10)* utilizes the terminal transferase activity of *Taq* polymerase to add deoxythymidine (T) residues to blunt ended restricted vectors (Fig. 1). The latter procedure has two general advantages. The first is that it can be used to generate T-vectors from many of the cloning vectors commonly found in molecular biology laboratories, and second, there is a very low background owing to self-ligation of the vector. It is this method that we describe in detail; covering primer design, setting up of a PCR reaction, product isolation, the steps involved in manufacturing T-vectors and then their use for cloning the PCR products. Finally, we describe a rapid PCR screening protocol for identifying colonies containing the desired clones.

2. Materials

All reagents should be prepared with sterile distilled water and stored at room temperature unless stated otherwise.

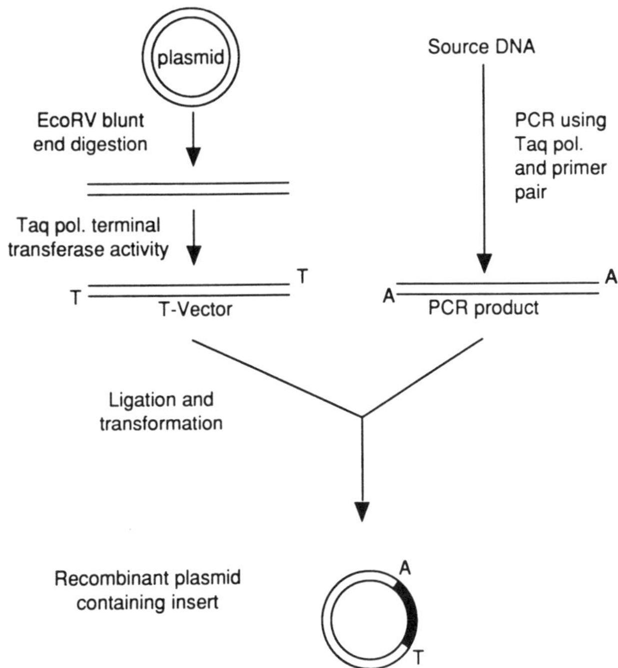

Fig. 1. Schematic diagram demonstrating the principle of cloning PCR products with plasmid T-vectors.

2.1. PCR Reaction and Product Isolation

1. PCR primers: Synthetic oligonucleotides diluted to 10 m*M* (*see* Note 1). The design of these is critical to the success of the PCR reaction (*see* Notes 2 and 3). Store at –20°C.
2. 10X PCR buffer: 100 m*M* Tris-HC1 (pH 8.3 at 25°C), 500 m*M* KCl, 15 m*M* $MgCl_2$, 0.1% w/v gelatin. This buffer should be incubated at 50°C to fully melt the gelatin and then filter sterilized. Store at –20°C.
3. dNTP mix: A mix in distilled water containing 2 m*M* of each dNTP. Liquid stocks of dNTPs are available from companies such as Pharmacia (Piscataway, NJ) and Boehringer (Indianapolis, IN) and we find these convenient and reliable. Store at –20°C.
4. Mineral oil.
5. *Taq* polymerase (5 U/mL). For T-vector extensions use the native form since it is not clear whether the cloned enzyme retains terminal transferase activity. Store at –20°C.
6. PCR Stop mix: 25% Ficoll 400, 100 m*M* EDTA, 0.1% w/v bromophenol blue.

7. Agarose gels: Melt electrophoresis grade agarose in either TAE (50X TAE: 242 g Tris, 57.1 mL glacial acetic acid, 100 mL 0.5*M* EDTA [pH 8.0]/L) or TBE (10X TBE: 121 g Tris, 55 g orthoboric acid, 7.4 g EDTA/L) by gentle boiling. (This can be done in a microwave.) Cool until hand warm and pour into prepared gel former. Run small gels at around 100 V using the dye in the stop mix as an indicator of migration (*see* Note 4).
8. Phenol: Ultra-pure phenol is buffer saturated in TE. Store at 4°C.
9. Chloroform: 29:1 mix of chloroform with isoamyl alcohol.

2.2. T-Vector Preparation, Ligation, and Transformation

10. Vector DNA: A suitable plasmid, such as pBluescript II (Stratagene, La Jolla, CA), prepared as a miniprep *(11)*. Store at –20°C
11. Restriction endonucleases: *Eco*RV, *Sma*I, or other blunt-end cutter as required. Store at –20°C.
12. dTTP: 100 m*M* dTTP stock. Store as small aliquots at –20°C.
13. 10X Ligation buffer: 0.5*M* Tris-HCl, pH 7.6, 100 m*M* $MgCl_2$, 500 mg/mL BSA. Store at –20°C.
14. DTT: 100 m*M* Dithiothreitol stock. Store as small aliquots at –20°C.
15. ATP: 10 m*M* adenosine triphosphate stock. Store as small aliquots at –20°C.
16. T4 DNA Ligase: as available from numerous commercial suppliers. The unit activity of T4 DNA ligase may be assessed in Weiss units (a pyrophosphate exchange assay), circle formation units, or by the suppliers own unit assay (such as Lambda/*Hind*III fragment ligation). It is therefore best to refer to product information, although generally 0.5 µL is sufficient as T4 DNA ligase is always added in excess. Store at –20°C.

2.3. PCR Screening

17. TBG medium: To make 1 L add 12 g of tryptone, 24 g of yeast extract, and 4 mL of glycerol in a total volume of 882 mL of distilled water and autoclave. When cool, add 100 mL of sterile 0.17*M* KH_2PO_4, 0.72*M* K_2HPO_4 solution, and 18 mL of 20% glucose solution. The 20% glucose solution should be prepared and autoclaved separately.
18. Sterile cocktail sticks.
19. Heat resistant 96-well microtiter plate (Techne, Cambridge, UK).

3. Methods

3.1. PCR Reaction and Product Isolation

It is difficult to define a single set of conditions that will ensure optimal specific PCR amplification of the target DNA sequence. We describe a basic protocol that has been successful in our hands for most applications and that should be used first before attempting any variations. However, we advise readers who are not familiar with PCR to refer to Notes 5–8.

Although aliquots from the PCR reaction mix can be used directly for T-vector ligation, we have found an improvement in efficiency if the DNA products are first purified, and there are a number of options available as described in steps 6 and 7.

1. Prepare the PCR reaction mix as follows. To a 0.5-mL Eppendorf tube add 5 μL of each PCR primer, 5 μL of 10X PCR buffer, 5 μL of dNTP mix, 1 μL of template DNA (*see* Note 9), and 29 μL of distilled water, giving a total volume of 50 μL. Overlay the mixed reaction mix with sufficient mineral oil to prevent evaporation.
2. Transfer the tube to a thermal cycler and heat at 95°C for 5 min. "Hot start" the reaction by the addition of 0.3 μL of *Taq* DNA polymerase.
3. Immediately initiate the following program for 30–35 cycles (*see* Note 10):

Denaturation	95°C for 0.5 min
Primer annealing	50–60°C for 0.5 min
Primer extension	72°C for 0.5–3 min

4. After the final cycle carry out an additional step of 72°C for 5 min; this will ensure that primer extension is completed to give full length double-stranded product.
5. Add 1 μL of PCR stop mix to 5 μL of PCR product and run on a 1.5% agarose gel to determine the yield and specificity of the PCR reaction. The anticipated yield of PCR products is 10–50 ng DNA/μL reaction.
6. Where a PCR produces a single band or a number of bands all of which could be the correct product, the entire reaction mix may be phenol/chloroform extracted and precipitated. Resuspend in a volume of 10 μL of TE or distilled water (*see* Note 11).
7. If unwanted bands are also present the whole reaction should be run on a low melting point agarose gel and the bands of interest excised with a clean scalpel blade. When excising gel slices it is preferable to use a UV box that emits longer wavelength UV light (365 nm) since this causes less damage to the DNA. The resulting gel slices may be purified in a variety of ways (*see* Note 11).

3.2. T-Vector Preparation, Ligation, and Transformation

Preparation of T-vectors involves first digestion of the cloning vector with a restriction enzyme that generates blunt ends and then addition of T-residues to the cut vector utilizing the inherent terminal transferase activity of *Taq* DNA polymerase. The simplicity of this system allows T-vectors to be prepared using any cloning vector with blunt end restriction sites, such as *Eco*RV or *Sma*I. The pBluescript II series of vectors (Stratagene) are excellent for this purpose since both these restriction sites are located within the polylinker. The T-vector is prepared in batch so it will last for a number of cloning reactions. For example, 5 µg of T-vector DNA is sufficient for 100 cloning experiments.

The PCR product isolated following in vitro amplification is directly ligated into the prepared T-vector using the same conditions as for sticky end cloning. The optimum amount of PCR product(s) to be added per ligation reaction is difficult to estimate because of variance in yield, complexity of the PCR product profile and the efficiency of A addition to the DNA products by the *Taq* polymerase catalyzed terminal transferase activity. To ensure that the PCR products are within a range that should ensure successful cloning, we usually set up two ligations, one of which involves a 1:10 dilution of the purified DNA.

1. Digest 5 µg of vector DNA with a restriction enzyme that generates a unique blunt ended site, for example *Eco*RV or *Sma*I, for 2 h at 37°C (*see* Note 12).
2. Run the digest on a 1% low melting point agarose gel. Excise the linear vector DNA under UV, phenol/chloroform extract, and ethanol precipitate (*see* Notes 11 and 13). Resuspend in 20 µL of water in a 0.5-µL Eppendorf tube.
3. Add 5 µL of 10X PCR buffer, 1 µL of 100 m*M* dTTP, 24 µL of distilled water, and 0.4 µL of *Taq* DNA polymerase. Overlay with 40 µL of mineral oil. Incubate at 72°C for 2 h. For convenience a thermal cycler set at 72°C may be used.
4. Purify the T-vector by phenol/chloroform extraction and ethanol precipitation. Resuspend the prepared T-vector in 100 mL of water or TE, giving a concentration of 50 ng/µL.
5. Set up three tubes containing 1.0 µL of either undiluted PCR product; a 1:10 dilution of the PCR product, or distilled water (a control that will indicate the background of vector self ligation). Add to each tube 1.0 µL of prepared T-vector, 1.0 µL of 10X ligation buffer, 1.0 µL of DTT, 1.0 µL of ATP, 4.5 µL of distilled water, and 0.5 µL of T4 DNA ligase. Incubate overnight at 16°C.
6. Transform 5.0 µL of each ligation reaction into a suitable competent strain of *E. coli*. (*see* Note 14).

Typically, up to several hundred colonies may be isolated following transformation (*see* Notes 15–17). If a blue/white colony selection has been used, we generally find a ratio of about 50%. The white colonies or, if no method of nonrecombinant differentiation has been used, random colonies, are then screened for inserts.

3.3. PCR Screening

Colonies may be rapidly screened for the presence and size of cloned inserts by carrying out PCR with primers that flank the vector cloning site *(12)*. The screen is rapid since colonies can be used directly in the PCR reaction (*see* Note 18). In the case of the pBluescript series of plasmids (Stratagene) the T3 and T7 promoter primers may be used as PCR primers. In the absence of any insert, the vector polylinker sequence is amplified, providing an internal control to show that the PCR reaction has worked. This protocol may be used with a microtiter thermal cycler, as described in this protocol, however the method can be adapted to the standard thermal cyclers with a block for 500-μL Eppendorf tubes.

1. Make up a stock PCR mix containing 1 μL of PCR primer, 1 μL of 10X PCR buffer, 1 μL of dNTP mix, 5.95 μL of distilled water, and 0.05 μL of *Taq* DNA polymerase for each colony to be screened.
2. Pipet 10 μL of the stock PCR mix into each designated well of the microtiter plate. At the same time pipet 100 μL of TBG medium, supplemented with the appropriate antibiotic selection, into a second sterile microtiter plate well for each colony to be analyzed. Carefully number the wells so that after analysis of the PCR reactions, a particular clone can be readily identified.
3. Pick each colony with a sterile cocktail stick and swirl first in the PCR reaction mix and then its corresponding aliquot of fresh culture broth. If a blue/white selection has been used on the final pick, stab a blue colony as a control. Place the culture microtiter plate at 37°C for at least 6 h.
4. Overlay each well of the PCR test plate with 40 μL of mineral oil to prevent evaporation.
5. Transfer the PCR test plate to a thermal cycler and heat at 95°C for 2 min to lyse the cells. Initiate the following program for 35 cycles:

Denaturation	95°C for 0.5 min
Primer annealing	50°C for 0.5 min
Primer extension	72°C for 0.5–3 min

After the final cycle carry out an additional step of 72°C for 5 min.

6. Stop the reaction by addition of 10 µL of a 1:5 dilution of PCR stop mix and run the PCR products on an agarose gel (*see* Note 19). PCR products on screening will be increased in size by the distance between the primers on the vector. This should be taken into account when estimating insert size. Where blue/white selection has been used, >90% of the white colonies should contain T-vector with insert; when no selection has been employed this figure is reduced to about 50%.
7. Positive clones can then be taken from the culture microtiter plate and used to inoculate cultures in order to prepare plasmid DNA.

The same universal primers used for PCR can be employed depending on their distance from the cloning site (between 10–50 bases is suitable) as initial primers for sequencing the cloned DNA fragments (*see* Note 20). Remember that on sequencing an additional T:A pair will be found on each of the vector arms flanking the blunt end restriction site. Typical results from a PCR cloning experiment are shown in Fig. 2.

4. Notes

1. We have found it unnecessary for primers of the size we have outlined to be gel or HPLC purified. We routinely precipitate the primers and resuspend them in 200–500 µL sterile distilled water or TE buffer. Their concentration can be obtained using a spectrophotometer (1 A_{260} = 20 µg of single-stranded oligonucleotide). The molarity is estimated, assuming an average mol wt/base of 325 daltons and an aliquot of each stock is diluted to a concentration of 10 µ*M*.
2. The selection of a pair of oligonucleotide primers is the first step in preparing a PCR reaction. Although there are no hard and fast rules one can follow to absolutely ensure a given pair of PCR primers will result in the isolation of a desired DNA fragment, we have outlined some guidelines that should be taken into account when designing your oligonucleotides, and that will enhance your chances of achieving successful amplification of the target DNA sequence.
   a. Some sequence information (17–25 bp) is generally necessary at either end of the region to be amplified (*see* Note 3). The length of this segment should not exceed 3–4 kb for practical purposes (*see* Note 10).
   b. For PCR from a complex genomic DNA source we have found 20–24-mers to be long enough to give specific amplification of the target region; when the template is less complex, for example from a plas-

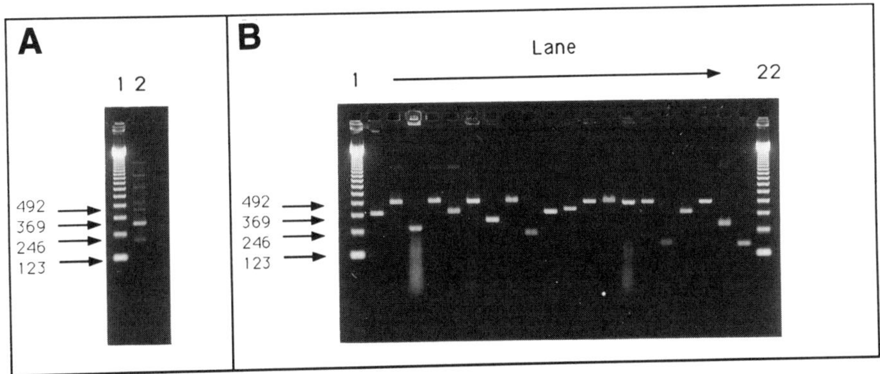

Fig. 2. A pair of degenerate primers were used to PCR genomic DNA from a complex organism. **(A)** Product profile on a 1.5% agarose gel. Lane 1, 123 bp ladder (BRL); Lane 2, PCR products. **(B)** PCR screen of nineteen white colonies (Lanes 2–20) and 1 blue control colony (Lane 21) with T3 and T7 universal primers. Lanes 1 and 22, 123-bp ladder. Analysis by agarose gel electrophoresis of an aliquot of the PCR reaction revealed a complicated series of products with a dominant band of around 340 bp (B). Since the anticipated size of the target sequence was 200–400 bp, the PCR reaction was purified as in Section 3.1. step 6 and then directly ligated into a prepared pBluescript II SK(+) T-vector. The ligation was transformed into *E. coli* strain XL1-Blue and selection was made on enriched media plates (with 100 µg/mL ampicillin) in the presence of X-gal and IPTG. Approximately 200 colonies were isolated with a blue/white ratio of around 50%. Nineteen white colonies and one blue were subjected to a PCR screen with flanking T3 and T7 universal primers to determine both the size of the products cloned and efficiency of the method. Agarose gel electrophoresis (1.5% gel) (B) revealed that all 19 white colonies carried cloned PCR products, with a predominant fragment of 500 bp. As the polylinker of Bluescript contributes an additional 166 bp, the actual insert size is approx 340 bp. This confirms cloning of the PCR products.

mid template, the primer length may be reduced to a 17mer. Longer primers can be prepared, but this is usually unnecessary and costly.

c. The primers should match their target hybridization sites well, especially at their 3' ends. If possible keep the G/C content of each primer to about 50% and try to avoid long stretches of the same base. Both primers should be approximately the same length.
d. Check that the two primers do not have significant complementarity to each other, particularly at their 3' ends, to avoid "primer dimers," where two primers hybridize to one another forming a very effective substrate for PCR that subsequently may become the dominant product *(13)*.

e. Where the aim of the PCR is to clone a known gene, or its homolog, check the primer sequences against the EMBL/Genbank database to try to ensure they are unique to the template DNA.

f. It is worth bearing in mind that both length and G/C content of primers determine the optimum annealing temperature in the PCR reaction. There are now a number of computer programs available that will design primers for specific target sequences.

One of the most up to date primer design programs is 'PRIMER Version 0.5' (1991), which can be obtained from: Stephen E. Lincoln, Mark J. Daly, and Eric S. Lander, MIT Center for Genome Research and Whitehead Institute for Biomedical Research, Nine Cambridge Center, Cambridge, MA 02142.

3. One of the difficulties in primer design when cloning from complex genomes is that the precise sequence of the target region is often unknown, and therefore must be predicted from reported sequences of the same gene in different organisms or from sequence information from similar genes. For structural genes, the known DNA sequence homologs should be first translated into protein and then aligned. There are a number of alignment programs available on mainframe computers, such as CLUSTAL *(14)*, that can assist in alignments and therefore allow identification of conserved regions. Within the conserved region, amino acids of low degeneracy are particularly sought after, such as those with one codon; methionine and tryptophan, and two codons; asparagine, aspartate, cysteine, glutamine, glutamate, histidine, lysine, phenylalanine, and tyrosine; which thereby lower the degeneracy of the primers. The amino acids arginine, leucine, and serine are encoded by 6 different codons and should be avoided if possible. When designing the primers one should also take into account the codon usage for the particular organism under study for which codon usage tables are available *(15)*. Finally, try to locate the most conserved sequence information available at the 3' ends of the primers.
4. When making agarose gels it is worth remembering that TAE gels are easier to use with kits such as GeneClean™. For products of 100 bp or less it may be necessary to pour gels of greater than 2%. Products above 300–400 bp can be run on 1.5% gels. It is useful to run PCR products against a low mol wt ladder, such as a 123-bp ladder (BRL), or *pBR*322/*Msp*I digest.
5. There are many modifications that can be made to the basic PCR reaction mix in an attempt to improve the specificity of the amplification reaction. However, even after taking precautions, multiple PCR products may be generated owing to nonspecific priming. The parameter

that will give the greatest variation in product yield and complexity is the annealing temperature. By increasing the annealing temperature the specificity of the reaction is increased because of more stringent binding of the primers. However, if the temperature of annealing is raised too high, the yield of product will decrease and eventually disappear. By experimenting with a number of different annealing temperatures, it is usually possible to select the optimum PCR conditions for a particular pair of primers. Improvements in specificity may also be achieved by altering the Mg^{2+} concentration in the PCR buffer over the range 0.5–5 m*M* final concentration. If all else fails a second set of primers, which are located internally to the first pair (nested), may be used to reamplify the products obtained from the initial amplification *(2)*.

6. Even after optimizing the PCR reaction there may be a number of seemingly specific bands that may not correspond to the desired product. We have found that this is most often caused by specific double-priming of one of the primer pair so as to produce a PCR product. We therefore routinely run parallel reactions that contain only one primer. By running these reactions side by side on a gel it is possible to determine which products are formed by the interaction of two primers and which are formed from one primer only.
7. *Taq* polymerase lacks a 3'–5' proofreading exonuclease activity *(16)*. Therefore, during enzymatic DNA amplification errors can accumulate in the PCR product at rates as high as 2×10^{-4} *(3)*. Errors occurring early in the PCR will be present in a substantial number of the amplified molecules and therefore in many of the clones. In situations where unknown sequences are being isolated we suggest determining the sequence of clones from at least three independent PCR reactions. An alternative is to use a different thermostable DNA polymerase that has a "proofreading" activity, such as "pfu" from Stratagene or "Vent" from New England Biolabs (Beverly, MA).
8. The exquisite sensitivity of the amplification reaction makes it liable to contamination by minute amounts of exogenous DNA. This is a particular problem for PCR reactions involving complex genomes when small amounts of template are present and that therefore need extra rounds of amplification. Measures can be taken to avoid this contamination, including the use of gloves, pipet tips with filters, UV treating solutions, and tubes, but obviously not DNA, and if necessary carrying out the preparation work in a lamina flow hood.
9. The source of template DNA can vary from genomic DNA, cDNA, and DNA prepared from YACs, cosmids, lambda, plasmids, and M13 phages, and may be single- or double-stranded. As a guide use more DNA as

the complexity of the source increases. For complex genomes use a DNA concentration of 10–100 μg/mL, whereas for plasmid DNA preparations use 1–10 μg/mL. The amplification efficiency is reduced in the presence of too much template. We have also successfully amplified by directly picking or aliquoting from single clone lambda phage stocks or plaques, plasmid colonies and glycerol stocks, and M13 plaques.

10. As a very rough guide to annealing temperature, allow 4°C/G or C and 2°C for every A or T. Thus, for an 18-mer with 10(G + C), annealing temperature is $(10 \times 4) + (8 \times 2) = 56°C$. The time allowed for the primer extension step is based on a rate of approx 1 kb/min. Although PCR products as large as 10 kb have been reported after detection by Southern blotting *(17)*, for practical purposes the maximum product size that can be visualized on a gel, purified, and cloned is around 3–4 kb.
11. To purify PCR product by extraction: Add an equal volume of phenol, mix, and centrifuge 12,000*g* for 10 min, transfer the aqueous phase (upper phase) to a fresh tube, and add an equal volume of chloroform. Mix, respin, and then add 1/10 volume 3*M* potassium acetate and 2.5 vol of ethanol to the transferred aqueous phase (top). Leave at either –20°C or on dry ice for 10 min, and then spin 12,000*g* for 10 min. Remove the supernatant and wash pellet with 200 μL of 70% ethanol. Dry the pellet in either air or under vacuum and resuspend in the desired volume of TE or water. If extracting from low melting point agarose; weigh the agarose slice, add an equal volume of water, and heat to 65°C to melt the agarose completely before adding the phenol. DNA binding matrices in kits, such as GeneClean (DNA >500 bp) and Mermaid (10–500 bp) from Bio 101 (La Jolla, CA), or gel filtration columns, such as Primerase from Stratagene, may also be used to purify the PCR products.
12. *Eco*RV is preferred to *Sma*I if available because it is more stable at 37°C and tends to give more efficient cleavage of vector DNA. If *Sma*I is the only blunt end cleavage site available then digestion should be carried out by incubation at 25°C.
13. By gel purifying the T-vector any uncut vector is removed, reducing the background of colonies that lack inserts.
14. Competent *E. coli* can be either prepared by the method of Hanahan *(18)* or purchased directly from commercial suppliers.
15. If the cloning vector of choice has β-galactosidase blue/white selection, inclusion of IPTG and X-Gal in the plating mix on transformation will result in recombinant clones, giving rise to white colonies. This is because of insertional inactivation of the β-galactosidase gene, thereby preventing it from cleaving the chromogenic lactose analog X-Gal,

which in turn prevents blue colony formation. White colonies do not definitively indicate recombinants since some may be caused by frameshift events or by misligation of noncompatible ends *(19)*. Conversely, some blue colonies may be recombinant clones that have left the β-galactosidase gene in frame in such a way that the enzyme still retains activity.

16. The lack of any colonies on a plate following transformation may be attributable to either low competency of the *E. coli* or failure of the PCR product to ligate to the T-vector. For each set of transformations a control plasmid should also be included to monitor the transformation efficiency of the *E. coli*. In addition, confirm the concentration of both T-vector and PCR product, and whether successful ligation has taken place, by running samples on an agarose gel.
17. An extremely high background of nonrecombinant colonies, either as blue colonies or as determined by the screening protocol, is most likely owing to failure of the T addition to the vector arms. Repeat this step ensuring all the components are included and that the *Taq* polymerase is active. An indicator of successful T-vector preparation can be ascertained by its ligation in the absence of PCR product, which following transformation should result in the presence of only a small number of nonrecombinant (blue) colonies.
18. The screening protocol can be adapted for PCR products cloned into M13. After picking plaques, place the cocktail stick into 100 µL of TBG medium containing 1/100 volume of an overnight culture of *E. coli*.
19. When running PCR products on an agarose gel wipe the tip of the pipet on a tissue before loading the sample to remove any oil that may have collected on its outside, since this may cause the sample to float out of the well into the buffer.
20. Even if a single band is visible after PCR, it is possible that there is more than one species of product present. It is always wise, therefore, to sequence a number of recombinant clones to ensure that the desired fragment is present. If a single band of known sequence is the product of the PCR reaction then further confirmation can be provided by restriction mapping of a small aliquot (2–5 µL) in 20–50 µL digestion mix.

Acknowledgments

We would like to thank Sydney Brenner for providing us with the opportunity to work on this project. M. K. Trower is supported by a fellowship from E. I. du Pont de Nemours and Co. Inc., Wilmington, DE. G. S. Elgar is supported by a grant from The Jeantet Foundation.

References

1. Saiki, R. K., Scharf, S. J., Faloona, F. A., Mullis, K. B., Horn, G. T., Erlich, H. A., and Arnheim, N. (1985) Enzymatic amplification of b-globin genomic sequences and restriction site analysis for diagnosis of sickle cell anemia. *Science* **230,** 1350–1354.
2. Mullis, K. B. and Faloona, F. A. (1987) Specific synthesis of DNA in vitro via a polymerase-catalyzed chain reaction. *Meth. Enzymol.* **155,** 335–350.
3. Saiki, R. K., Gelfand, D. H., Stoffel, S., Scharf, S. J., Higuchi, R., Horn, G. T., Mullis, K. B., and Erlich, H. A. (1988) Primer-directed enzymatic amplification of DNA with a thermostable DNA polymerase. *Science* **239,** 487–491.
4. Erlich, H. A., Gelfand, D., and Sininsky, J. J. (1991) Recent advances in the polymerase chain reaction. *Science* **252,** 1643–1651.
5. Scharf, S. J., Horn, G. T., and Erlich, H. A. (1986) Direct cloning and sequence analysis of enzymatically amplified genomic sequences. *Science* **233,** 1076–1078.
6. Kaufman, D. L. and Evans, G. A. (1990) Restriction endonuclease cleavage at the termini of PCR products. *Biotechniques* **9,** 304–305.
7. Clark, J. M. (1988) Novel non-templated nucleotide reactions catalyzed by procaryotic and eucaryotic DNA polymerases. *Nucleic Acids Res.* **18,** 9677–9686.
8. Hemsley, A., Arnheim, N., Toney, M. D., Cortopassi, G., and Galas, D. J. (1989) A simple method for site-directed mutagenesis using the polymerase chain reaction. *Nucleic Acids Res.* **17,** 6545–6551.
9. Mead, D. A., Pey, N. K., Herrnstadt, C., Marcil, R. A., and Smith, L. (1991) A universal method for the direct cloning of PCR amplified nucleic acid. *Biotechnology* **9,** 657–663.
10. Marchuk, D., Drumm, M., Saulino, A., and Collins, F. S. (1991) Construction of T-vectors, a rapid and general system for direct cloning of unmodified PCR products. *Nucleic Acids Res.* **19,** 1154.
11. Sambrook, J., Fritsch, E. F., and Maniatis, T. (1989) Plasmid vectors, in *Molecular Cloning. A Laboratory Manual*, 2nd ed. (Ford, N., ed.), Cold Spring Harbor Laboratory, Cold Spring Harbor, NY, pp. 1.25–1.28.
12. Güssow, D. and Clackson, T. (1989) Direct clone characterisation from plaques and colonies by the polymerase chain reaction. *Nucleic Acids Res.* **17,** 4000.
13. Mullis, K. B. (1991) The polymerase chain reaction in an anemic mode: how to avoid cold oligodeoxyribonuclear fusion. *PCR Methods and Applications* **1,** 1–4.
14. Higgins, D. G. and Sharp, P. M. (1988) CLUSTAL: a package for performing multiple sequence alignment on a microcomputer. *Gene* **73,** 237–244.
15. Wada, K.,Wada, Y., Doi, H., Ishibashi, F., Gujubori, T., and Ikemura, T. (1991) Codon usage tabulated from the GenBank genetic sequence data. *Nucleic Acids Res.* **19 (Suppl.)** 1981–1986.
16. Tindall, K. R. and Kunkel, T. A. (1988) Fidelity of DNA synthesis by the *Thermus aquaticus* DNA polymerase. *Biochemistry* **27,** 6008–6013.

17. Jeffreys, A. J., Wilson, V., Neumann, R., and Keyte, J. (1988) Amplification of human minisatellites by the polymerase chain reaction: towards DNA fingerprinting of single cells. *Nucleic Acids Res.* **16,** 10953–10971.
18. Hanahan, D. (1985) Techniques for transformation of *E. coli,* in *DNA Cloning: A Practical Approach,* vol. 1 (Glover, G. M., ed.), IRL, Oxford, pp. 109–135.
19. Wiaderkiewicz, R. and Ruiz-Carillo, A. (1987) Mismatch and blunt to protruding-end joining by DNA ligases. *Nucleic Acids Res.* **15,** 7831–7848.

CHAPTER 4

Thermal Cycle Dideoxy DNA Sequencing

Barton E. Slatko

1. Introduction

Since DNA sequencing has rapidly become standard practice in many laboratories, a large variety of new cloning vectors, sequencing strategies, and techniques have been developed to allow more efficient sequencing of a large variety of DNA templates. Despite a current focus on the automation of DNA sequencing procedures, other methods are still required until automated DNA sequencing is in more general use. One recent addition to this repertoire of sequencing methods is termed thermal cycle sequencing.

Thermal cycle DNA sequencing protocols are based on the dideoxynucleotide chain termination method of Sanger et al. *(1)*. In the reaction, an appropriate primer DNA molecule is annealed to a complementary single-stranded stretch of DNA. This primer template complex is incubated with a highly thermostable DNA polymerase, such as the exonuclease deficient DNA polymerase from *Thermococcus litoralis*, $\text{Vent}_\text{R}^{\text{TM}}$ (exo^-) DNA polymerase *(2,3)*, or from *Thermus aquaticus Taq* polymerase *(4–7)* in the presence of deoxynucleotide triphosphates (dNTPs) and dideoxynucleotide triphosphates (ddNTPs). Four separate reaction mixes, each with all four dNTPs and one of the four ddNTPs, generate four different sets of fragments, each set corresponding to terminations at specific nucleotide residues. Repetitive cycles of denaturation, annealing, and chain extensions from small amounts of template molecules in the pres-

From: *Methods in Molecular Biology, Vol. 31: Protocols for Gene Analysis*
Edited by: A. J. Harwood

ence of primer excess, $Vent_R^{TM}$ (exo^-) DNA polymerase, dNTPs, and ddNTPs achieve a linear amplification of reaction products and a strong sequencing signal (Fig. 1). In each cycle, the reaction is raised to 95°C to denature double-stranded templates or secondary structure regions of single-stranded DNA templates, lowered to 55°C for the annealing of the primer to the template, and subsequently raised to 72°C for the elongation step of enzymatic synthesis. Subsequent cycles of denaturation, annealing, and extension occur in which the excess primer anneals to the identical denatured template molecules as in the first cycle.

Thermal cycle sequencing offers several important advantages over previously developed techniques *(4–8)*.

1. The reactions are rapid, easy to perform, efficient, and useful for both manual and automated DNA sequencing.
2. The method requires much less template than does a standard reaction, owing to the linear amplification of labeled product.
3. There is no need to denature ("collapse") double-stranded DNA templates before initiating the sequencing reactions.
4. There is no separate annealing step preceding the reactions.
5. The use of a highly thermostable DNA sequencing enzyme allows sequencing at high temperatures, an advantage for obtaining DNA sequence information from DNA templates that have high degrees of secondary structure and that may be recalcitrant with lower temperature sequencing methods.

The method can be used to sequence single-stranded DNA templates, such as those derived from M13, fl, fd phage, or from phagemid vectors, and double-stranded templates, such as plasmid DNA, PCR products, or large linear double-stranded DNA, such as bacteriophage λ. In addition, sequencing directly from phage plaques and bacterial colonies is also feasible *(8–10)*.

In order to visualize the dideoxy terminated chains on the gel, various means of labeling have been developed. Conventionally, ^{32}P and ^{35}S have been used for incorporation into the nascent chain by using α-labeled deoxynucleotide triphosphates in the reaction mixture. Alternatively, a second approach utilizes end-labeled primers, wherein T4 polynucleotide kinase can be used to transfer a γ-[^{32}P] (or γ-[^{33}P]) from rATP to the 5' end of a primer. Similarly, primers can be

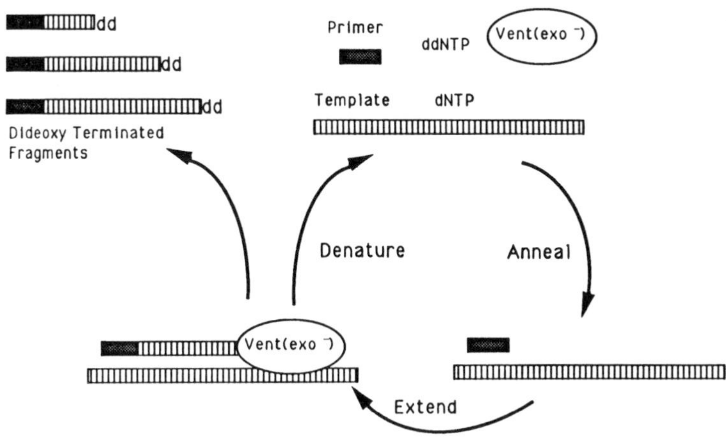

Fig. 1. Diagram of the thermal cycle sequencing reaction, demonstrating the annealing, extension, and denaturation steps that generate the dideoxy-terminated DNA fragments. Reprinted with permission from New England Biolabs, Inc.

biotinylated for chemiluminescent DNA sequencing *(11)*, or end-labeled with fluorescent dyes for automated DNA sequencing *(8,12)*. 5' end-labeling with ^{32}P or ^{33}P is used in thermal cycle sequencing when using lesser amounts of template DNA (*see* Section 3.2.). 5' end-labeled primers are also recommended for sequencing large templates (such as λgt11) or for templates that yield less than optimal results with incorporated label techniques.

The sequencing reaction products, DNA strands of varying lengths, are separated on a denaturing polyacrylamide gel. The gel is subsequently processed for exposure to X-ray film, when utilizing radioactively labeled DNA molecules or when using chemiluminescent (nonradioactive) detection. The resultant autoradiogram is analyzed for sequence information. The data is collected "in real time" when using automated DNA sequencers and no "postprocessing" of the gel is required.

This chapter provides a set of methods for sequencing nanogram amounts of single-stranded and double-stranded DNA templates by thermal cycle sequencing, using 5' end-labeled primers (^{32}P, ^{33}P, or biotin for chemiluminescent detection) or by ^{35}S, ^{32}P, or ^{33}P [dATP] radiolabel incorporation.

2. Materials

1. 10X Vent$_R$™ (exo$^-$) DNA polymerase sequencing buffer: 100 m*M* $(NH_4)_2SO_4$, 100 m*M* KCl, 200 m*M* Tris-HCl, 50 m*M* $MgSO_4$, pH 8.8 (room temperature). Store at –20°C.
2. Vent$_R$™ (exo$^-$) DNA polymerase deoxy/dideoxy sequencing mixes: Make up in 1X Vent$_R$™ (exo$^-$) DNA polymerase sequencing buffer (*see* Note 1).

	μ*M* Concentrations			
	A Mix	C Mix	G Mix	T Mix
ddATP	900	—	—	—
ddCTP	—	400	—	—
ddGTP	—	—	360	—
ddTTP	—	—	—	720
dATP	30	30	30	30
dCTP	100	37	100	100
dGTP	100	100	37	100
dTTP	100	100	100	33

Store at –20°C.

3. 30X Triton X-100: 3% Triton X-100. Store at –20°C.
4. Radiolabel: (*see* Note 2). Use α-[^{35}S]-dATP, 500 Ci/mmol; α-[^{32}P]-dATP, 400 Ci/mmol, or α-[^{33}P]-dATP, 3000 Ci/mmol for labeled dATP incorporation as described in Section 3.1. Use γ-[^{32}P]-rATP, 3000 Ci/mmol or γ-[^{33}P]-rATP, 3000 Ci/mmol for end-labeled primers as described in Section 3.2.). Store at –20°C.
5. Vent$_R$™ (exo$^-$) DNA polymerase: 2 U/μL (New England Biolabs, Inc., Beverly, MA). Store at –20°C.
6. Stop/loading dye solution: Deionized formamide containing 0.3% xylene cyanol, FF 0.3% bromophenol blue and 0.37% EDTA (pH 7.0). Store at –20°C.

3. Methods

3.1. Thermal Cycle Sequencing with Labeled dATP Incorporation

1. Label four microcentrifuge tubes A, C, G, T. Using the Vent$_R$™ (exo$^-$) DNA polymerase deoxy/dideoxy sequencing mixes, add 3 μL of A mix to the bottom of the tube A, and 3 μL of the C, G, and T mixes to the bottoms of the tubes C, G, and T, respectively.
2. Mix together the following in a 0.5-mL microcentrifuge tube: 0.04 pmol of single-stranded template DNA or 0.1 pmol of double-stranded template DNA (*see* Notes 3 and 4), 0.6 pmol primer for a single-stranded

template DNA or 1.2 pmol primer for a double-stranded template DNA, 1.5 μL of 10X $Vent_R$™ (exo$^-$) DNA sequencing buffer, 1 μL of 30X Triton X-100 solution and distilled water to a total vol of 12.0 μL. Mix the solution by gentle pipeting.

3. Individually process each template/primer tube through this step. When all sets of reactions are complete, proceed to step 4. To the tube containing the template, primer, buffer, Triton X-100, and water, add 2 μL of radiolabel. Add 2–4 U of $Vent_R$™ (exo$^-$) DNA polymerase and mix the solution by gentle pipeting. Immediately distribute 3.2 μL of this reaction to the deoxy/dideoxy tube labeled A, and mix the solution by gentle pipeting. Changing pipet tips each time, repeat this addition to the C, G, and T tubes.
4. Overlay each reaction with one drop of sterile mineral oil. Place the tubes in the thermal cycler, which has been preset for reaction times and temperatures (*see* Note 5). Start the thermal cycler.
5. After completion of the thermal cycling, add 4 μL of stop/loading dye solution to each tube, beneath the mineral oil. The reactions are now complete and ready to be electrophoresed in appropriate denaturing sequencing gels (*see* Note 6). The reactions may be stored at –20°C.

3.2. Thermal Cycle Sequencing with 5' End-Labeled Primers

1. Label four microcentrifuge tubes A, C, G, and T. Using the $Vent_R$™ (exo$^-$) DNA polymerase deoxy/dideoxy sequencing mixes, add 3 mL of A mix to the bottom of the tube A, and 3 μL of the C, G, and T mixes to the bottoms of the tubes C, G, and T, respectively.
2. Mix together the following in a 0.5-mL microcentrifuge tube: 0.004 pmol of single-stranded template DNA or 0.01 pmol of double-stranded template DNA (*see* Notes 3, 4, and 7), 0.6 pmol end-labeled primer for single-stranded template DNA or 1.2 pmol end-labeled primer for double-stranded template DNA (*see* Note 8), 1.5 μL 10X $Vent_R$™ (exo$^-$) DNA polymerase sequencing buffer, 1 μL 30X Triton X-100 solution, and distilled water to a total volume of 14.0 μL. Mix the solution by gentle pipeting.
3. Individually process each template/primer tube through this step. When all sets of reactions are complete, proceed to step 4. Add 2–4 U of $Vent_R$™ (exo$^-$) DNA polymerase to the tube containing the template, primer, buffer, Triton X-100, and water and mix the solution by gentle pipeting. Immediately distribute 3.2 μL of this reaction to the deoxy/dideoxy tube labeled A, and mix the solutions by gentle pipeting. Changing pipet tips each time, repeat this addition to the C, G, and T tubes.

4. Overlay each reaction with a drop of sterile mineral oil. Place the tubes in the thermal cycler that has been preset for reaction times and temperatures (*see* Note 5). Start the thermal cycler.
5. After completion of the thermal cycling, add 4 µL of stop/loading dye solution to each tube beneath the mineral oil. The reactions are now complete and ready to be electrophoresed in appropriate denaturing sequencing gels (*see* Note 6). The reactions may be stored at –20°C.

Typical results are shown in Fig. 2. Potential problems are described in Notes 9 and 10.

4. Notes

1. Improved sequencing of high secondary structure regions may be accomplished by substituting an equal molar amount of 7-deaza dGTP (7-deaza-2'-deoxyguanosine-5'-triphosphate) for the deoxyguanosine triphosphate or by substituting an equal molar amount of 7-deaza dATP (7-deaza-2'-deoxyadenosine-5'-triphosphate) for the deoxyadenosine triphosphate. dITP (2'-deoxyinosine-5'-triphosphate) is not recommended as a substitute for deoxyguanosine triphosphate because it is not incorporated as efficiently and because it tends to show shadow banding (premature terminations) in all four sequencing lanes.
2. For sequencing small amounts of DNA template (0.004–0.01 pmol) or for sequencing directly from bacterial colonies or phage plaques, use ^{33}P or ^{32}P end-labeled primers. For all other applications, including sequencing PCR products, ^{35}S, ^{33}P, or ^{32}P can be effectively utilized. Because ^{35}S is not efficiently incorporated in the kinase reaction, it is not recommended for sequencing smaller amounts of template DNA. When preparing end-labeled primers by the kinase reaction, it is recommended to use the higher specific activity (3000 Ci/mmol) rATP, since it provides a stronger signal. In all protocols it is recommended that all radioactive material be used within two radioactive decay half-lives.
3. The following template and preparations have given good results:
   a. Plasmid templates should be purified by CsCl gradient, standard minipreparation methods *(13)*, mini-column chromatography procedures, or by more recent methods, such as Insta-Prep™ (5 prime→3 prime, Inc., Boulder, Co).
   b. Single-stranded M13 templates should be purified by PEG precipitation methods *(13)* or solid support purification procedures *(14)*.
   c. λ (λgt10, 11, and so forth) and cosmid templates should be purified by CsCl gradient protocols or by minipreparation methods *(13)*.

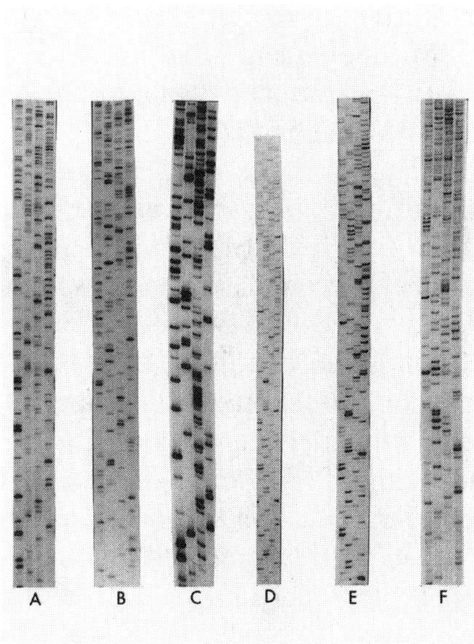

Fig. 2. Autoradiograms of Vent$_R$™ (exo$^-$) DNA polymerase thermal cycle sequencing reactions, as described in the text. Panel **A**: DNA sequence obtained directly from a bacterial colony containing pUC19 double-stranded plasmid DNA with a ^{32}P 5' end-labeled (kinased) M13 forward sequencing primer (NEB, New England Biolabs, Inc. #1224); Panel **B**: DNA sequence obtained directly from an M13mp18 phage plaque with a ^{32}P 5' end-labeled (kinased) M13 forward sequencing primer (NEB #1224); Panel **C**: DNA sequence obtained from 0.01 pmol double-stranded PCR template with a ^{32}P 5' end-labeled (kinased) 20-mer primer; Panel **D**: DNA sequence obtained from 0.004 pmol (10 ng) M13mp18 single-stranded DNA template (NEB #404C) sequenced with the M13 forward universal primer (NEB #1224) and α-[^{35}S] dATP incorporation; Panel **E**: DNA sequence obtained from 0.01 pmol (20 ng) double-stranded pUC19 double-stranded DNA template (NEB#304-1) sequenced with the M13 forward universal primer (NEB #1224) and α-[^{35}S] dATP incorporation; Panel **F**: DNA sequence obtained from a minipreparation of double-stranded cosmid DNA template with a ^{32}P 5' end-labeled primer. All lanes loaded left to right: G, C, A, T. All exposures overnight without an intensifying screen, except for panels D and E, which were 36 h exposures.

d. Double-stranded and single-stranded (asymmetric) PCR products must have the excess primers and nucleotides removed after completion of the PCR reaction. Phenol/chloroform extract and alcohol precipitate the PCR product from the reaction. Resuspend the dried pellet in 50 µL water. Drop dialyze the reaction *(15)* using a PVDF membrane (0.025 µ) (Millipore Corp., Bedford, MA) (1 h against 200 mL of water), or purify the product by gel purification, glass bead procedures, or by exclusion methods (i.e., "spin" columns).

e. For direct sequencing from bacterial colonies *(8,10)* resuspend the colony in 12 µL of freshly prepared colony/plaque lysis solution (10 m*M* Tris-HCl, pH 7.5, 1 m*M* EDTA, 50 mg/mL proteinase K; prepare fresh each time. A proteinase K stock solution may be prepared and aliquots stored at –20°C.). Briefly vortex the solution, incubate 15 min at 55°C, and then 15 min at 80°C. Place on ice 1 min and microcentrifuge 3 min. Use 9 µL of the supernatant as the template in the end-labeled primer sequencing reaction.

f. For direct sequencing of M13 or λ plaques *(8–10),* transfer with a toothpick the top agar containing a plaque, to a 1.5-mL centrifuge tube containing 12 µL of freshly prepared colony/plaque lysis solution (10 m*M* Tris-HCl, pH 7.5, 1 m*M* EDTA, 50 mg/mL proteinase K; prepare fresh each time. A proteinase K stock solution may be prepared and aliquots stored at –20°C). Briefly vortex the solution, incubate 15 min at 55°C, and then 15 min at 80°C. Place on ice 1 min and microcentrifuge 3 min. Use 9 µL of the supernatant as the template in the end-labeled primer sequencing reaction.

4. Molar ratio calculations: The following formulae can be used to calculate pmol amounts of primer and template:

Single-stranded DNA (M13 phage, ss PCR products, oligonucleotide primers, and so on.)*

$$\text{\# pmol} = (\text{\# µg DNA} \times 10^6\ \text{pg/µg})/(\text{\# bases} \times 330\ \text{pg/pmol})$$

Double-stranded DNA (ds PCR products, plasmids, λ DNA, restriction fragments, and so on.)**

$$\text{\# pmol} = (\text{\# µg DNA} \times 10^6\ \text{pg/µg})/(\text{\# base pairs} \times 660\ \text{pg/pmol base pair})$$

Useful numbers:

0.04 pmol of a 7.25-kb single-stranded DNA template = ~100 ng
0.004 pmol of a 7.25-kb single-stranded DNA template = ~10 ng

0.1 pmol of a 3-kb double-stranded DNA template = ~200 ng
0.01 pmol of a 3-kb double-stranded DNA template = ~20 ng

*40 µg = 1 OD_{260}; # µg oligonucleotide = # OD_{260} × 40 µg/OD_{260}
**50 µg = 1 OD_{260}; # µg oligonucleotide = # OD_{260} × 50 µg/OD_{260}

5. Thermal cycler parameters: For most thermal cyclers, including the Techne (Duxford, UK) PHC-2 and the Perkin-Elmer-Cetus (Norwalk, CT) 480 thermal cycler, use 20 cycles:

   95°C 20 s
   55°C 20 s
   72°C 20 s

   It is possible to alter the cycling conditions of the reaction and still achieve excellent results. One may reduce the number of cycles in the reaction, especially if the amount of template DNA being used exceeds the recommendations of the protocol guideline. However, increasing the number of cycles to more than 30 has not been successful in sequencing smaller amounts of DNA; it often increases the shadow banding on the resultant autoradiograms. It is also possible to alter the cycle temperatures and/or times for the reaction. Templates may be sequenced using only a two-step cycle, a 95°C denaturation step plus a 72°C annealing and extension step, using primers with apparent melting temperatures (T_m values) of as low as 55°C. The signal intensities with this approach may be weaker than with the more standard three-step recommendation. It is also possible to lower the annealing/extension temperature, corresponding to the primer T_m, to perform sequencing cycles consisting of two steps, for example, a 95°C denaturation step and a 55°C annealing and extension step. If weak sequence signal is observed when using the standard three-step method, the recommended annealing temperature may be too high for the primer being used. The annealing step can be lowered to correspond to the individual primer T_m. Successful use of some thermal cyclers requires longer reaction times (increasing each step from 20 s to 1 min in each cycle).
6. A brief (5 s) microcentrifugation of the completed reaction helps ensure the aqueous layer fully separates from the oil layer. Immediately before loading the sequencing gel, heat the completed samples at 80°C for 2 min. For each loading, insert the pipet tip underneath the oil covering the reaction and remove a 2.5 µL sample. It may be necessary to briefly touch the pipet tip to a tissue to remove residual oil, before loading the sample on the gel. Alternatively, one can remove the oil before loading the reaction on a gel by using silicone oil to cover the reactions followed by precipitation of the reaction products from the oil *(16)* or by using Parafilm™ to remove the oil *(17)*. Another approach is to use a "hot-top" apparatus, set 15°C hotter than the warmest step in the cycle reaction (~115°C), which eliminates evaporation and precludes the requirement for oil on the reactions. It should be emphasized that we

have never observed any sequence aberration caused by overlay oil contaminating the reaction in the sequencing gel lane.

7. As much as 10-fold greater quantity of template DNA can be successfully used in this protocol.
8. For end-labeling with γ-[^{32}P] or γ-[^{33}P] rATP, the following protocol works well:
   a. Mix the following in a 0.5-mL microcentrifuge tube: 13.5 µL water, 2.5 µL 10X T4 polynucleotide kinase buffer (10X buffer = 500 m*M* Tris-HCl, pH 7.6, 100 m*M* $MgCl_2$, 50 m*M* DTT), 10.5 pmol primer (1 µL of stock primer solution; stock = 2.5 mg 24-mer primer resuspended in 30 µL of water), 7 µL γ-[^{32}P] rATP (11.5 pmol) and 1 µL (10 U) T4 polynucleotide kinase (10,000 U/mL).
   b. Allow the reaction to proceed at 37°C for 30 min.
   c. Terminate the reaction at 95°C for 5 min. Briefly microcentrifuge at room temperature. Store at –20°C.

   This protocol provides 10.5 pmol of 5' end-labeled primer in a 25 µL reaction. The final concentration of primer is thus 0.4 pmol/µL.
9. The following sequencing patterns are characteristic of $Vent_R$™ (exo^-) DNA polymerase:
   a. The first C of a run of Cs is darker than the following Cs.
   b. The second A of a run of As is darker than the preceding (and/or following) As.
   c. G following an A tends to be darker than other Gs.
   d. T following an A tends to be darker than other Ts.
10. When DNA sequencing results are less than optimal, several factors might be considered. Often the problem lies with template preparation. A phenol/chloroform extraction step or alcohol precipitation followed by a 70% alcohol rinse will often eliminate the problem. A second common problem is the use of too much or too little primer and/or template in the reaction; the pmol amounts in the protocols have been empirically determined to provide optimal results.

    Another potential problem involves suboptimal performance of thermal cyclers. Some may require longer reaction times, may also cause problems, and are often indicated by one or two lanes of a reaction that are lighter than the others or by failure of one lane. If one or two lanes of a reaction are lighter than the others, or if failure of one lane occurs, a problem with the cycler is indicated.

References

1. Sanger, F., Nicklen, S., and Coulson, A. R. (1977) DNA sequencing with chain inhibitors. *Proc. Natl. Acad. Sci. USA* **74,** 5463–5467.

2. Kong, H., Kucera, R., and Jack, W. (1992) Characterization of a DNA polymerase from the hyperthermophile Archaea *Thermococcus litoralis*. *J. Biol. Chem.* **268,** 1965–1975.
3. Perler, F., Comb, D., Jack, W., Moran, L., Qiang, B.-Q., Kucera, R., Benner, J., Slatko, B., Nwankwo, D., Hempstead, K., Carlow, C., and Jannasch, H. (1991) Novel intervening sequences in an Archaea DNA polymerase gene. *Proc. Natl. Acad. Sci. USA* **89,** 5577–5581.
4. Adams, S. and Blakesley, R. (1991) Linear amplification sequencing. *Focus (BRL)* **13(2),** 56–57.
5. Carothers, A. M., Urlaub, G., Mucha, J., Grunberger, D., and Chasin, L. A. (1989) Point mutation analysis in a mammalian gene: rapid preparation of total RNA, PCR amplification of cDNA and *Taq* sequencing by a novel method. *BioTechniques* **7,** 494–499.
6. Craxton, M. (1991) Linear amplification sequencing: a powerful method for sequencing DNA. *Methods*, a companion to *Meth. Enzymol.* **3,** 20–26.
7. Murray, V. (1989) Improved double-stranded DNA sequencing using the linear polymerase chain reaction. *Nucleic Acids Res.* **17,** 8889.
8. Sears, L. E., Moran, L. S., Kissinger, C., Creasey, T., Sutherland, E., Perry-O'Keefe, H., Roskey, M., and Slatko, B. (1992) Thermal cycle sequencing and alternative manual and automated DNA sequencing protocols using the highly thermostable $Vent_R$™ (exo^-) DNA polymerase. *BioTechniques* **13,** 626–633.
9. Krishnan, B. R., Blakesly, R. W., and Berg, D. E. (1991) Linear amplification DNA sequencing directly from single phage plaques and bacterial colonies. *Nucleic Acids Res.* **19,** 1153.
10. Young, A. and Blakesley, R. (1991) Sequencing plasmids from single colonies with the dsDNA cycle sequencing system. *Focus (BRL)* **13(4),** 137.
11. Creasey, A., D'Angio, L., Dunne, T. S., Kissinger, C., O'Keeffe, T., Perry-O'Keefe, H., Moran, L. S., Roskey, M., Schildkraut, I., Sears, L. E., and Slatko, B. (1991) Application of a novel chemiluminescence-based DNA detection method to single-vector and multiplex DNA sequencing. *BioTechniques* **11,** 102–109.
12. *Biosystems Reporter* (Applied Biosystems, Inc., Foster City, CA) (1991) **13,** 1–2.
13. Slatko, B., Heinrich, P., Nixon, B. T., and Eckert, R. (1991) Preparation of templates for DNA sequencing unit 7.3, in *Current Protocols in Molecular Biology* (Ausubel, F., Brent, R., Kingston, R., Moore, D., Seidman, J., Smith, J., and Struhl, K., eds.), Wiley, New York.
14. Hultman, T., Bergh, S., Moks, T., and Uhlen, M. (1991) Bidirectional solid phase sequencing of in vitro amplified plasmid DNA. *BioTechniques* **10,** 84–93.
15. Silhavy, T., Berman, M., and Enquist, L. (1984) *Experiments With Gene Fusions.* Cold Spring Harbor Laboratory, Cold Spring Harbor, NY.
16. Ross, J. and Leavitt, S. (1991) Improved sample recovery in thermocycle sequencing protocols. *BioTechniques* **11,** 618–619.
17. Whitehouse, E. and Spears, T. (1991) A simple method for removing oil from cycle sequencing reactions. *BioTechniques* **11,** 616–618.

PART II

In Vitro Mutagenesis

Chapter 5

Ordered Deletions Using Exonuclease III

Denise Clark and Steven Henikoff

1. Introduction

An important manipulation in molecular genetics is to make ordered deletions into a cloned piece of DNA. The most widely used application of this method is in DNA sequencing. Ordered deletions can also be used in delineating sequences that are important for the function of a gene, such as those required for transcription. The principle behind using deletions for sequencing is that consecutive parts of a fragment cloned into a plasmid vector are brought adjacent to a sequencing primer site in the vector. Deletions are generated by digesting DNA unidirectionally with *Escherichia coli* exonuclease III (ExoIII) *(1)*. ExoIII digests one strand of double-stranded DNA by removing nucleotides from 3' ends if the end is blunt or has a 5' protrusion. A 3' protrusion of 4 bases or more is resistant to ExoIII digestion.

We present two procedures, outlined in Fig. 1, for preparation of the template for ExoIII digestion. Procedure I begins with double-stranded plasmid DNA. The DNA is digested with restriction endonucleases A and B, where A generates a 5' protrusion or blunt end next to the target sequence and B generates a 3' protrusion next to the sequencing primer site. The linearized plasmid DNA is digested with ExoIII, with aliquots taken at time points that will yield a set of deletions of the desired lengths. Procedure II *(2)* begins with single-stranded phagemid DNA. A nicked double-stranded circle is generated by annealing a primer to the phagemid so that its 5' end is adjacent to the insert and synthesizing the second strand with T4 DNA polymerase.

From: *Methods in Molecular Biology, Vol. 31: Protocols for Gene Analysis*
Edited by: A. J. Harwood

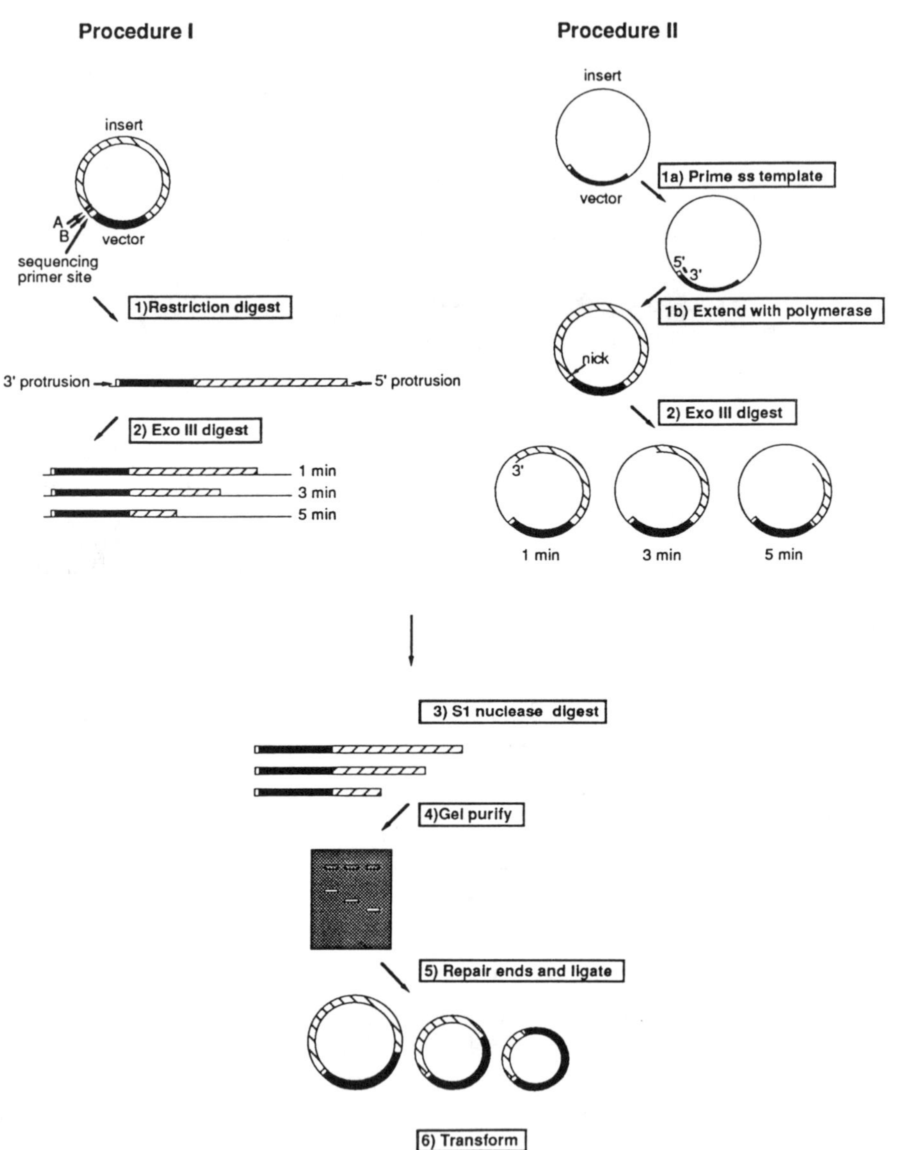

Fig. 1. Outline of the method.

The 3' end of the nicked strand is then digested with ExoIII. The remaining undigested single strands are digested with S1 nuclease. The deletion reactions are examined by electrophoresis in a low melting point agarose gel and the desired deletion products are isolated in gel slices

(3,4). This step serves to eliminate DNA products that do not contain the desired deletion. Repair of ends with Klenow DNA polymerase and ligation to circularize are performed on the DNA in diluted agarose. Following transformation into *E. coli*, clones are selected for sequencing.

An advantage of Procedure I is that deletions can be made into the insert from either end, so that both strands of the insert can be sequenced starting with a single plasmid clone. However, the requirement for multiple unique restriction sites in the vector polylinker cannot always be met, and this is where Procedure II is advantageous, since no restriction sites are necessary. This also means that, with Procedure II, a set of nested deletions can be made from any fixed point using a custom primer. A disadvantage of Procedure II is that, to sequence both strands of an insert, two clones with the insert in either orientation are required. Nevertheless, the two procedures can be used together to obtain the sequence of both strands from a single parent clone, reducing the need for multiple unique restriction sites. When planning a sequencing strategy, it is important to realize that although either single- or double-stranded templates for sequencing are possible for Procedure I, only double-stranded templates are possible for Procedure II. The reason is that for most phagemid vectors the primer for synthesis of the second strand of the nicked circle anneals to a site on the other side of the insert from the primer site that would be used for sequencing a single-stranded template.

2. Materials

2.1. General

Solutions are made with distilled water that when used for enzyme buffers is sterilized by autoclaving. Store solutions at room temperature, unless otherwise specified. Store enzymes as directed by the manufacturer.

2.1.1. Restriction Enzyme Digestion

1. 10 μg of double-stranded DNA: The fragment should be cloned into the multiple cloning site (polylinker) of a plasmid or phagemid vector such as Bluescript (Stratagene, La Jolla, CA). DNA can be prepared using an alkaline lysis procedure (*5*; *see* Note 1). A 10-mL liquid culture incubated overnight in LB media with the appropriate selection (e.g., ampicillin) should yield a sufficient amount of DNA.
2. Restriction enzymes and buffers: as supplied by manufacturer. Select a restriction enzyme (A in Fig. 1) that cuts the plasmid only in the

polylinker adjacent to the insert and yields a 5' protrusion or blunt end. Restriction enzyme B must also cut at a unique site in the polylinker and yield a 4 base 3' protrusion to make the end of the linearized fragment nearest the sequencing primer site resistant to ExoIII. If such a site is not available, a 5' protrusion can be generated that is filled in with α-phosphorothioate nucleotides using Klenow fragments *(6)*.
3. Phenol:Chloroform solution: 1:1 mix; store at 4°C.
4. 10*M* Ammonium actetate.
5. Ethanol: both 95% and 70% (v/v).
6. 10X ExoIII buffer: 660 m*M* Tris-HCl, pH 8.0, 6.6 m*M* $MgCl_2$.

2.1.2. Procedure II: Synthesis of a Nicked Circular Plasmid

7. Template: 2.5 µg single-stranded phagemid DNA, 0.2–2.0 µg/µL. A small-scale preparation (2–3 mL culture) should yield a sufficient amount of DNA *(5)*.
8. Primer: α-T3 20-mer (5' CCCTTTAGTGAGGGTTAATT 3'), or the equivalent, e.g., reverse hybridization 17-mer (5' GAAACAGCTATGACCAT 3') at 4 pmol/µL; store at –20°C.
9. T4 DNA polymerase: 1–10 U/µL, cloned or from T4-infected cells.
10. 10X TM: 660 m*M* Tris-HCl, pH 8.0, 30 m*M* $MgCl_2$.
11. 2.5 m*M* dNTPs: 2.5 m*M* each of the 4 deoxynucleoside triphosphates; store at –20°C.
12. BSA: 1 mg/mL bovine serum albumin (nuclease-free); store at –20°C.

2.2. Exonuclease III and S1 Nuclease Digestion

13. *E. coli* exonuclease III: 150–200 U/µL (1 U = amount of enzyme required to produce 1 nmole of acid-soluble nucleotides in 30 min at 37°C).
14. S1 buffer concentrate: 2.5*M* NaCl, 0.3*M* potassium acetate (titrate to pH 4.6 with HCl, not acetic acid), 10 m*M* $ZnSO_4$, 50% (v/v) glycerol.
15. S1 nuclease: e.g., 60 U/µL from Promega (1 U = amount of enzyme required to release 1 µg of perchloric acid-soluble nucleotides per minute at 37°C). Mung bean nuclease may also be used.
16. S1 mix: 27 µL S1 buffer concentrate 173 mL of H_2O, 1 µL (60 U) of S1 nuclease. Prepare just before use and store on ice.
17. S1 stop: 0.3*M* Tris base (no HCl), 50 m*M* EDTA.

2.3. Gel Purification, End Repair, Ligation, and Transformation

18. Agarose: Low melting point agarose. Use a grade appropriate for cloning, e.g., Seaplaque GTG agarose (FMC).
19. Ethidium bromide: 10 mg/mL; store in a dark container.

20. 50X TAE gel running buffer: 1 L = 242 g Tris base, 57.1 mL glacial acetic acid, 100 mL 0.5*M* EDTA (pH8). When diluted this gives a 1X buffer of 40 m*M* Tris-acetate, 1 m*M* EDTA.
21. 10X Gel-loading buffer: 0.25% bromophenol blue, 50% (v/v) glycerol in distilled water.
22. 10X Klenow buffer: 20 m*M* Tris-HCl, pH 8.0, 100 m*M* $MgCl_2$.
23. Klenow enzyme: the large fragment of the *E. coli* DNA polymerase (2 U/µL). Store at –20°C.
24. Klenow mix: For every 10 µL 10X Klenow buffer, add 1 U of Klenow enzyme. Prepare just before use and store on ice.
25. dNTPs: 0.125 m*M* of each deoxyribonucleoside; store at –20°C.
26. 10X Ligase buffer: 0.5*M* Tris-HCl, pH 7.6, 100 m*M* $MgCl_2$, 10 m*M* ATP; store at –20°C.
27. PEG: 50% (w/v) polyethylene glycol 6000–8000 fraction, store at 4°C.
28. DTT: 0.1*M* dithiothreitol; store at –20°C.
29. T4 DNA ligase: (1 U/µL). Store at –20°C.
30. Ligation cocktail (1 mL): 570 µL H_2O, 200 µL ligase buffer, 200 µL PEG, 20 µL 0.1*M* DTT, 10 U of T4 ligase. Prepare just before use and store on ice.
31. Host *E. coli* cells: Competent recA$^-$ cells, e.g., DH5α.
32. Growth media: LB medium *(5)*.

3. Methods

3.1. Preparation of Exonuclease III Substrate

3.1.1. Procedure I: Restriction Enzyme Digestion

1. Digest 10 µg of plasmid to completion in a 100 µL volume. At least 200 ng is required for each ExoIII digestion time point (*see* Note 1).
2. To extract the DNA, vortex in 100 µL of phenol:chloroform and remove the aqueous layer to a fresh tube.
3. Add 25 µL of 10*M* ammonium acetate and 200 µL of 95% ethanol. Chill on ice for 15 min and pellet the DNA in a microfuge (12,000*g*) for 5 min.
4. Wash pellet with 70% ethanol and air dry. Resuspend the DNA pellet in 90 µL of distilled H_2O and 10 µL of 10X ExoIII buffer (i.e., 100 ng/µL).

3.1.2. Procedure II: Synthesis of a Nicked Circular Phagemid

1. Mix in a volume of 22 µL: approx 2.5 µg single-stranded phagemid DNA, 4 µL of 10X TM, 1 µL (4 pmole) of α-T3 primer (or equivalent).
2. Heat to 75°C for 5 min, then cool slowly over 30–60 min. Evaporation and condensation can be minimized by placing a piece of insulating

foam over the tube in an aluminum tube-heating block. Remove a 2 µL aliquot for subsequent gel analysis.
3. Mix in a volume of 20 µL: 2 µL of DTT, 4 µL of 2.5 m*M* dNTPs, 4 µL of BSA, and 5 U of T4 DNA polymerase.
4. Prewarm to 37°C and add to the primed DNA at 37°C. Incubate 2–8 h. An example of a nicked circle is shown in Fig. 2. When the plasmid is as large as this one, we recommend supplementing the extension reaction with an additional 0.4 m*M* dNTPs and 0.1 U/µL polymerase after 4–6 h, and then incubation overnight. Extension can be monitored by removing a 0.5-µL aliquot and electrophoresing it on an agarose gel.
5. Inactivate polymerase by heating for 10 min at 70°C and store on ice.

3.2. Exonuclease III Digestion

Deletions separated by 200–250 bases are required for obtaining contiguous sequence with some overlap, therefore to sequence a 4-kb insert, 16–20 time points are desirable (*see* Note 2). The following protocol is written to give 16 time points.

1. Prepare 16 7.5-µL aliquots (*see* Note 3) of S1 nuclease mix on ice, one for each time point.
2. Place 40 µL of DNA (from Section 3.1.4. or 3.2.5.) into a single tube. Warm the DNA to 37°C in a heating block (*see* Note 3). Add 1–2 µL of ExoIII (5–10 U/µL), mix rapidly by pipeting up and down.
3. Remove aliquots of 2.5 µL at 30-s intervals and add into the S1 mix tubes on ice. Mix by pipeting up and down and hold on ice until all aliquots are taken.
4. Remove samples from ice and incubate at room temperature for 30 min.
5. Add 1 µL of S1 stop and heat to 70°C for 10 min to inactivate the enzymes.

3.3. Gel Purification, End Repair, Ligation, and Transformation

1. Add 1 µL of 10X gel loading buffer to each sample and load onto a 0.7% low melting point agarose gel in 1X TAE gel running buffer containing 0.5 µg/mL ethidium bromide. Electrophorese for about 2 h at 3–4 V/cm until the deletion time points can be resolved (*see* Note 4). Using a 300-nm wavelength ultraviolet light source, remove the desired bands from the gel with a small spatula or scalpel.
2. Dilute gel slices in approx 2 vol of distilled water and melt at 68°C for 5 min. Transfer 1/3 of the diluted DNA to a new tube containing 0.1 vol Klenow mix at room temperature and mix by pipeting up and down. Incubate at 37°C for 3 min. Add 0.1 vol of dNTPs and mix. Incubate for 5 min at 37°C (*see* Note 5).

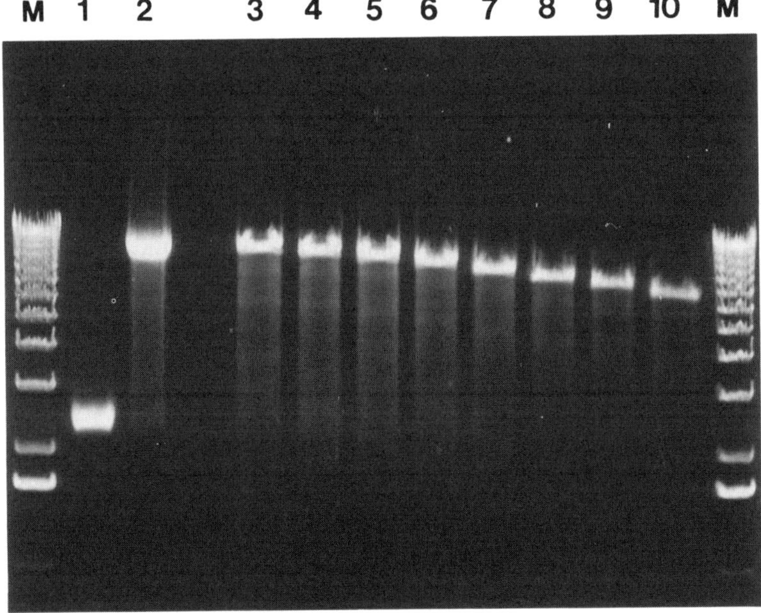

Fig. 2. Low melting point agarose gel electrophoresis of a set of deletions prepared following Procedure II. The size of the phagemid is 9.5 kb. Lane 1: Single-stranded phagemid. Lane 2: Double-stranded nicked circle after overnight extension with T4 DNA polymerase. Depending on the gel, one occasionally sees a minor band below the nicked circle that may be the result of extension of single-stranded linear molecules that are primed by their own 3' end *(2)*. Lanes 3–10: ExoIII digestion of the material in Lane 2 for 1, 2, . . ., 8 min. Lane M: 1-kb ladder (BRL), with the 7-kb band comigrating with the 8 min time point.

3. Add an equal volume of ligation cocktail, mix, and incubate at room temperature for at least 1 h. The yield of transformants is greatest if ligations are incubated overnight.
4. Transform *E. coli* by combining 1/2 of each ligation with 2–3 vol of competent cells that have been thawed on ice. Incubate for 30 min on ice, heat shock at 42°C for 90 s and further incubate the cells in 0.2 mL LB medium for 60 min at 37°C. Plate using the appropriate selection and incubate at 37°C overnight. The yield from each transformation can be up to 500 colonies, depending on ligation time and competence of the cells. Store the remaining ligation mixtures at –20°C.
5. Set up cultures to prepare plasmid DNA from one colony from each time point and verify that each one contains a deletion of expected size before sequencing.

4. Notes

1. ExoIII digestion occurs at nicks in the DNA as efficiently as at ends. This gives rise to a background of undesired clones. Nicks can be introduced by impurities in the DNA and by restriction enzymes. Gel purification eliminates this background. As a result of the gel purification step, plasmid DNA can be prepared with a standard alkaline lysis procedure, including a phenol:chloroform extraction *(4)*. If nicking proves to be a problem, supercoiled DNA may be column purified using the premade columns from Qiagen, Inc. (Chatsworth, CA). Restriction enzymes can also occasionally have excessive nicking or "star" activity. Often this problem can be overcome by trying different suppliers.
2. At 37°C, ExoIII digests 400–500 bases/min, with the rate changing directly with temperature about 10%/1°C in the 30–40°C range *(7)*. It is advisable to perform a test run of the ExoIII digestion on your DNA, picking a few time points. The three reasons for this are:
   a. You can confirm that both enzymes cut to completion if using Procedure I (*see also* Note 4);
   b. You can get an accurate rate of digestion for your particular batch of ExoIII; and
   c. You can confirm that there is not excessive nicking of your DNA (*see* Note 3).
3. When processing a large number of time points, it is much more efficient to use a conical bottom microtiter plate with a lid rather than microfuge tubes. These plates can be used for the S1 nuclease step, collecting and diluting gel slices, end repair, ligation, and transformation.
4. Occasionally, an ExoIII resistant fragment is seen. This results from incomplete digestion by enzyme A (Fig. 1). Provided this is not a dominant band, the gel isolation of the digested fragment should eliminate this as a problem. Alternatively, a second fragment that digests at a higher rate may be present. This results from incomplete digestion by enzyme B and subsequent ExoIII digestion of the fragment from both ends. Again, if this is not a dominant product, gel isolation should alleviate this problem.
5. An alternative method for in-gel ligation incorporates end repair and ligation into one step. For 1 mL of cocktail, mix 560 µL of water, 200 µL of ligation buffer, 200 µL PEG, 20 µL DTT, 20 µL each of dNTP (2.5 m*M*), 1 U of T4 DNA polymerase, and 10 U of T4 DNA ligase. Mix 1/3 of the DNA in the diluted gel slice with an equal volume of the cocktail. Incubate at room temperature 1 h to overnight. Transformation is done following step 6.

References

1. Henikoff, S. (1984) Unidirectional digestion with exonuclease III creates targeted breakpoints for DNA sequencing. *Gene* **28,** 351–359.
2. Henikoff, S. (1990) Ordered deletions for DNA sequencing and in vitro mutagenesis by polymerase extension and exonuclease III gapping of circular templates. *Nucleic Acids Res.* **18,** 2961–2966.
3. Nakayama, K. and Nakauchi, H. (1989) An improved method to make sequential deletion mutants for DNA sequencing. *Trends Genet.* **5,** 325.
4. Steggles, A. W. (1989) A rapid procedure for creating nested sets of deletions using mini-prep plasmid DNA samples. *Biotechniques* **7,** 241–242.
5. Sambrook, J., Fritsch, E. F., and Maniatis, T. (1989) *Molecular Cloning: A Laboratory Manual.* Cold Spring Harbor Laboratory, Cold Spring Harbor, NY.
6. Putney, S. D., Benkovic, S. J., and Schimmel, P. R. (1981) A DNA fragment with an α-phosphorothioate nucleotide at one end is asymmetrically blocked from digestion by exonuclease III and can be replicated in vivo. *Proc. Natl. Acad. Sci. USA* **78,** 7350–7354.
7. Henikoff, S. (1987) Unidirectional digestion with exonuclease III in DNA sequence analysis. *Meth. Enzymol.* **155,** 156–165.

CHAPTER 6

Site-Directed Mutagenesis Using a Double-Stranded DNA Template

Stéphane Viville

1. Introduction

It is now technically possible to create almost any desired mutation in a given DNA sequence. So called site-directed mutagenesis allows the introduction of designed mutations into specific locations. This approach is invaluable for studying gene regulation as well as for functional assessment of proteins and their interactions. Several protocols have been successfully employed to generate such mutants. Here I describe a "linker-scanning" method that I have used to systematically mutate the murine Major Histocompatibility Complex (MHC) class II Eα gene promotor (Fig. 1, ref. *1*).

1.1. General Considerations

All site-directed mutagenesis is based on the use of synthetic oligonucleotides that contain the desired mutations. Although single-stranded DNA vectors (generally M13; *2*) and the polymerase chain reaction (PCR; *3,4*) have been most widely employed, the linker scanning method offers several advantages. Because mutations are introduced directly into a double-stranded plasmid template, the amount of subcloning required is minimal and PCR errors are completely eliminated. In addition, double-stranded templates allow cloning of larger target fragments and are much more stable than single-stranded templates.

In principal, this protocol is similar to those involving single-stranded DNA templates. Basically, a frame of single-stranded DNA encompassing only the region of interest is created within a double-

From: *Methods in Molecular Biology, Vol. 31: Protocols for Gene Analysis*
Edited by: A. J. Harwood

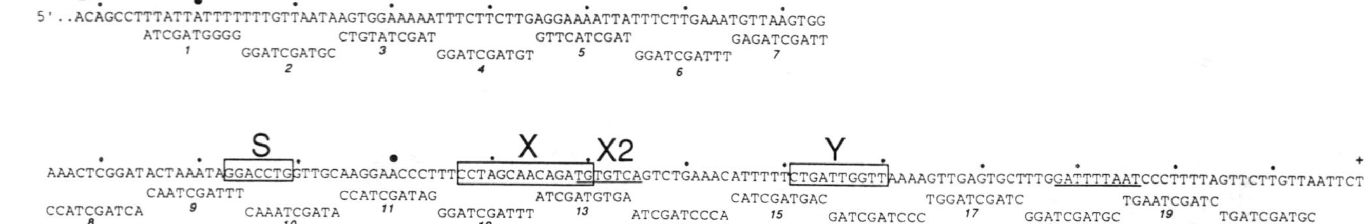

Fig. 1. Linker-scanning mutation of the Eα promoter. The sequence of each of the 10-bp linker-scanning mutations (numbered 1–20) is shown under the wild-type Eα promoter sequence. Class II regulatory motifs are underlined or boxed (reproduced from ref. *1*).

stranded plasmid. The mutagenic oligonucleotide is used as a primer for the Klenow fragment of DNA polymerase I (Klenow) that synthesizes the second strand of the target region, and free ends are joined by T4 DNA ligase. To generate the single-stranded target sequence, two distinct forms of the template plasmid are produced by digestion with appropriate restriction enzymes. For the first, the target sequence is completely excised from the plasmid; for the second, the plasmid is linearized outside the target sequence. The two forms are mixed, denatured, and allowed to reanneal. During this process four hybrids are produced: two correspond to the original forms and two are heteroduplexes, each containing a stretch of single-stranded DNA encompassing the target sequence (Fig. 2). The mutagenic oligonucleotide will anneal to one of these heteroduplexes. Incubating a mixture of hybrid plasmids and mutagenic oligonucleotides with Klenow and T4 ligase will engender two types of viable plasmids: wild type and those carrying the desired mutation. Using the mutagenic oligonucleotide as a probe, it is fairly straightforward to screen colonies for the mutated plasmid. In practice, 1–20% of the transformants carry the mutation. The protocol described below is based on the previously described "gapped heteroduplex" method *(5)* that we have simplified at several steps.

2. Materials

This protocol requires at least two or three restriction endonucleases, the Klenow fragment of DNA polymerase I, T4 ligase, polynucleotide kinase, and calf intestinal phosphatase (CIP). These enzymes are commonly used in a variety of molecular biology protocols and are commercially available. Complete descriptions of their activities are outlined in vol. 16 of this series. Storage, working conditions, and often appropriate buffers are supplied by the manufacturers. Solutions sould be prepared using deionized H_2O and molecular biology-grade reagents. Use high quality phenol chloroform and ethanol.

2.1. Preparation of the Plasmid Fragment

1. Phenol: Phenol must be equilbrated to pH 7.0–7.5 with Tris-HCl prior to use. (Follow the protocol recommended by the supplier or in vol. 16 of this series.) Equilibrated phenol can be stored (under H_2O) at 4°C for 1 or 2 mo, but should be kept in aliquots at –20°C for long-term storage.
2. 3*M* Na acetate, pH 5.6: Sterilize by autoclaving; store at room temperatuture.

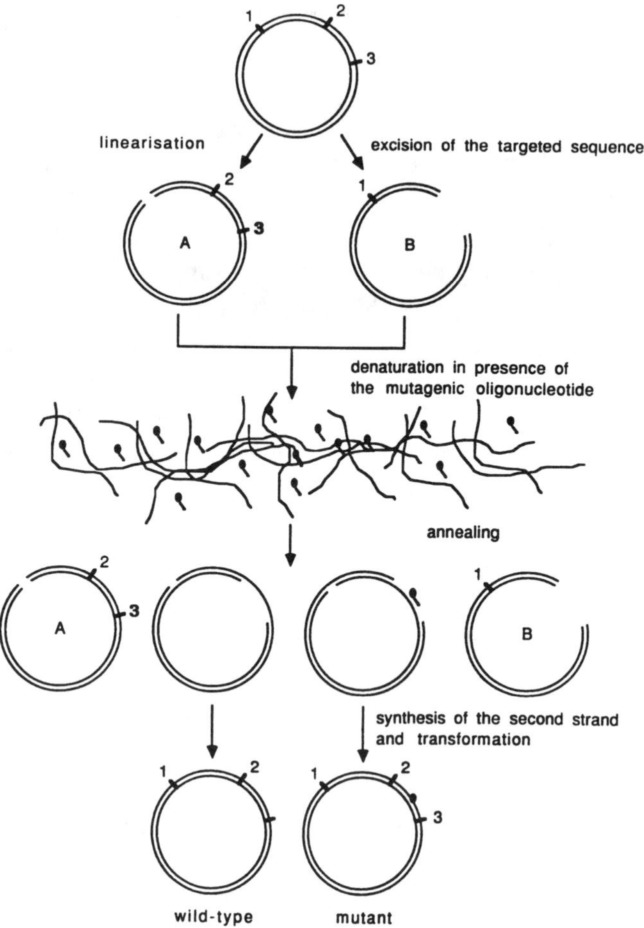

Fig. 2. Site directed mutagenesis using a double-stranded DNA plasmid. This technique is based on the creation of a stretch of single-stranded DNA within double-stranded plasmids. To accomplish this, two forms of the plasmid are produced: **(A)** the plasmid is linearized outside of the target sequence, and **(B)** the target sequence is excised. When these two DNA fragments are denatured and allowed to reanneal in the presence of the mutagenic oligonucleotide, four distinct hybrids are formed. Two are heteroduplexes containing a stretch of single-stranded DNA. The mutagenic oligonucleotide hybridizes to one of these and serves as a primer synthesis of the second strand. In theory, 33% of the transformed bacteria should contain the mutant plasmid.

3. TE buffer: 10 m*M* Tris-HCl, pH 7.4, 1 m*M* EDTA. Store at room temperature.
4. GeneClean™ Kit: Available from Bio 101, Box 2284, La Jolla, CA 92038-2284.

2.2. Phosphorylation of the Mutagenic Oligonucleotide

5. Mutagenic oligonucleotide: Appropriate design of the mutagenic oligonucleotides is critical; each must contain an informative mutation along with sufficient homology to facilitate hybridization to the target plasmid (*see* Note 1). If you do not have access to an oligonucleotide synthesizer, oligonucleotides are commercially available from a variety of companies. Adjust stock concentration to 10 pmole/μL.
6. ATP: 10 m*M* adenosine triphosphate in TE buffer. Aliquot and store at –20°C.
7. 10X nick translation NT buffer: 0.5*M* Tris-HCl, pH 7.5, 0.1*M* $MgSO_4$, 1 m*M* dithiothreitol, 500 μg/mL bovine serum albumin (Fraction V, Sigma, St. Louis, MO). Store in small aliquots at –20°C.

2.3. Site-Directed Mutagenesis

8. 1*M* NaCl: Sterilize by autoclaving; store at room temperature.
9. dNTP mix: 5 m*M* of each deoxynucleotide (dGTP, dATP, dCTP, and dTTP) in TE buffer. Store in small aliquots at –20°C.
10. Competent *E. coli*: We have sucessfully used both strains MC1061 and DH5 (*see* Note 2).
11. 2X YT medium: To make 1 L dissolve 16 g of bacto-tryptone, 10 g yeast extract, and 10 g NaCl. Sterilize by autoclaving; store in a dark place.
12. 15% 2X YT agar: To 1 L of 2X YT medium add 15 g of agar just prior sterilization; pour into sterile Petri dishes (30 mL/90 mm dish) after it has cooled to 50°C. For antibiotic plates add the amount of antibiotic required immediately before pouring the medium (*see* Note 3).

3. Methods

3.1. Preparation of the Plasmid Fragments

3.1.1. Linearizing the Plasmid

As stated above, the plasmid containing the target must be linearized at a unique restriction site located outside of the target sequence. Subsequent treatment of the linearized plasmid with alkaline phosphatase, to eliminate the two 5' phosphates that are required for religation of the plasmid, markedly decreases the efficiency of religation.

1. Digest 10 μg of plasmid with the chosen enzyme; generally incubation with 25 units of enzyme (2.5 U/μg) for 2 h under appropriate conditions (in a volume of 100 μL) is sufficient. A 2–3 μL aliquot should be run on an agarose gel to ensure that digestion is complete (*see* Note 4).

2. Increase the volume of the digestion reaction to 200 μL with sterile, deionized H_2O and add 0.5 U of CIP. Incubate for 30 min at 37°C. Inactivate the enzyme by heating the mixture for 15 min at 65°C.
3. Extract once with an equal volume of phenol:chloroform (1:1) and then with an equal volume of chloroform. Add 1/10 vol (20 μL) of 3*M* Na acetate (pH 5.6) and precipitate the DNA with 2.5 vol (500 μL) of 100% ethanol.
4. Pellet the DNA in a microfuge for 10 min and wash once with 500 μL of 70% ethanol. Dry, and resuspend in 20 μL of TE.
5. To estimate the DNA concentration run a 2-μL sample on an agarose gel with standard fragments of known concentration. The concentration should be approx 400–500 ng/μL. Hereafter, I will refer to this fragment as "A" (DNA frgt. A, Fig. 2).

3.1.2. Excision of the Target Locus

The target region should be between 400 bp and 1 kb in length and can be excised by one enzyme for which two sites exist (one on each side of the sequence of interest) or by two enzymes (which cut at two unique restriction sites that flank the target locus, *see* Note 5). If one enzyme is used, CIP treatment will be required; if two are used this will not be necessary unless they generate cohesive ends (which is rare).

1. Digest 10 μg of the plasmid as in Section 3.1.1.
2. Separate the two fragments on an appropriate agarose gel and cut out slice of agarose containing the plasmid with the targeted region excised.
3. Purify using the GeneClean kit (*see* Note 6).
4. Evaluate the concentration of the DNA solution by electophoresis as in Section 3.1.1. It is preferable to run frgt. A alongside. This fragment will be referred to as DNA fragment B (DNA frgt. B, Fig. 2).

3.2. Phosphorylation of the Mutagenic Oligonucleotide

To serve as a substrate for T4 DNA ligase, the mutagenic oligonucleotide must be phosphorylated.

1. Add 7 μL of oligonucleotide (10 pmole/μL), 1 μL of 10 m*M* ATP, 1 μL of 10X NT buffer, 1 U of T4 polynucleotide kinase, and sterile dH_2O to final volume of 10 μL.
2. Incubate for 30 min at 37°C. To inactivate the enzyme, incubate further at 65°C for 10 min (*see* Note 7).

No further purification is required.

3.3. Site-Directed Mutagenesis

1. To create the heteroduplex plasmid and allow hybridization of the mutagenic oligonucleotide, assemble 200–400 ng of DNA frgt. A, 200–400 ng of DNA frgt. B, 2.5 μL of 10X NT buffer, 2.5 μL of 1*M* NaCl, 3 μL of 5'-phosphorylated oligonucleotide (21 pmole) and sterile dH_2O to final volume of 24 μL. Remove 8 μL to use as a control later.
2. Denature-anneal the fragments by heating the mixture to 100°C for 3 min. Put at room temperature for 20 min and then cool on ice for 30 min. Centrifuge for 30 s in a microfuge.
3. Remove 8 μL and run on a 1% agarose gel with the control aliquot from Section 3.3., step 1.

   If heteroduplex plasmids have formed, an additional band (larger than the two initial fragments) will be apparent in the denatured/annealed DNA sample (*see* Fig. 3). If this band is not visible, do not proceed.
4. Mix 8 μL of annealed fragments, 2 μL of 10 m*M* ATP, 2 μL of 5 m*M* dNTP, 0.5 μL of Klenow enzyme (4 U/mL), 0.2 μL of T4 DNA ligase (0.6 U/μL), and sterile dH_2O to a final volume of 15 μL. Incubate this mixture overnight (or greater than 8 h) at 15°C.
5. Mix 7 μL of the elongation-ligation mixture with 100 μL of competent *E. coli* (*see* Note 8) and incubate for 20 min at 0°C in ice/water.
6. Follow by a heat shock for 3 min at 37°C. Dilute cells with 700 μL of 2X YT and incubate for 45 min at 37°C with shaking.
7. Plate 200 μL of the bacteria onto each of four agar plates containing an appropriate antibiotic (*see* Notes 9 and 10).

Screening of transformants can be carried out in a number of ways, but one of the most reliable is that described by Hanahan and Meselson *(6)*. Here, the colonies are tranferred to a membrane (nitrocellulose or nylon), the bacteria lysed, and DNA denatured with NaOH. The DNA is fixed to the membrane and can then be hybridized with the mutagenic oligonucleotide. Detailed, reliable protocols can also be found in vol. 16 of this series.

Once the clones of interest have been identified, it is a good idea to confirm the mutation by sequencing (*see* Note 11). Double-stranded sequencing can be carried out on DNA from minipreps following the method described by Stephen et al. *(7)*.

4. Notes

1. In general, the length of homologous sequence should be proportional to the mismatches that create the mutation, and each should be between

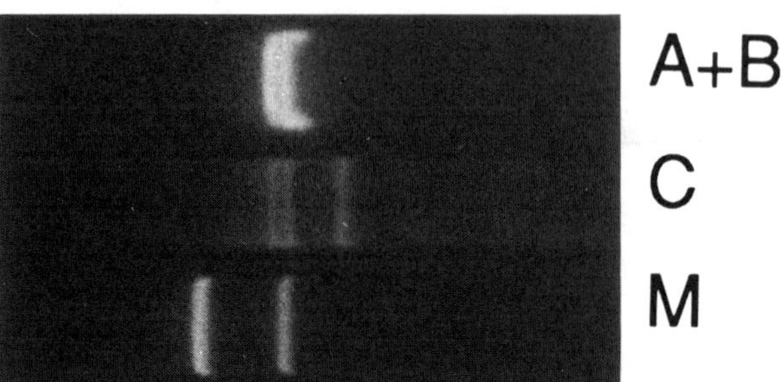

Fig. 3. Assessment of the denaturation/annealing step. If denaturation/annealing of the two DNA fragments A and B results in the formation of heteroduplexes, an additional band C should be apparent when the annealed mixture is run on an agarose gel. Lane 1, fragments A and B prior to denaturation/annealing. Lane 2, fragments A and B after denaturation/annealing. Lane 3, size markers (M).

5 and 15 bp. Long oligonucleotides (>35 bp) are expensive and occasionally difficult to synthesize. It is possible to design the oligonucleotides such that a restriction site is either created or eliminated in the mutant. This strategy simplifies the characterization of the mutants and can facilitate further manipulation of the DNA.

2. Many bacterial strains are available and can be used.
3. For ampicilin add 1 mL of a 50 mg/mL solution of ampicillin to 1 L of 2X YT agar.
4. Avoid loading samples directly from the refrigerator or freezer since in some cases the DNA may form some secondary structures that are stable at low temperature. Heat samples 5 min at 65°C to eliminate this effect.
5. If the enzymes require different NaCl concentrations, digest first at the lower concentration, then increase the volume and adjust the NaCl concentration for the second enzyme.
6. Any method for purifying DNA from agarose gels can be used.
7. This reaction can be monitored by adding 1 µL of (γ-[^{32}P])ATP (3000 Ci/mmole) to the reaction. If the reaction works properly, the ^{32}P ATP will be incorporated; this can be assessed by TCA precipitation or by running a small sample on a polyacrylamide gel and exposing it to film.
8. Competent bacteria with transformation efficiency greater than 10^6 colonies/µg of DNA should be used.
9. All transfections should include at least two controls: one sample in which no DNA is added and one in which 1 ng of an intact plasmid is present.

10. There could be many reasons for the lack of colonies after the transfection: Bacteria are not competent; Klenow and/or T4 DNA ligase could be deficient.
11. Two main problems can occur with site-directed mutagenesis. First, it is possible not to get the mutation, which is generally owing to the poor quality of the DNA fragments or of the oligonucleotide used. Repurification of the DNA is recommended before repeating the experiment. The oligonucleotide can be purified by loading it on an acrylamide gel. Second, it is possible to get the mutation at the wrong site. In this case it is because the homologous sequences of the mutagenic oligonucleotide are not long enough to be site specific. New oligonucleotide should be prepared.

Acknowledgments

I am very grateful to S. Gilfillan and R. Mantovani for their helpful discussions and critical reading of the manuscript.

References

1. Viville, S., Jongeneel, V., Koch, W., Mantovani, R., Benoist, C., and Mathis, D. (1991) The Eα promoter: a linker-scanning analysis. *J. Immunol.* **146,** 3211–3217.
2. Wallace, R. B., Schold, M., Johnson, M. J., Dembek, P., and Itakura, K. (1981) Oligonucleotide directed mutagenesis of the human β-globin gene: a general method for producing specific point mutations in cloned DNA. *Nucleic Acids Res.* **9,** 3647–3656.
3. Nelson, R. M. and Long, G. L. (1989) A general method of site-specific mutagenesis using a modification of the *Thermus aquaticus* polymerase chain reaction. *Anal. Biochem.* **180,** 147–151.
4. Hemsley, A., Arnheim, N., Toney, M. D., Cortopassi, G., and Galas, D. J. (1989) A simple method for site directed mutagenesis using the polymerase chain reaction. *Nucleic Acids Res.* **16,** 6545–6551.
5. Inouye, S. and Inouye, M. (1987) Oligonucleotide-directed site-specific mutagenesis using double-stranded DNA, in *Synthesis and Application of DNA and RNA* (Narang, S. A., ed), Academic, Orlando, FL, pp. 181–206.
6. Hanahan, D. and Meselson, M. (1980) Plasmid screening at high colony density. *Gene* **10,** 63–67.
7. Stephen, D., Jones, C., and Schofield, J. P. (1990) A rapid method for isolating high quality plasmid DNA suitable for DNA sequencing. *Nucleic Acids Res.* **18,** 7463–7464.

CHAPTER 7

Site-Directed Mutagenesis Using a Uracil-Containing Phagemid Template

Michael K. Trower

1. Introduction

Site-directed mutagenesis is a powerful technique by which specific changes can be generated in a target DNA molecule for either structure–function investigations or for creating and deleting endonuclease restriction sites. There have been several oligonucleotide-directed mutagenesis procedures described in the literature that allow for selection against the wild-type DNA, thereby giving enrichment in those sequences carrying the desired mutation *(1–4)*. Collectively, these methods suffer from a number of drawbacks, including multi-enzymatic steps, the necessity for DNA precipitation, the need for convenient restriction sites, and a requirement for specialized vectors. An elegant method for achieving site-directed mutagenesis has been described by Kunkel et al. *(5,6)* that avoids these problems and at the same time produces a high percentage of DNA molecules with the designed mutation. This process is schematically outlined in Fig. 1. It relies on the observation that an *E. coli* strain lacking two key enzymes, dUTPase *(dut)* and uracil *N*-glycosylase *(ung)*, will synthesize DNA with a small number of uracil bases substituted for thymine. Uracil-containing single-stranded DNA (ssDNA) prepared from this *dut*$^-$, *ung*$^-$ host strain can be used as a template for a mutagenic oligonucleotide to prime in vitro DNA synthesis of a complementary strand. When this heteroduplex is transformed into a wild type *(dut*$^+$,

From: *Methods in Molecular Biology, Vol. 31: Protocols for Gene Analysis*
Edited by: A. J. Harwood

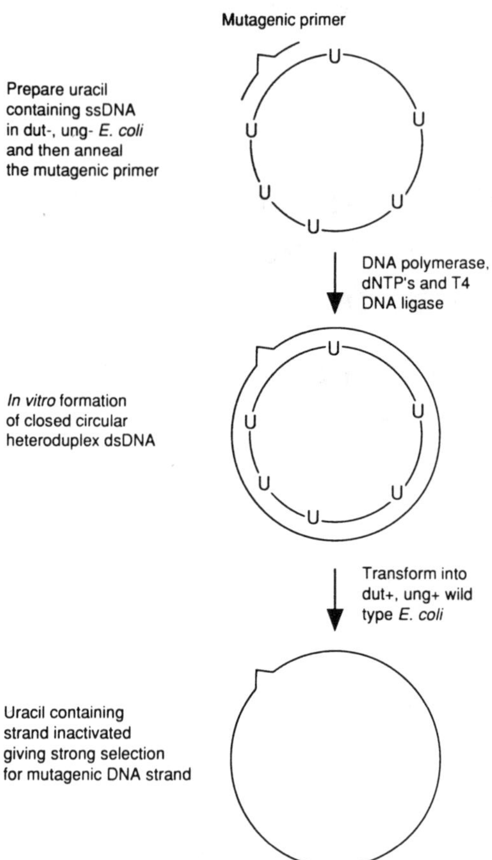

Fig. 1. Site-directed mutagenesis using the Kunkel method. A uracil-containing ssDNA is prepared by passage through a *dut*⁻, *ung*⁻ *E. coli* host strain. The template is primed in vitro with the mutagenic oligonucleotide and then a dsDNA heteroduplex is generated following the addition of a DNA polymerase and T4 DNA ligase. The heteroduplex is transformed into a *dut*$^+$, *ung*$^+$ wild-type *E. coli* that selects for the newly synthesized mutagenic DNA strand through preferential inactivation of the uracil-containing DNA strand.

ung$^+$) *E. coli* the uracil template strand is selectively inactivated resulting in a majority of the progeny having the designed mutation. Described below is a protocol based on the Kunkel method, which can avoid the necessity of two rounds of DNA subcloning; the first into a vector for the mutagenesis reaction, usually M13, and then from there into a suitable vector to investigate the effects of the changes made in the target DNA sequence. This is achieved by using plasmid vectors that contain the f1 origin of replication *(7)*. These so called phagemids propagate as normal plasmids in *E. coli,* unless the host cell is infected with a "helper" phage (*see* Note 1). In such circumstances the helper phage interacts with the f1 origin of replication so that ssDNA copies of one strand of the phagemid DNA are produced and nonlytically secreted from the cell where they may be recovered from the supernatant *(8)*. Phagemid ssDNA generated in this way by passage through a *dut*$^-$, *ung*$^-$ *E. coli* host will incorporate uracil residues and therefore can be used in the Kunkel method as templates for oligonucleotide-directed mutagenesis.

2. Materials

Solutions are stored at room temperature unless stated otherwise.

2.1. Preparation of the Uracil-Containing ssDNA Template

1. *E. coli* strains:
   a. CJ236, a *dut*$^-$, *ung*$^-$ strain containing the F' episome or other host with the same genotypic features.
   b. Wild type *dut*$^+$, *ung*$^+$ strain such as DH5α, JM109, NM522, TG1 or XL1-Blue. The wild type host need not have a F' episome.
2. Helper phage: M13K07 or VCSM13. These typically have a titer of 10^{11}–10^{12} phage particles/mL.
3. TBG medium: To make 1 L, add 12 g of tryptone, 24 g of yeast extract, and 4 mL of glycerol in a total volume of 882 mL of distilled water and autoclave. When cool add 100 mL of sterile 0.17*M* KH_2PO_4, 0.72*M* K_2HPO_4 solution, and 18 mL of 20% glucose solution. The 20% glucose solution should be prepared and autoclaved separately.
4. Kanamycin: a 75 mg/mL (i.e., 1000×) stock in water. Store at –20°C.
5. PEG/NaCl solution: 20% Polyethylene glycol 6000 or 8000, 2.5*M* NaCl.
6. TE buffer: 10 m*M* Tris-HCl, pH 7.4, 1 m*M* EDTA. Autoclave.
7. Phenol: double-distilled and saturated with TE buffer. Store at 4°C.
8. 3*M* Sodium acetate, pH 4.8–5.2.

2.2. Primer Phosphorylation, DNA Duplex Formation, and Transformation

9. T4 polynucleotide kinase (10 U/μL).
10. 10X T4 polynucleotide kinase buffer: 0.5*M* Tris-HCl, pH 7.5, 0.1*M* $MgCl_2$, 50 m*M* DTT, 0.5 mg/mL nuclease free BSA. Store at –20°C.

11. Klenow: The large fragment of DNA polymerase I (5 U/µL).
12. 10X Klenow polymerase buffer: 0.5*M* Tris-HCl, pH 7.4, 0.1*M* $MgCl_2$, 10 m*M* DTT, 0.25 mg/mL nuclease free BSA. Store at –20°C.
13. dNTP stock: 4 m*M* each of dATP, dCTP, dGTP, and dTTP. Store at –20°C.
14. ATP: 10 m*M* adenosine triphosphate solution. Store at –20°C.
15. T4 DNA ligase: 400 U/µL.
16. Mutagenic primer (*see* Notes 2 and 3): a synthetic oligonucleotide at 0.5 m*M*.

3. Methods

3.1. Preparation of the Uracil-Containing ssDNA Template

For mutagenesis a uracil-containing ssDNA template must be prepared. This can only be achieved if the target sequence is located in a cloning vector with an f1 filamentous phage origin of replication to allow recovery of the phagemid ssDNA template. Furthermore, the *dut*$^-$, *ung*$^-$ *E. coli* strain used must contain the F' episome to allow the helper phage to infect the host cells. The general protocol for template rescue is as follows:

1. Transform 10–50 ng of plasmid DNA, containing the insert with the target DNA sequence, into the *dut*$^-$, *ung*$^-$, F' *E. coli* strain CJ236 and select on an appropriate medium (*see* Note 4).
2. From the fresh transformation plate pick an isolated colony into 2 mL of TBG medium. Add approx 10^7–10^8 pfu/mL of either VCSM13 or M13KO7, helper phage. This gives a multiplicity of infection [moi] of 10–20 (*see* Note 5 and 6). If the plasmid can be selected with an antibiotic, such as ampicillin (100 µg/mL), this should also be included (*see* Note 7). Grow with vigorous aeration at 37°C for 2 h.
3. Add kanamycin to 75 µg/mL, since both VCSM13 and M13KO7 confer resistance against this antibiotic. This selects for cells infected with helper phage. Growth should then be continued overnight. Since the helper phage does not lyse the infected cells, the overnight culture should be saturated.
4. Transfer 1.4 mL of the overnight culture to a 1.5-mL Eppendorf tube and centrifuge at 12,000*g* for 5 min. Pour the supernatant into a fresh tube taking care not to transfer any cells. Respin for 5 min once more and again transfer the supernatant to a fresh tube. If the ssDNA cannot be processed at this time the supernatant may be stored for 24 h at 4°C, although it must be then recentrifuged before continuing with the protocol. The pellet can be discarded.

5. Add 200 µL of PEG/NaCl solution, mix and stand for 20 min on ice, and then centrifuge for 5 min in a microfuge. At this stage a small pellet should be visible. Discard the supernatant and then recentrifuge at 12,000*g* for 2 min. Carefully remove any residual PEG/NaCl solution with a micropipet tip.
6. Resuspend the pellet in 150 µL of TE buffer and add an equal volume of phenol. Vortex for 10–15 s and then stand the tubes for 5 min. Revortex and centrifuge at 12,000*g* for 5 min. Remove the upper phase and transfer to a fresh tube. Do not be concerned if the upper phase is very cloudy; this is quite common with these preparations.
7. Add 150 µL of chloroform, vortex for 10–15 s, and stand for 5 min. Revortex and then spin for 3 min. Remove the upper phase once more and again transfer to a fresh tube.
8. Add 15 µL of 3*M* sodium acetate, pH 4.8–5.2, and 300 µL of ethanol. Place the tube in either dry ice for 10 min or at –20°C overnight. Centrifuge at 12,000*g* for 10 min and pour off the liquid. Recentrifuge for 2 min, remove any remaining liquid with a micropipet tip, and then either vacuum or air dry. A small pellet should be visible. Redissolve the pellet in 50 µL of water or TE. If not used immediately store at –20°C. The yield of ssDNA from the CJ236 strain is usually around 10 µg DNA/mL of broth.

   Confirmation of template isolation and yield can be achieved by running a small aliquot (2–5 µL) on a 1% agarose gel. Two major bands, the helper phage ssDNA and the phagemid ssDNA, should be visible following ethidium bromide staining. The helper phage ssDNA should not interfere with the mutagenesis reaction.

3.2. Primer Phosphorylation, DNA Duplex Formation, and Transformation

The next stage involves phosphorylation of the mutagenic oligonucleotide, primer extension by the DNA polymerase with deoxynucleotides, and subsequent ligation of the nicked DNA to form in vitro a covalently closed circular relaxed heteroduplex dsDNA. An aliquot of this extension/ ligation mix is used to transform a wild type *dut*$^+$, *ung*$^+$ *E. coli* strain. This allows preferential selection for the newly synthesised DNA strand that incorporates the oligonucleotide directed mutations.

The mutagenic primer must be phosphorylated at the 5' position to ensure circularization of the heteroduplex, by ligation of the newly polymerized DNA strand to the mutagenic oligonucleotide. This reduces the extent of mismatch repair by *E. coli* host enzymes.

1. Mix 1.0 µL of 10X T4 polynucleotide kinase buffer, 1.0 µL of ATP, 1.0 µL of mutagenic primer, 6.5 µL of distilled water, and 0.5 µL of T4 polynucleotide kinase. Incubate for 60 min at 37°C and then inactivate the kinase at 70°C for 5 min.
2. While the phosphorylated oligonucleotide is still at 70°C, add 2 µL of the uracil-containing template (approx 20–50 ng DNA) and place the reaction tube into 200–300 mL of water at 70°C. Allow to cool to room temperature.
3. Add 2.0 µL of 10X Klenow polymerase buffer, 1.0 µL of dNTP stock, 2.0 µL of ATP, 3.0 µL of distilled water, 1.0 µL of Klenow enzyme, and 1.0 µL of T4 DNA ligase. Incubate at 16°C overnight (*see* Note 8).
4. Transform 1–10 µL of the extension/ligation reaction into a competent wild type *dut*$^+$, *ung*$^+$ *E. coli* host strain and plate out on selective media.

Following transformation it is common to find ten to several hundred colonies present on the plates. Typically the efficiency of mutagenic site incorporation is >50% (*see* Notes 9–12), which allows direct screening of a small number of clones by DNA sequencing to identify those carrying the designed mutation. Moreover, if the site directed changes create or delete a restriction site then these can initially be investigated by endonuclease restriction analysis, as shown in Fig. 2. All site-directed changes, including those involving restriction sites, should be confirmed by DNA sequencing (*see* Note 13).

4. Notes

1. Phagemids include the pBluescript II (Stratagene, La Jolla, CA), pGEMZf (Promega, Madison, WI) and pUC118/9 series of cloning vectors.
2. A mutagenic primer should consist of a region of mismatch flanked by sequences complementary to the template to ensure the oligonucleotide anneals efficiently to the specified target. Mismatches can involve insertions, deletions, and substitutions, or a combination of these. If a single or double change is introduced, 10 flanking bases on either side of the mutation site are usually sufficient. This can be extended up to 15 bases and more if the flanking regions are either rich in A + T residues or if substantial alterations to the target sequence are to be incorporated. Whenever possible include one or more G or C residues at the 5' terminus of the primer to protect the oligonucleotide from displacement following extension by the DNA polymerase, even if this extends the primer length. Avoid primers that include inverted repeats since this can lead to primer duplex formation preventing their hybridization to the target sequence. Do not be afraid to design primers that

loop out significant regions of DNA. For instance, using the protocols described, a 237-bp region of the human cytomegalovirus IE_1 coding sequence was specifically deleted with a 30-mer containing 15 bp flanking each side of the deletion site (Stephen Walker, personal communication).

3. The orientation of the f1 origin of replication determines which DNA strand of the phagemid is generated following helper phage infection. This is usually denoted for a given phagemid by the designation (+) or (–) and in the pBluescript II phagemids refers to the recovery of the sense and antisense strands of the *lac*Z gene, respectively. Such information is clearly critical when designing the mutagenic primer to ensure that it will hybridize to the rescued uracil containing ssDNA template.
4. Competent cells of CJ236 and wild type host strain can be prepared by the Frozen Storage Buffer (FSB) method of Hanahan *(9)*. This method is convenient since it enables aliquots of competent cells to be stored at –70°C until required.
5. R408 helper phage may also be used, but the antibiotic addition in Section 3.1. step 3 should not be carried out since R408 lacks resistance against kanamycin.
6. Multiplicity of infection (moi) refers to the ratio of helper phage to cells present in the culture. For example, a moi of 10 means there are 10 helper phages for each bacterial cell. Generally a moi of 10–20 generates good yields of the phagemid ssDNA.
7. Kunkel *(5,6)* incorporates a uridine supplement (0.25 μg/mL) in his M13 protocol, but I have found this addition to be unnecessary with phagemid preparations.
8. Klenow is the DNA polymerase used for primer extension in this protocol. It lacks ssDNA 5' exonuclease activity and therefore is unable to degrade the mutagenic primer. T4 and T7 DNA polymerases may also be used. Both have the advantage of higher processivities compared to Klenow, and their use can considerably reduce the primer extension reaction time period. In my hands, however, I have found Klenow to be a much more reliable enzyme in ultimately isolating clones with the desired site-directed changes and it is this enzyme that I recommend.
9. A lack of any colonies present on the plate after transformation is usually attributable to low competency of the host strain. For all transformations a control plasmid should be transformed at the same time to confirm that the *E. coli* strain used is competent.
10. Lack of colonies may also arise from failure to form heteroduplex DNA. In these circumstances, first check that dsDNA is being generated from the ssDNA template by comparing a sample of the oligonucleotide extension/ligation to a control sample of the ssDNA template on a 1% agarose

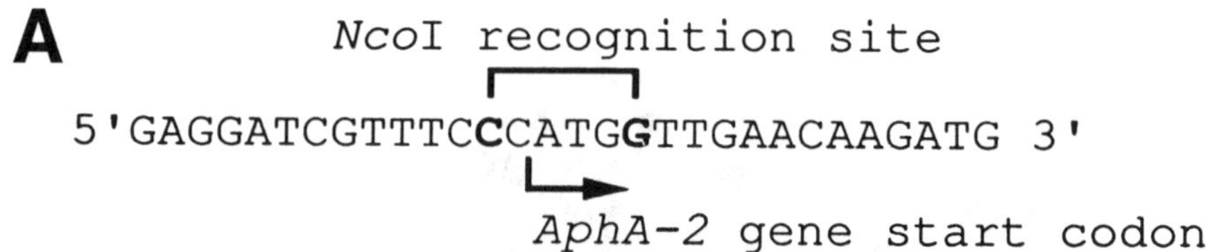

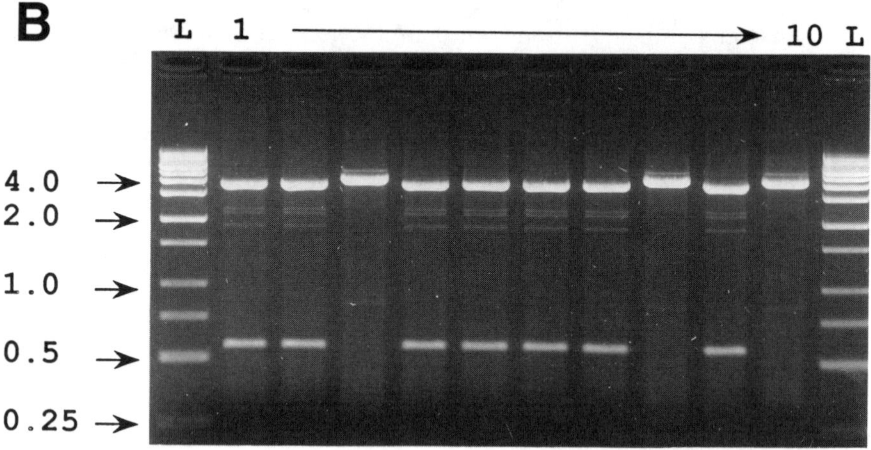

gel. The phagemid ssDNA band should have disappeared from the extension/ ligation sample and a new band of greater intensity formed that runs significantly slower then the ssDNA template. This band should be either the closed circular relaxed dsDNA (form IV) heteroduplex, or through failure of the ligation reaction, the nicked circular dsDNA (form III).

The presence of only a phagemid ssDNA band in the extension/ligation reaction sample indicates that no heteroduplex DNA has been formed. This may be because of one of a number of problems: inactive DNA polymerase or the presence of inhibitory contaminants; failure of the mutagenic primer to anneal; or the presence of secondary structures, such as hairpin loops, in the template. In the first instance use a new batch of Klenow, check the dilution of the mutagenic primer, and reprecipitate the ssDNA template to remove any contaminants. Confirmation that the mutant oligonucleotide is hybridizing to the correct site on the template can also be demonstrated by using it as a primer in a sequencing reaction with the uracil-containing phagemid ssDNA containing the cloned target sequence. The DNA sequence obtained should be located just downstream of the mutagenic priming site. Lastly, problems with secondary structure that cause a block to DNA polymerase

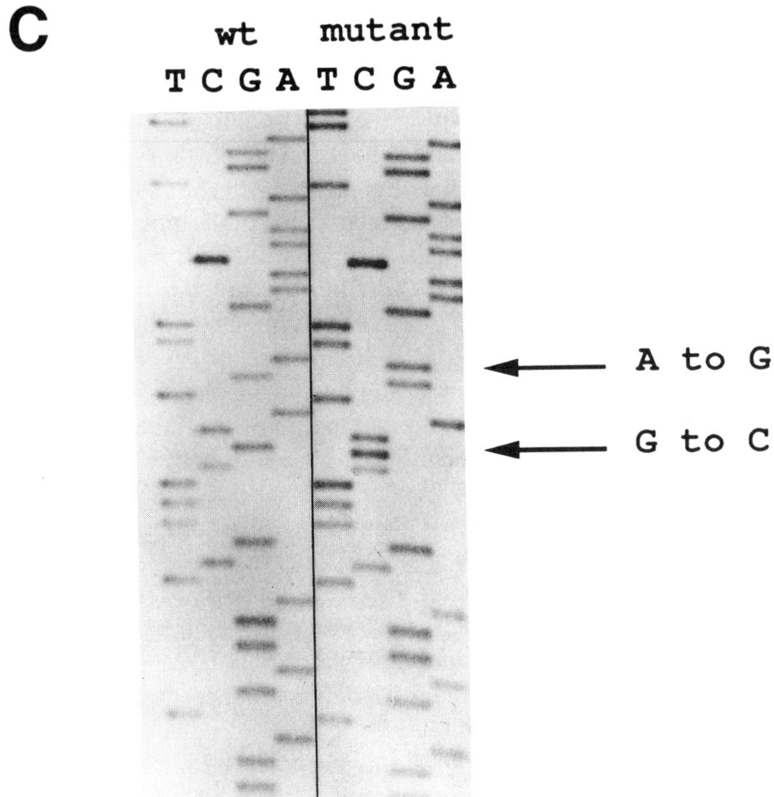

Fig. 2. (**A**) Design of the mutant oligonucleotide to introduce a *Nco*I restriction enzyme site (CCATGG) at the start codon of the Tn5 *AphA-2* kanamycin phosphotransferase gene. This generates two changes from the wild-type sequence (highlighted in bold); a G to C at –2 and A to G at +4 with respect to the *AphA-2* start codon. Reading 5' to 3', the 30-mer has 12 bases flanking each side of the two point mutations. The arrow on the figure indicates direction of transcription of the *AphA-2* structural gene. (**B**) Site-directed mutagenesis was carried out and plasmid DNA was prepared from ten of the several hundred colonies on the transformation plates. Seven of the ten digested with *Nco*I to generate two fragments of 3.75 kbp and 560 bp (Lanes 1,2,4–7,9). This is consistent with the introduction of a second *Nco*I recognition sequence into the vector; in clones lacking the designed mutation a single linearized 4.3 kbp band of the plasmid DNA is formed (Lanes 3,8,10). L, 1-kbp ladder (Stratagene) (**C**) Sequencing of wild-type and mutated plasmid DNA. A 17-mer oligonucleotide that hybridizes 55 bases upstream of the *AphA-2* start codon was used as a primer to obtain the sequences for both wild-type and *Nco*I mutated vectors.

extension may be overcome by addition of *E. coli* single-strand DNA binding protein to the mix. This is helix-stabilizing protein that can denature DNA secondary structures.

11. Mutant colonies should still be isolated, although probably at a lower efficiency, in a situation where the T4 DNA ligase is working inefficiently and not removing the nick between the newly synthesized strand and the 5' end of the mutagenic oligonucleotide. The DNA repair systems of the host strain will complete circularization of the mutant strand in vivo.
12. A low frequency of mutagenesis is likely to result from either lack of uracil in the ssDNA template so that it is not inactivated following transformation into the wild type host strain, mutant oligonucleotide displacement by the advancing DNA polymerase, or DNA impurities in the phagemid ssDNA preparation that prime the template for DNA synthesis.

    The incorporation of uracil into the template can be tested by transformation of 10 ng of the ssDNA template into both CJ236 and a wild-type *dut*$^+$, *ung*$^+$ *E. coli.* A greater than 10^5-fold reduction would be expected between CJ236 and the wild-type strains, since a template containing uracil should be selectively degraded.

    Displacement of the primer may be reduced by increasing the number of template complementary G/C base pairs in the 5' end of the primer. Alternatively, either T4 or native T7 DNA polymerase (not Sequenase) may be used as a replacement for Klenow since these enzymes lack strand displacement activities.

    Priming on the second strand owing to DNA contaminants can be identified by carrying out the extension/ligation step in the absence of mutagenic oligonucleotide. If the phagemid ssDNA is converted to a dsDNA product, fresh template should be prepared.
13. The whole site-directed mutagenesis procedure, including sequencing to identify clones with the designed mutation, can be completed in a week. Furthermore since only 1/25 of the uracil-containing ssDNA is consumed per reaction, the same template can be used to generate other mutations in the cloned insert, thereby reducing the time window by a day.

Acknowledgments

I would like to thank Sydney Brenner for providing me with the opportunity to work on this project and E. I. duPont de Nemours Co., Inc. (Wilmington, DE) for a Fellowship.

References

1. Vandeyar, M. A., Weiner, M. P., Hutton, C. J., and Batt, C. A. (1988) A simple and rapid method for the selection of oligodeoxynucleotide-directed mutants. *Gene* **65,** 129–133.

2. Stanssens, P., Opsomer, C., McKeown, Y. M., Kramer, W., Zabeau, M., and Fritz, H.-J. (1989) Efficient oligonucleotide-directed construction of mutations in expression vectors by the gapped duplex DNA method using alternating selectable markers. *Nucleic Acids Res.* **17,** 4441–4454.
3. Lewis, M. K. and Thompson, D. V. (1990) Efficient site directed in vitro mutagenesis using ampicillin selection. *Nucleic Acids Res.* **18,** 3439–3443.
4. Sayers, J. R., Schmidt, W., Wendler, A., and Eckstein, F. (1988) Strand specific cleavage of phosphorthioate containing DNA by reaction with restriction endonucleases in the presence of ethidium bromide. *Nucleic Acids Res.* **16,** 803–814.
5. Kunkel, T. A. (1985) Rapid and efficient site-specific mutagenesis without phenotypic selection. *Proc. Natl. Acad. Sci. USA.* **82,** 488–492.
6. Kunkel, T. A., Roberts, J. D., and Zakour, R. A. (1987) Rapid and efficient site-specific mutagenesis without phenotypic selection. *Meth. Enzymol.* **154,** 367–382.
7. Dotto, G. P., Enea, V., and Zinder, N. D. (1981) Functional analysis of bacteriophage f1 intergenic region. *Virology* **114,** 463–473.
8. Vieira, J. and Messing, J. (1983) Production of single-stranded plasmid DNA. *Meth. Enzymol.* **153,** 3–11.
9. Hanahan, D. (1985)Techniques for transformation of *E. coli*, in *DNA Cloning: A Practical Approach*, vol. 1 (Glover, G. M., ed.), IRL, Oxford, pp. 109–135.

CHAPTER 8

Construction of Linker-Scanning Mutations by Oligonucleotide Ligation

Grace M. Hobson, Patricia P. Harlow, and Pamela A. Benfield

1. Introduction

The purpose of linker-scanning mutagenesis is to create a series of mutant molecules in which individual sections are sequentially replaced with the same "neutral" mutant sequence (*see* Fig. 1). Typically the technique has been applied to DNA and has most commonly been used to determine the *cis*-regulatory roles of upstream flanking regions of genes. The technique has several advantages over more random mutational analysis:

1. The approach is systematic and implies no previous prejudice about the location of functional elements.
2. The same mutated segment is substituted in each mutant, so that potential contributions to activity from the mutated segment are minimized when comparing mutants, and
3. Most importantly, the spacing and torsional orientation of remaining functional elements remains unchanged in the mutants.

Linker-scanning mutagenesis was first promoted by McKnight and his colleagues *(1)*, and the name arises from the fact that they used a replacement segment containing an endonuclease restriction site generated by the addition of synthetic linker DNA to appropriate DNA fragments generated by deletion. As a result, the replaced segment DNA was small (approx 10 bp) and contained a restriction site that could then be further used to generate mutations (e.g., internal deletions).

From: *Methods in Molecular Biology, Vol. 31: Protocols for Gene Analysis*
Edited by: A. J. Harwood Copyright ©1994 Humana Press Inc., Totowa, NJ

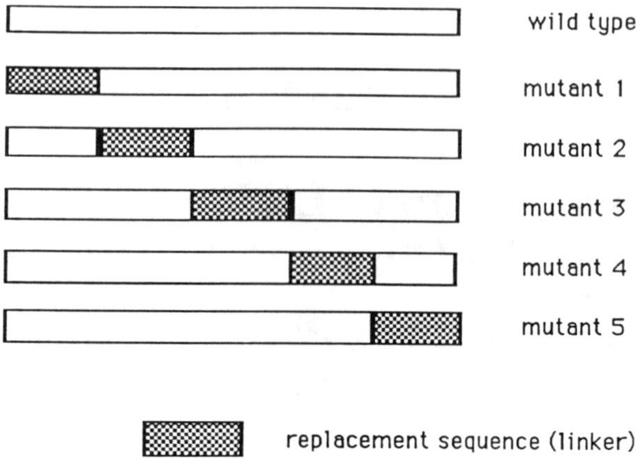

Fig. 1. A schematic representation of a series of linker-scanning mutants in a biopolymer such as DNA or protein. The open boxes represent the wild-type sequence and the hatched boxes represent a replacement mutant segment.

For analysis of gene regulatory function the technique is usually used to examine relatively short stretches of DNA (200–300 bp) in more detail, having established the location of potential regulatory sites by more gross deletion experiments. Technically, however, the principle can be applied to analysis of other biopolymers, e.g., proteins. Alanine scanning mutagenesis *(2)* represents an extension of linker scan technology to analysis of functionally important residues in proteins.

This chapter describes a modification of the linker-scanning method described by Karlsson et al. *(3)*. A series of synthetic oligonucleotides is produced to cover both strands of the DNA sequence under investigation. Each oligonucleotide is designed to partially span the two adjacent complementary strands as outlined in the sample scheme in Fig. 2. The entire set of oligonucleotides anneal together to assemble a complete DNA region. Sufficient oligonucleotides to create a totally wild-type sequence are synthesized, but by exchanging pairs of wild-type for mutant oligonucleotides, changes can be introduced at any position in the DNA region. By ensuring that the terminal oligonucleotides possess the cohesive ends of restriction enzyme sites, the synthetic region is then used to replace a corresponding restriction enzyme fragment in the piece of DNA under investigation. Although obviously the efficiency of the method will decrease as the length of the segment

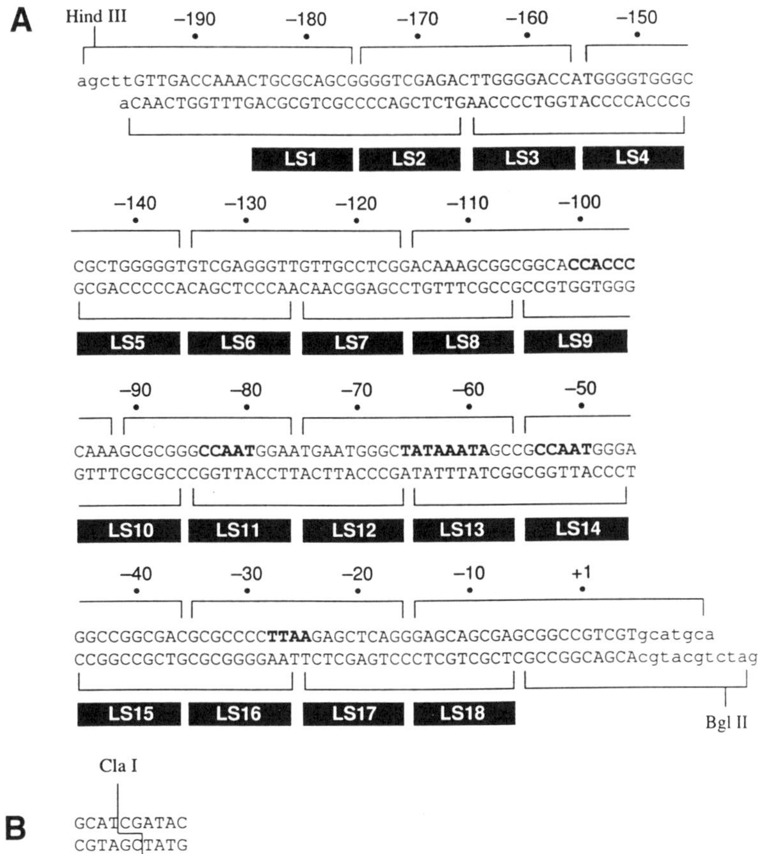

Fig. 2. Sample strategy *(4)* for construction of linker scan (LS) mutant promoter fragments for the rat brain creatine kinase (*ckb*) promoter. **(A)** Sequence of the wild type *ckb* promoter DNA insert showing the location of overlapping oligonucleotides used for its construction. Black boxes labeled LS1–LS18 indicate the positions of the linkers in the LS mutants. Lower case letters indicate non-*ckb* sequences. *Hind*III and *Bgl*II overlaps used to insert synthetic promoter regions into the vector are indicated. **(B)** The sequence of the *Cla*I linker showing the *Cla*I restriction site. The sequences of mutant oligonucleotides used to construct the LS mutants can be deduced by replacing the wild-type sequences shown in A with the *Cla*I linker for each LS mutant.

to be constructed increases, we have used this method successfully to prepare linker-scanning mutations through a 200-bp section of the brain creatine kinase (*ckb*) promoter *(4)*.

2. Materials

1. Synthetic oligonucleotides: These may be generated using a variety of commercially available synthesis machines using the β-cyanoethyl phosphoramite chemistry with a dimethoxytrityl blocking group (*see* Note 1). Oligonucleotides (not gel purified before use) were resuspended to a concentration of 1 mg/mL. The design of oligonucleotides is discussed in Notes 2–5.
2. 10X kinase buffer: 0.5*M* Tris-HCl, pH 7.5, 0.1*M* $MgCl_2$, 50 m*M* DTT, 1 m*M* spermidine, 1 m*M* EDTA. Store at –20°C.
3. ATP: 42 m*M* adenosine triphosphate (ATP) solution. Store at –20°C.
4. T4 polynucleotide kinase: Store as 20 U/µL stock at –20°C.
5. 10*M* Ammonium acetate.
6. Ethanol: 100% and 70% (v/v) solutions.
7. TE: 10 m*M* Tris-HCl, pH 8.0, 1 m*M* EDTA.
8. 100 m*M* $MgCl_2$.
9. 3*M* Na acetate.
10. 10X ligase buffer: 0.5*M* Tris-HCl, pH 7.5, 0.1*M* $MgCl_2$, 100 m*M* DTT, 10 m*M* ATP. Store at –20°C.
11. T4 DNA ligase: Store as 6 U/µL stock at –20°C.
12. Appropriate restriction endonucleases and their buffers.
13. Agarose: Low melting point agarose. Use a grade appropriate for cloning, e.g., Seaplaque GTG agarose (FMC).
14. TBE gel running buffer: 89 m*M* Tris, 89 m*M* borate, 2.5 m*M* EDTA, pH 8.3.

3. Methods

1. Kinase all oligonucleotides, except the terminal ones (*see* Note 6), separately as follows. Mix 40 µL of oligonucleotide, 6 µL of 10X kinase buffer, 1.5 µL of 42 m*M* ATP, 10 µL of dH_2O and 1µL of T4 polynucleotide kinase in a microfuge tube (*see* Note 7).
2. Incubate at 37°C for 1 h, then heat to 65°C for 15 min to stop the reaction. At this point oligonucleotides may be stored at –20°C.
3. Mix appropriate oligonucleotides together. Use 3.3 µL of the two unphosphorylated terminal oligonucleotides and 5 µL of each kinased internal oligonucleotide stock. For mutants leave out the appropriate wild-type pair and substitute with the corresponding mutant pair.
4. Heat to 95°C and slowly cool to room temperature in a water bath. This is usually best performed in a 1-L beaker. Boil 300 mL of water in the beaker; remove from heat. Float tubes containing the hybridization reactions in the water bath and allow to cool to room temperature.
5. Add 1/5 vol of 10*M* ammonium acetate to each tube and ethanol precipitate by adding 2 vol of 100% ethanol. Cool in dry ice/ethanol for 5–

10 min. Pellet by spinning for 15 min in a microfuge. Wash the pellet with 70% ethanol and respin. Remove the supernatant and dry pellet. Resuspend in 200 µL TE.

6. Add 25 µL of 100 m*M* $MgCl_2$ and 25 µL of 3*M* Na acetate. Ethanol precipitate as described in step 5. Resuspend in 42.5 µL of H_2O.
7. Add 5 µL of 10X ligase buffer and 2.5 µL of T4 DNA ligase. Incubate at 14°C overnight. Ligated oligonucleotides may be stored at 4°C (*see* Note 8).
8. Double digest 1 µg of vector DNA with appropriate restriction enzymes as directed by the supplier. Electrophorese on a 0.6% low melting temperature agarose gel in TBE running buffer. Visualize DNA by staining with ethidium bromide. Excise the vector band and melt in 4 vol of 10 m*M* Tris-HCl, pH 8.0. This results in a total volume of approx 300 µL (*see* Note 9).
9. Set up the following ligations: 16 µL of vector DNA, 1 µL of oligonucleotide insert, 2 µL of 10X ligase buffer, and 1 µL of T4 DNA ligase (*see* Note 10). Incubate overnight at 14°C.
10. Transform into a suitable *E. coli* host strain (*see* Note 11). Pick colonies and determine which colonies contain appropriately sized inserts by restriction and sequence (*see* Note 12).

4. Notes

1. Oligonucleotides synthesized using the methyl phosphite protecting group contain up to 30% methylated thymidine residues: These lead to mutations if read as cytidine *(5)*.
2. We have found it useful to design the terminal oligonucleotides so that the overhang constitutes a restriction site. Preferably these restriction sites should be different to allow insertion of the final synthetic fragment into an appropriately doubly-digested vector. The remaining oligonucleotides should be made in two forms, one that allows generation of the wild-type sequence and one that allows replacement of each segment with a section of DNA that contains the linker restriction site of choice. Oligonucleotides should be designed to minimize any potential redundancy in the overlap regions. In this way the correct annealing scheme should be greatly favored and unwanted deletions or mutations can be avoided.
3. It is preferable to pick unique and different restriction sites for the terminal oligonucleotides and for the linker. In choosing these sites it is advisable to consider further manipulations that might be desired and avoid restriction sites that might appear elsewhere in the DNA constructs. Try to avoid any sites that might contain sequences of potential functional relevance, since the aim is to make the linker as "neutral" as possible.

4. Oligonucleotides do not have to be the same size. We have used oligonucleotides between 16 and 27 bases in length successfully. This allows some flexibility in the design scheme—for example, to avoid redundant overlaps and repeated sequences.
5. Try to avoid oligonucleotides with internal repeats that might allow hairpin bend formation.
6. Terminal oligonucleotides are not kinased to avoid multimerization of the synthetic fragment.
7. This procedure produces a generous amount of kinased oligonucleotides. Considerably less material could be kinased for the mutant sequences that are only used in a limited number of ligations compared with the wild-type sequences.
8. Presence of the appropriately sized band could now be checked by agarose gel electrophoresis and gel purified for subsequent ligation into the vector. We have always seen multiple bands and even a smear interrupted by discrete bands when checking the ligation reaction by electrophoresis. It is probable that most of the material smaller than the appropriate insert size does not have the appropriate restriction site ends for ligation into the vector. Although in some cases we have gel purified the insert we have found this to be unnecessary. In general, we do not check ligation mixtures by electrophoresis or gel purify the inserts; nevertheless, we have had a high success rate in isolating appropriate clones after ligation and transformation.
9. Alternatively, the restriction enzyme can be inactivated following restriction digestion by heating to 65°C for 10 min (for heat sensitive enzymes). The vector can then be used without gel purification if the restriction enzyme digestion was efficient.
10. Different vector insert ratios can be tried if problems are encountered.
11. Usually we use pUC based vectors as recipients for our linker scanned segments and transform into *E. coli* [DH5] by electroporation *(6)* to achieve high transformation efficiencies.
12. Plasmid DNA has been prepared from drug resistant colonies by the boiling preparation described by *(7)*. Plasmid DNA containing appropriately sized inserts has been sequenced using the USB Sequenase kit using the conditions suggested by the manufacturer.

References

1. McKnight, S. L. and Kingsbury, R. (1982) Transcriptional control signals of a eukaryotic protein-coding gene. *Science* **217,** 316–324.
2. Cunningham, B. C. and Wells, J. A. (1989) High-resolution epitope mapping of hGH-receptor interactions by alanine-scanning mutagenesis. *Science* **244,** 1081–1085.

3. Karlsson, O., Edlund, T., Moss, J. B., Rutter, W. J., and Walker, M. D. (1987) A mutational analysis of the insulin gene control region: expression in beta cells is dependent on two related sequences within the enhancer. *Proc. Natl. Acad. Sci. USA* **84,** 8819–8823.
4. Hobson, G. M., Molloy, G. R., and Benfield, P. A. (1990) Identification of cis-acting regulatory elements in the promoter region of the rat brain creatine kinase gene. *Mol. Cell. Biol.* **10,** 6533–6543.
5. Bell, L. D., Smith, J. C., Derbyshire, R., Finlay, M., Johnson, I., Gilbert, R., Slocombe, P., Cook, E., Richards, H., Clissold, P., Meredith, D., Powell-Jones, C. H., Dawson, K. M., Carter, B. L., and McCullagh, K. G. (1988) Chemical synthesis, cloning and expression in mammalian cells of a gene coding for human tissue-type plasminogen activator. *Gene* **63,** 155–163.
6. Dawes, W. J., Miller, J. F., and Ragsdale, C. W. (1988) High efficiency transformation of *E. coli* by high voltage electroporation. *Nucleic Acids Res.* **16,** 6127–6145.
7. Sambrook, J., Fritsch, E. F., and Maniatis, T. (1989) In vitro amplification of DNA by the polymerase chain reaction. *Molecular Cloning: A Laboratory Manual,* 2nd ed., Chapter 14. Cold Spring Harbor Laboratory, Cold Spring Harbor, NY, pp. 14.1–14.35.

CHAPTER 9

Construction of Linker-Scanning Mutations Using the Polymerase Chain Reaction

Patricia P. Harlow, Grace M. Hobson, and Pamela A. Benfield

1. Introduction

In the previous chapter we described linker scanning mutagenesis by oligonucleotide ligation. In this chapter we describe a more recent and versatile procedure that makes use of amplification by the polymerase chain reaction (PCR). For a description of the traditional method of generating linker scanning mutants by combination of deletion fragments, *see* ref. *1*.

Use of PCR provides an economical approach to linker scanning since fewer oligonucleotides are required to generate a set of mutants. Three general methods have been described for mutagenesis: the overlapping, asymmetric, and circular methods. The overlapping method *(2,3)* involves four oligonucleotides and three PCR reactions to create point, substitution, insertion, or deletion mutations at any point within a DNA fragment. Asymmetric methods *(4–6)* reduce the number of oligonucleotides to three and the number of PCR reactions to two for each mutation, but require purification of the PCR fragment from the first PCR reaction before proceeding to the second PCR reaction. Both the overlapping and asymmetric methods limit the amount of DNA sequencing required, but both require unique restriction sites appropriately placed near the ends of the fragment to be mutagenized. These methods have also been applied to the mutagenesis of proteins

From: *Methods in Molecular Biology, Vol. 31: Protocols for Gene Analysis*
Edited by: A. J. Harwood

(4–9). The circular method *(10,11)* reduces the number of PCR reactions and obviates the requirement for appropriately placed restriction sites. However, if appropriate restriction sites are not available, the circular method increases the amount of DNA sequencing to be performed. Recently, Jayaraman et al. *(12)* have combined the oligonucleotide ligation and a PCR method to synthesize DNA encoding horseradish peroxidase.

To create 18 linker-scan mutations in a 200-bp promoter, the oligonucleotide ligation method (*see* Chapter 8) requires 56 oligonucleotides (36 mutant and 20 wild-type). The PCR methods require fewer: 38 for the overlapping method, 20 for the asymmetric method, and 36 for the circular method. Although the asymmetric PCR method minimizes both the number of oligonucleotides and the amount of DNA sequencing required to create a series of linker-scan mutations, it requires purification of the fragment produced in the first PCR reaction.

We have successfully used the overlapping method without any intervening purification of PCR fragments to create mutations in the *ckb* promoter, and it is this method we describe in this chapter. The method can be adapted to the asymmetric PCR method by carrying out only one of the initial PCR reactions using a mutant and one common oligonucleotide. The resulting fragment, however, must be purified before proceeding to the second PCR reaction (Fig. 1).

2. Materials

1. Synthetic oligonucleotides: These may be generated using a variety of commercially available synthesis machines using the β-cyanoethyl phosphoramite chemistry with a dimethoxytrityl blocking group (*see* Note 1). Oligonucleotides (not gel purified before use) were resuspended to a concentration of 1 mg/mL. The oligonucleotides used in this protocol are named in accordance with Fig. 1. For a discussion of oligonucleotide design, *see* Note 2.
2. Template DNA: The template DNA is generally plasmid DNA that contains the region to be mutagenized (*see* Note 3).
3. 10X PCR reaction buffer: 500 m*M* KCl, 15 m*M* $MgCl_2$, 100 m*M* Tris-HCl, pH 8.3, 0.1% gelatin (*see* Note 4).
4. 2 m*M* dNTP mix: 2 m*M* of each dNTP, neutralized to pH 7.0. Store in aliquots at –20°C.
5. Mineral oil.

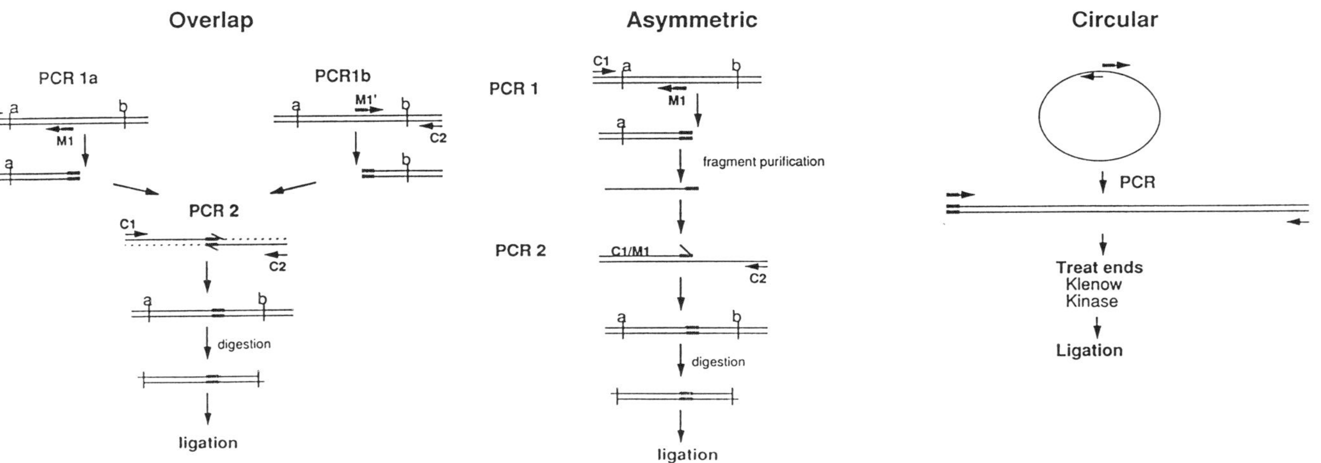

Fig. 1. Site-directed mutagenesis by PCR. In the overlapping and asymmetric methods the common oligonucleotides are C1 and C2, whereas the mutant oligonucleotides are M1 and M1'. The unique restriction sites are labeled a and b. In the asymmetric method the fragment produced from the first PCR reaction is labeled C1/M1. In the circular method only one oligonucleotide is required to be a mutant (M1); the other oligonucleotide may be a wild-type or mutant oligonucleotide.

6. *Taq* DNA polymerase: DNA *Thermus aquaticus* DNA polymerase or AmpliTaq™ (Perkin Elmer Cetus, Norwalk, CT). Store according to manufacturer's instructions.
7. Chloroform.
8. Appropriate gels, either agarose or acrylamide, and DNA size markers for checking reaction products and possibly purifying appropriate sized fragments.
9. Appropriate restriction endonucleases and their buffers for digesting the PCR fragments and the recipient vector.
10. Phenol/chloroform: A 1:1 (v/v) mix of Tris-HCl, pH 7.4 buffered phenol and chloroform.
11. 10X ligase buffer: 0.5*M* Tris-HCl, pH 7.5, 0.1*M* $MgCl_2$, 100 m*M* DTT, 10 m*M* ATP. Store at –20°C.
12. T4 DNA ligase: Store as 6 U/µL stock at –20°C.
13. Vector DNA: Appropriately digested plasmid DNA, e.g., pUC. Phenol extracted, ethanol precipitated, and resuspended at 3 ng/µL. Store at –20°C.

3. Methods

3.1. PCR Reaction 1

For each mutation set up two PCR reactions, 1a and 1b, in appropriately sized microfuge tubes.

1. In tube 1a place 2 µL of 10 µ*M* (20 pmol) common oligonucleotide 1 (C1); 2 µL of 10 µ*M* (20 pmol) mutant oligonucleotide 1 (M1); 5 µL of 10X PCR reaction buffer; 2.5 µL of 2 m*M* dNTP mix; 0.5–1 ng (0.15–0.3 fmole) of template DNA; and sterile water to 50 µL (*see* Notes 4 and 5).
2. In tube 1b place 2 µL of 10 µ*M* (20 pmol) common oligonucleotide 2 (C2); 2 µL of 10 µ*M* (20 pmol) complementary mutant oligonucleotide (M1'); 5 µL of 10X PCR reaction buffer; 2.5 µL of 2 m*M* dNTP mix; 0.5–1.0 ng of template DNA, and sterile water to 50 µL.
3. Overlay each reaction with a drop (about 50 µL) of mineral oil. Heat the reaction to 95°C for 5 min. Allow the reaction to cool to 70°C (*see* Note 6), then add below the oil layer 0.5–1.0 U of *Taq* polymerase that has been diluted into 1X PCR reaction buffer, and mix with the pipet tip.
4. Perform a PCR reaction in a programmable thermal cycler of 25–35 cycles of denaturation at 94°C for 30 s; annealing at an appropriate temperature for 1 min; extension at 72°C for 30 s; or an appropriate time for the length of fragment expected (*see* Note 7).
5. Extract the two reactions with chloroform to remove the mineral oil (*see* Note 8).

Remove an aliquot (1/10) from each reaction and electrophorese on an appropriate gel (agarose or acrylamide) to determine whether a single PCR fragment of the appropriate size has been produced. If there is only a single fragment, proceed to PCR reaction 2 (Section 3.2.). If there are multiple fragments, decide on the basis of number and intensity whether it would be advantageous to purify the appropriate sized fragment, repeat the reaction using different conditions, or proceed (*see* Note 9). If no PCR fragment has been produced, then repeat the PCR reaction using different conditions (*see* Note 4).

3.2. PCR Reaction 2

1. In an appropriately sized microfuge tube place: 0.5–5.0 µL of PCR reaction 1a (*see* Note 10); 0.5–5.0 µL of PCR reaction 1b; 2 µL of 10 µ*M* (20 pmol) common oligonucleotide 1 (C1); 2 µL of 10 µ*M* (20 pmol) common oligonucleotide 2 (C2); 5 µL of 10X PCR reaction buffer; 2.5 µL of 2 m*M* dNTP mix; and sterile water to 50 µL.
2. Overlay with mineral oil and heat to 95°C for 5 min. Allow the reaction to cool to 70°C and add below the mineral oil 0.5–1.0 U of *Taq* polymerase diluted in 1X reaction buffer.
3. Perform a PCR reaction in a programmable thermal cycler of 25–35 cycles of denaturation at 94°C for 30 s; annealing at an appropriate temperature for 1 min; extension at 72°C for 30 sec.
4. Extract the reaction at least once with chloroform to remove the mineral oil.

Remove an aliquot (1/10), and electrophorese the sample on an appropriate gel (agarose or acrylamide) to determine whether a single PCR fragment of the appropriate size has been produced. If there is a single fragment, proceed to digest the fragment with appropriate restriction endonucleases. The choice of restriction endonuclease will determine whether the fragment can be digested without an intervening ethanol precipitation step (*see* Note 11). If multiple fragments are produced, decide whether it would be advantageous to purify the appropriate sized fragment, repeat the reaction using different conditions, or proceed directly.

3.3. Restriction Endonuclease Digestion and Ligation

1. Add the appropriate restriction endonucleases and digest for 2 h. Remove an aliquot (1/10), and electrophorese on an appropriate gel to determine if the expected sized fragments are present (*see* Note 12).

2. Extract the digested fragments with phenol/chloroform and then chloroform (*see* Note 13). Precipitate the fragments with ethanol. Estimate the concentration of the fragment by electrophoresis of an aliquot on an appropriate gel in comparison to a known amount of marker DNAs.
3. Add 1 µL of the digested fragments to 16 µL (50 ng) of an appropriately digested vector, 2 µL of 10X ligase buffer, and 1 µL of T4 DNA ligase. Incubate overnight at 14°C (*see* Note 14).
4. Transform the ligated DNA into an appropriate *E. coli* host (*see* Note 15).
5. Screen the mutant clones either by restriction digestion of miniprep DNA (*see* Note 16) or by hybridization with the mutant oligonucleotide (*see* Note 17).
6. Sequence the mutant clones using the common oligonucleotides, C1 and C2, to confirm that only the expected mutation was generated (*see* Note 18).

4. Notes

1. Oligonucleotides synthesized using the methyl phosphite protecting group contain up to 30% methylated thymidine residues that can cause mutations if read as cytidine. Generally the oligonucleotides do not need to be purified by acrylamide electrophoresis, although batches of nucleotides with a high proportion of short products may produce lower amounts of PCR reaction products or require alteration of PCR conditions (e.g., temperature).
2. The design of the oligonucleotides to be chemically synthesized is critical. The number of changes from the wild-type template affect the length of oligonucleotide required. When making a cluster of changes from the wild-type sequence, as is the case when making linker-scan mutations, the changes should be confined to the 5' end of the oligonucleotide. The length and sequence of the 3' portion of the oligonucleotide that hybridizes to the template should be such that the melting temperature, T_m, of hybridization to the template is within a reasonable annealing temperature range (40–60°C). The T_m is estimated from the equation: $T_m = 4(G + C) + 2(A + T)$. Since the template DNA for PCR 2 is the annealed product of the PCR fragments from reactions 1a and 1b, it is critical when designing the mutant oligonucleotides, M1' and its complement (*see* Fig. 1), to have overlapping sequences with a reasonable T_m (40–60°C). The common oligonucleotides C1 and C2 (*see* Fig. 1) should be chosen so that their T_m is similar to the mutant oligonucleotides. It helps for subsequent cloning of the fragment to incorporate into C1 or C2 at least one restriction enzyme site that cleaves to produce a 5' or 3' overhang.

3. We have successfully utilized miniprep DNA prepared by the alkaline lysis or boiling preparation procedures as template DNA. The plasmid DNA may be either circular or linearized at a unique restriction site located away from the region to be mutagenized. If the DNA is linearized, it need not be purified from the restriction digestion reaction when the concentration is high (>1 ng/µL). Because PCR can amplify very small amounts of template DNA, precautions against contamination are sometimes required. In the method described here, relatively large amounts of template DNA are used. Therefore, extraordinary precautions against contamination are not required. However, reasonable care to prevent contamination should be exercised (*see* ref. *13*).
4. As in any PCR reaction, the $MgCl_2$ concentration may need to be adjusted for a particular set of oligonucleotides *(14)*. The concentration used here has been used successfully for many oligonucleotide pairs. However, if no PCR fragment is observed, optimization of the $MgCl_2$ concentration is suggested.
5. Since the amount of template DNA can influence the frequency of unexpected mutations, Ho et al. *(3)* recommended a high initial template concentration. However, a high concentration may necessitate that the PCR fragments from PCR 1a and 1b be purified. We have found that a lower initial template concentration (0.5–1.0 ng) allowed us to recover mutant clones without having to purify the fragments from PCR 1.
6. After the initial denaturation of the DNA, maintenance of the reaction above the annealing temperature prevents spurious hybridization of the oligonucleotides that can result in the production of multiple PCR fragments.
7. The exact annealing temperature used for PCR will depend on the oligonucleotides chosen. Since *Taq* DNA polymerase extends nucleotides at the rate of 60 nucleotides/s *(15)* at 70°C, a 30-s extension time is sufficient to prepare PCR fragments of 1 kb or less. A longer extension time is necessary to prepare larger fragments. In order to reduce extraneous mutations, the number of cycles should be kept as low as possible. Usually 25–35 cycles are sufficient to produce a detectable amount of fragment starting with 1 ng of initial template.
8. The chloroform extraction step can be eliminated if an aliquot is carefully removed from the oil layer. This is sometimes easier if the reaction mix is dispensed onto a piece of parafilm.
9. Even if multiple bands result from the PCR reactions, we may proceed to PCR 2 if the expected sized band represents at least 30–50% of the ethidium bromide stained fragments. In some cases, however, we have purified the PCR fragments from PCR 1, especially when generating point mutations.

10. To dilute the wild-type template and eliminate purification of PCR fragments 1a and 1b, only 0.5–5.0 µL of PCR reactions 1a and 1b should be used.
11. We have successfully digested the PCR fragments from reaction 2 with *Hind*III/*Bgl*II without an intervening ethanol precipitation step by adjusting the pH, NaCl, and $MgCl_2$ concentrations before adding the enzymes. Some restriction endonucleases have different requirements and an intervening ethanol precipitation step may be necessary.
12. Although we have noted multiple bands for some PCR reactions, digestion with restriction endonucleases results primarily in the expected sized fragment.
13. A phenol extraction is used to inactivate the restriction endonucleases. If the restriction endonuclease utilized can be inactivated by heating to 70°C, the phenol/chloroform extractions can be eliminated.
14. Different vector insert ratios can be tried if problems are encountered.
15. Usually we use pUC based vectors as recipients for our linker-scanned segments and transform them into *E. coli* [DH5] by electroporation to achieve high transformation efficiencies (*16*; and Chapter 1).
16. Plasmid DNA may be prepared from drug resistant colonies by the boiling preparation described in ref. *13*.
17. If linker-scan mutations are generated by replacement with a unique restriction site, mutants may be screened by restriction digestion of miniprep DNA from a limited number of clones, however, when generating point mutations it may be necessary to screen many more clones by hybridization with a labeled mutant oligonucleotide.
18. Sequencing of potential positive mutants is essential before functional studies are initiated. We sequence using the USB Sequenase kit using the conditions suggested by the manufacturer. Oligonucleotides used in mutant preparation can often be used as sequencing primers. Not only is it necessary to confirm that the expected mutation was generated, but it is also necessary to rule out spurious mutations in the region of interest. *Taq* DNA polymerase has a poor fidelity, so unexpected mutations can occur generally at a rate of 1/400–1/4000 bases per cycle *(17)*. We have also noted unexpected mutations, which apparently were present in the initial mutant oligonucleotides. Additionally, *Taq* DNA polymerase can add an additional nucleotide, generally an adenine residue, to the 3' end of a fragment *(18)*. Kuipers et al. *(19)* have suggested that this potential error may be eliminated by appropriately choosing the first 5' nucleotide of the mutant oligonucleotide to follow a thymidine residue in the same strand of the template.

References

1. McKnight, S. L. and Kingsbury, R. (1982) Transcriptional control signals of a eukaryotic protein-coding gene. *Science* **217,** 316–324.
2. Higuchi, R., Krummel, B., and Saiki, R. K. (1988) A general method of in vitro preparation and specific mutagenesis of DNA fragments: study of protein and DNA interactions. *Nucleic Acids Res.* **16,** 7351–7367.
3. Ho, S. N., Hunt, H. D., Horton, R. M., Pullen, J. K., and Pease, L. R. (1989) Site-directed mutagenesis by overlap extension using the polymerase chain reaction. *Gene* **77,** 51–59.
4. Sarkar, G. and Sommer, S. S. (1990) The "megaprimer" method of site-directed mutagenesis. *Biotechniques* **8,** 404–407.
5. Nelson, R. M. and Long, G. L. (1989) A general method of site-directed mutagenesis using a modification of the *Thermus aquaticus* polymerase chain reaction. *Anal. Biochem.* **180,** 147–151.
6. Perrin, S. and Gilliland, G. (1990) Site-specific mutagenesis using asymmetric polymerase chain reaction and a single mutant primer. *Nucleic Acids Res.* **18,** 7433–7438.
7. Horton, R. M., Cai, Z., Ho, S. N., and Pease, L. R. (1990) Gene splicing by overlap extension: tailor–made genes using the polymerase chain reaction. *Biotechniques* **8,** 528–535.
8. Horton, R. M., Hunt, H. D., Ho, S. N., Pullen, J. K., and Pease, L. R. (1989) Engineering hybrid genes without the use of restriction enzymes: gene splicing by overlap extension. *Gene* **77,** 61–68.
9. Vallette, F., Mege, E., Reiss, A., and Adesnik, M. (1989) Construction of mutant and chimeric genes using the polymerase chain reaction. *Nucleic Acids Res.* **17,** 723–733.
10. Hemsley, A., Arnheim, N., Toney, M. D., Cortopassi, G., and Galas, D. J. (1989) A simple method for site-directed mutagenesis using the polymerase chain reaction. *Nucleic Acid Res.* **16,** 6545–6551.
11. Jones, D. H. and Howard, B. H. (1990) A rapid method for site-specific mutagenesis and directional subcloning by using the polymerase chain reaction to generate recombinant circles. *Biotechniques* **8,** 178–183.
12. Jayaraman, K., Fingar, S. A., Shah, J., and Fyles, J. (1991) Polymerase chain reaction-mediated gene synthesis: synthesis of a gene coding for isozyme c of horseradish peroxidase. *Proc. Natl. Acad. Sci. USA* **88,** 4084–4088.
13. Sambrook, J., Fritsch E. F., and Maniatis T. (1989) In vitro amplification of DNA by the polymerase chain reaction, in *Molecular Cloning: A Laboratory Manual,* 2nd ed., Chapter 14, Cold Spring Harbor Laboratory, Cold Spring Harbor, NY, pp. 14.1–1435.
14. Saiki, R. K. (1989). The design and optimization of the PCR, in *PCR Technology: Principles and Applications for DNA Amplification* (Erlich, H. A., ed.), Stockton, New York, pp. 7–16.

15. Gelfand, D. H. (1989) *Taq* DNA polymerase, in *PCR Technology: Principles and Applications for DNA Amplification* (Erlich, H. A., ed.), Stockton, New York, pp. 7–122.
16. Dawes, W. J., Miller J. F., and Ragsdale, C. W. (1988) High efficiency transformation of *E. coli* by high voltage electroporation. *Nucleic Acids Res.* **16,** 6127–6145.
17. Dunning, A. M., Talmud, P., and Humphries, S. E. (1988) Errors in the polymerase chain reaction. *Nucleic Acids Res.* **16,** 10393.
18. Clark, J. M. (1988) Novel non-templated nucleotide addition reactions catalyzed by prokaryotic and eukaryotic DNA polymerases. *Nucleic Acids Res.* **16,** 9677–9686.
19. Kuipers, O. P., Boot, H. J., and de Vos, W. M. (1991) Improved site-directed mutagenesis method using PCR. *Nucleic Acids Res.* **19,** 4558.

CHAPTER 10

Localized Random Polymerase Chain Reaction Mutagenesis

Claude G. Lerner and Masayori Inouye

1. Introduction

Since the invention of the polymerase chain reaction (PCR) for in vitro amplification of DNA using thermostable DNA polymerase *(1)*, an impressive assortment of powerful modifications of the method have been further developed *(2,3)*. Until recently, surprisingly little attention has been paid to application of PCR as a tool for localized random mutagenesis. *Taq* DNA polymerase (from *Thermus aquaticus*), the most commonly used thermal stable DNA polymerase in PCR, exhibits a relatively high single base misincorporation rate of 10^{-4} since it lacks detectable 3'–5' proofreading exonuclease activity *(4,5)*. Depending on the conditions, including pH and relative concentrations of $MgCl_2$ and deoxynucleotide triphosphates, the base substitution error rates of *Taq* DNA polymerase can vary between 1/3200 and 1/300,000 *(5)*. Base substitution error rates on the order of 1/10,000 are common using standard PCR conditions (PE Cetus, Norwalk, CT kit, *4*). Other factors that influence misincorporation events during PCR include the length of the sequence, the number of amplification cycles, and the initial number of template copies *(6)*. Thus, PCR products obtained using *Taq* DNA polymerase contain substitution mutations at low but appreciable frequencies. Although it is generally desirable to minimize the misincorporation activity in conventional PCR amplification, the mutagenic activity of *Taq* DNA polymerase can be successfully exploited to rapidly isolate random mutations through-

From: *Methods in Molecular Biology, Vol. 31: Protocols for Gene Analysis*
Edited by: A. J. Harwood

out an entire or defined region of a gene. The unique aspect of this procedure is that random mutations can be easily generated in vitro over large but defined sequences of gene-sized DNA molecules. Several methods utilizing chemicals *(7–9)*, mixtures of spiked oligonucleotides *(10–14)*, or enzymatic misincorporation of nucleotides *(15–19)* have been used to generate targeted random mutations in vitro. Although these methods have been successfully used to mutate a variety of genes, low mutagenesis rates, cost, or difficulty in focusing the mutagenesis over large but defined regions of DNA have limited the breadth of their application. Localized random PCR mutagenesis (PCRM), on the other hand, has been found to be a powerful mutagenic method by which mutagenized gene libraries can be constructed relatively easily, rapidly, and inexpensively *(20–27)*. One of the most desirable features of PCRM is that specified regions of DNA molecules can be randomly mutagenized while nontarget site mutations are essentially eliminated. Another advantage of PCRM is that ssDNA templates are not required, unlike the highly efficient, yet costly, random method of Hermes et al. *(11)*. The disadvantage of PCRM is that it is not truly random; a strong bias exists for T–C/A–G transitions and certain target sequences contain regions more or less prone to mutation *(4,5)*. Currently no model for the template sequence dependence of *Taq* infidelity has been proposed *(2)*.

Figure 1 diagrams the basic steps in the localized random PCR mutagenesis protocol. The mutagenesis is localized between restriction sites that are chosen or constructed to flank the target region of interest. Restriction sites with noncohesive ends facilitate forced cloning of the mutagenized fragment into a plasmid vector and eliminate the possibility of vector religation or incorrect insert orientation and so limit the number of false positive mutant candidates. Oligonucleotides are designed such that they flank the restriction sites and give rise to a PCR product in which the sites are located approx 50 bp from the product's ends. This allows clean resolution of the completely digested restriction fragment during preparative gel electrophoresis. After PCR amplification for 25 cycles and subsequent restriction endonuclease digestion, the fragment is gel purified. The whole target plasmid used in the PCR reaction is digested with the same restriction endonucleases and the vector fragment lacking the region

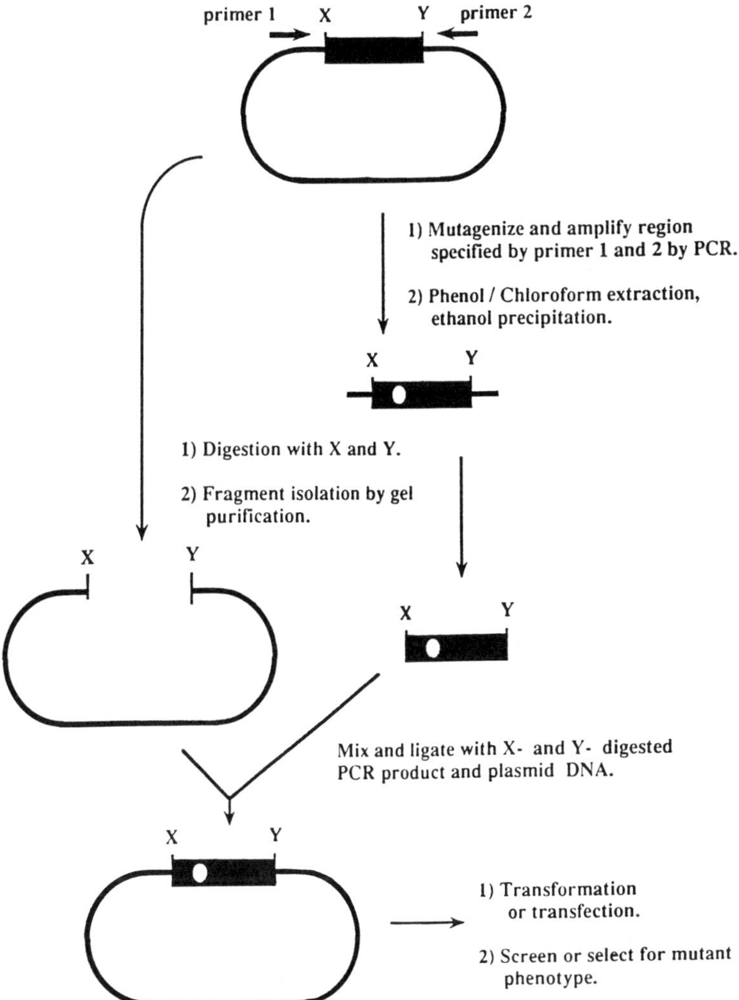

Fig. 1. Schematic representation of localized polymerase chain reaction mutagenesis (PCRM). Primers 1 and 2 define the region over which PCRM will occur; restriction enzymes X and Y are used to excise the mutagenized target region and force the correct orientation during ligation with the similarly digested plasmid vector. The ligation mixture consists of a mutagenized DNA library in the original, but unmutagenized, plasmid vector. The ligation mixture is transformed or transfected into the appropriate host cells. The transformants are then exposed to selective conditions for mutant isolation.

targeted for PCRM is similarly isolated by preparative gel electrophoresis. Thus, isolation of completely double-digested vector is assured. The PCRM target and plasmid vector fragments are mixed in 1:1 molar ratios and ligated. The ligation mixture is then used to transform an appropriate *E. coli* strain and the culture is plated to screen or select for colonies with mutants of the desired phenotype. Plasmid DNA is isolated from primary mutant candidates and retested for the ability to confer the mutant phenotype. DNA sequencing of the mutagenized target region in plasmids conferring validated mutant phenotypes ultimately characterizes the sequence alterations that give rise to the mutant phenotypes. To confirm that the PCRM target region is the only part of the plasmid responsible for conferring the mutant phenotype, the fragment can be cut out of the plasmid and recloned into a fresh vector background. The reconstructed plasmid is retested to confirm the location of the mutation in the target region, not the vector.

Recently PCRM has been applied to isolate mutants in a variety of genes (Table 1). These include mutations within the prosequence of subtilisin *(21)* and mature region *(22)* of subtilisin (an alkaline serine protease form *Bacillus subtilus*); the *Escherichia coli recA (20)* and *crp (27)* genes; the 2A proteinase of human rhinovirus 2 *(24)*; the kringle 1 domain of human tissue plasminogen activator *(26)*; and the light (V_L) and heavy (V_H) chain variable regions of mouse IgG *(25)*. As in all random mutagenesis experiments, the key to success lies in a clear screen or strong selection for mutants of the desired phenotype. In the case of subtilisin prosequence mutants, defective protease activity is easily scored as lack of halo formation around *E. coli* colonies growing on casein plates. Halos fail to form around colonies in which cells secrete plasmid encoded prosubtilisin (inactive) that is unable to be processed to mature (active) subtilisin owing to PCR-induced prosequence mutations *(21)*.

In addition to the four mutations initially reported, more than 25 additional prosequence mutations have subsequently been isolated. PCRM was further applied to isolate second-site suppressor mutations in the mature region that allow formation of active protease and thus compensate for a prosequence mutation *(22)*. Thus, PCRM can help identify important domain interactions required for proper protein folding. Regions of the *E. coli* RecA protein involved in self-assembly or dsDNA interaction were also identified by PCRM. The PCR mutagenized *recA* gene was cloned into a λgt11 expression vector. Mutant RecA proteins were identified by immunoscreening for loss of epitope reactivity with monoclonal antibodies that block self assembly or dsDNA binding *(20)*. Mutations in the *E. coli crp* gene encoding the catabolite gene activator protein were identified by PCRM using a simple screen consisting of a functional plate assay for the Crp^- phenotype on indicator agar *(26)*.

Mutations within the 2A proteinase of human rhinovirus 2 were isolated by PCRM using a unique screen based on a gene fusion construct in which a portion of the *E. coli lacZ* gene that encodes the α-fragment of β-galactosidase was fused in frame with the 2A proteinase coding region. The 2A protease is naturally produced as a polyprotein precursor that cleaves itself out of the polyprotein. A *lacZ*::2A-proteinase hybrid was constructed such that active protease would liberate the *lacZ* α-fragment and allow α-complementation (formation of active β-galactosidase). Protease activity was thus scored as blue M13 plaques on plates containing the chromogenic substrate X-gal. Defective 2A proteinase mutants, scored as faint blue or white plaques, were found to contain the unprocessed α-fragment trapped in the α-fragment::2A proteinase hybrid *(23)*. PCRM was also used to identify mutations in the human tissue plasminogen activator (t-PA) kringle 1 domain that result in failure of the protein to bind its clearance receptor. The mutations were isolated by fluorescence-activated cell sorting (FACS) of cells transiently transfected with a plasmid expressing a hybrid t-PA fusion protein that contains a membrane glycophospholipid anchor sequence. The hybrid t-PA is thus bound to the cell surface. FACS was used to select a population of cells expressing mutant t-PA that was unable to bind to a monoclonal antibody that inhibits binding of t-PA to its receptor *(25)*. PCRM was also used to generate monoclonal antibodies in a combinatorial Fab library with enhanced affinity for a progesterone. The combinatorial Fab library displayed on the surface of filamentous phage was constructed with PCR-mutagenized IgG light (V_L) and heavy (V_H) chain variable regions from nonimmune mice. PCRM mutagenesis allowed isolation of high affinity mutant antibodies from an otherwise nonimmune library, thus bypassing the need for immunization steps. This is especially useful for haptens, such as progesterone, that elicit restricted immune responses *(24)*.

Table 1
Applications of Localized Random PCR Mutagenesis

Reference	21	22	27	20	24	26	25
Target (gene/protein)	Pro-sequence of subtilisin, a secreted protese from *B. subtilus*	Subtilisin pro-sequence and mature region in gene containing pro-sequence mutation	*Crp* gene encoding the *E. coli* catabolite gene activator protein	*E. coli recA* gene	Human rhinovirus 2A proteinase	Human tissue plasmin-ogen activator kringle 1 domain	Mouse IgG light (V_L) and heavy (V_H) chain variable regions
Target size	460 bp	460 bp pro-sequence and 930 bp mature region	633 bp	ca. 273 bp	450 bp	330 bp	ca. 324 bp
PCR reaction conditions	Standard	Standard	Standard	Error prone (+dITP)	Standard with additional *Taq* and cycles	Error prone (ref. *23)*	Error prone (ref. *23)*
% Phenotypic mutants	4.3%	4.2% in pro-sequence and 0.7% in mature region	11%	0.5% Maximal	ca. 6%	ND	ND
Mutation bias	71% A–G/T–C transitions	ND (Only aa changes discussed)	Transitions greater than trans-versions	70% A–G/T–C transitions, 15% transversions	ND (Only aa changes discussed)	ND (Only aa changes discussed)	68% A–G/T–C transitions

Hot spots	None	Yes, in domains	None	ND	ND	ND	Yes
% Multiple mutations; avg. no/ clone	14%; 1.4	1.8%; 1.03	ND	20%; 1.3	46%; 1.5	74%; 2.9	100%; 14.5
% Deletions	14%	ND	ND	ND	12.5%	ND	ND
Screen selection method	Plate assay for protease activity detected by halo formation	Plate for assay for loss or recovery of protease activity detected by halo formation	Plate assay for Crp phenotype on indicator agar	Epitope loss detected by immunoblotting of RecA in λgt11 expression vector	Plate assay for protease activity detected by cleavage of protease-*LacZ* fusion	FACS enrichment for loss of epitope expressed on transfected cell surface	Affinity selection of combinatorial filamentous phage Fab library to isolate high affinity species

To date, mutations isolated by PCRM in six genes have been reported. A compilation of information describing the mutagenesis conditions, screening systems, and results are listed. bp = base pairs, aa = amino acids, hot spots = higher than average mutational frequency, ND = not discussed.

2. Materials

Standard procedures for working with nucleic acids should be followed. Solutions should be prepared with distilled and deionized water. Other than electrophoresis buffers, all solutions as well as disposable plasticware should be sterilized by autoclaving. Microfuge tubes for PCR reactions or dilution of *Taq* DNA polymerase should be siliconized. To do so, flush the tubes several times with a 5% solution of dichlorodimethylsilane in chloroform using a Pasteur pipet. Rinse the tubes in water several times and autoclave. Be sure to wear goggles, a lab coat, and two pairs of gloves and to work in a fume hood since the solution is toxic and highly volatile. Alternatively, siliconized PCR reaction tubes can be obtained commercially (e.g., tube no. RA0650-GST, National Scientific Supply Co. Inc., San Rafael, CA).

2.1. Generating the PCRM Product

1. Target plasmid DNA template: Purify by conventional means (*see* Note 1), resuspend in TE, and dilute according to usage.
2. PCR primers: Depending on their quality, the oligonucleotide primers can be used without further purification. Store as a 2-µ*M* (10 pmole/µL) stock at –20°C (*see* Note 2).
3. PCR reagents: 10X PCR buffer (500 m*M* KCl, 100 m*M* Tris-HCl, pH 9.0 25°C, 0.1% gelatin); 1.25 m*M* deoxynucleotide triphosphates (dNTP), 25 m*M* $MgCl_2$, *Taq* DNA polymerase (1.25 U/µL) and mineral oil. These can be obtained as a kit (Perkin Elmer/Cetus, Norwalk, CT).
4. 1X TE: 10 m*M* Tris-HCl, pH 8.0, 1 m*M* EDTA.
5. Phenol:chloroform: Prepare buffer-saturated phenol by mixing equal volumes melted phenol (65°C) and 0.5*M* Tris-HCl, pH 8.0 in a separatory funnel. After phase separation the lower phenol phase is collected and the procedure is repeated with 0.1*M* Tris-HCl, pH 8.0 until the pH of the phenolic phase is >7.8 (as measured with pH paper). Mix equal volumes of phenol and chloroform and 1/25 vol of isoamyl alcohol. For a chloroform extract mix 1 vol of chloroform with 1/25 vol of isoamyl alcohol.
6. 3*M* Sodium acetate (NaOAc): Bring to pH 7.0 with glacial acetic acid and sterilize by autoclaving.
7. Restriction enzymes and buffers. Use as according to manufacturer's instructions. Store as 5–40 U/µL stocks at –20°C.
8. 10X DNA loading dye: 0.25% bromophenol blue, 0.25% xylene cyanol FF, 30% glycerol, 5 m*M* EDTA, pH 8.0, 10 m*M* Tris-HCl, pH 8.0. Prepare by dissolving 0.125 g of bromophenol blue, 0.125 g of xylene

cyanol FF in 20 mL of water. Add 15 mL of glycerol, 0.5 mL of EDTA, pH 8.0, 0.5 mL of 1*M* Tris-HCl, pH 8.0, and sterile water to a final volume of 50 mL.

10. 1X TBE: Dissolve 10.8 g of Tris base and 5.5 g of boric acid in H_2O. Add 4 mL of 0.5*M* EDTA (pH 8.0). Add H_2O to 1 L.
11. PAGE: A preparative 5% polyacrylmide gel with 1-mm thick spacers. For 60 mL of gel solution, mix 10 mL of acrylamide-*bis* acrylamide (29:1) stock solution, 6 mL of 10X TBE, 4 mL of 50% glycerol (v/v in water), 40 mL of water, and 85.5 mg of ammonium persulfate. Add 25 μL of TEMED (*N,N, N', N'*, tetramethylethylenediamine) just before pouring. Caution must be taken since unpolymerized acrylamide is a neurotoxin: Always wear gloves when handling it.
12. Nuclease-free dialysis tubing: To prepare, cut the tubing (Spectra/Por, 25 mm 2 mL/cm, mwco: 12,000–14,000, Spectrum Medical Industries, Inc., Los Angeles, CA) into strips 8–10 cm in length. Place the strips in a large volume of 2% (w/v) sodium bicarbonate and 1 m*M* EDTA (pH 8.0) preheated to near boiling in a microwave. After 10 min, rinse thoroughly in distilled H_2O. Transfer the strips to 1 m*M* EDTA in a 1-L beaker, cover with aluminum foil, and autoclave for 10 min on liquid cycle. Allow the tubing to cool and store at 4°C. Use with dialysis tubing closure no. 132736, available from Spectrum.
13. 1X TAE: Dissolve 4.84 g of Tris base in H_2O. Add 1.14 mL of glacial acetic acid and 2 mL of 0.5*M* EDTA (pH 8.0). Add H_2O to 1 L.

2.2. Cloning of PCRM Product

14. 10X ligase buffer: For 1 mL, mix 0.35 mL of H_2O, 0.5 mL of 1*M* Tris-HCl, pH 7.5, 50 μL of 2*M* $MgCl_2$, and 31 mg of dithiothreitol.
15. T4 DNA ligase: Store stock at –20°C. Just prior to use dilute between 1:10 to 1:20 with 1X ligase buffer to a concentration of 0.1–0.3 U (Weiss)/μL.
16. High transformation efficiency competent cells: These can be obtained commercially (e.g., Gibco BRL, Gaithersburg, MD) or prepared according to the method of Hanahan *(28)* as described in Perbal (*29*; *see* Note 3).

3. Methods

3.1. Generating the PCRM Product

1. In a siliconized microfuge tube mix 30 μL of 10X PCR buffer, 18 μL of $MgCl_2$, 10 μL of oligonucleotide primer 1, 10 μL of oligonucleotide primer 2, 25 ng of plasmid template containing target gene, and sterile H_2O to 270 μL. Heat mixture at 95°C for 10 min and then quickly cool in an ice-water bath.

2. Dispense 90 µL/tube into a siliconized 0.5-mL microfuge tube. Place the tubes in a thermal cycler set at 73°C. Add 10 µL of a freshly prepared mixture consisting of 2.0 µL of *Taq* DNA polymerase and 8 µL of dNTP. Overlay the solution with 50 µL of mineral oil.
3. Run the PCR in a thermal cycler set to incubate at 73°C for 3 min and then cycle 25 times at 94°C for 1 min; 50°C, 1 min; 72°C, 1 min.
4. Remove lower aqueous phase to a fresh tube. Analyze 2.5 µL of the reaction products by standard agarose gel electrophoresis.
5. Add 100 µL of TE to 90 µL of the PCR reaction. Mix 100 µL of phenol:chloroform by tapping the tube repeatedly with your finger. Separate the phases by centrifugation for 10 min in a microcentrifuge at room temperature and remove the upper aqueous phase to a fresh tube containing 100 µL of chloroform. Mix, and centrifuge for 5 min as above.
6. Precipitate the DNA by adding 20 µL of 3*M* NaOAc (pH 7) and 440 µL of absolute EtOH, mixing by inversion, and chilling at –70°C for 10 min. Collect the precipitate by centrifugation for 15 min in a microcentrifuge at 4°C. Resuspend the PCR product in 63.5 µL of sterile deionized water.
7. Add 7.5 µL of 10X concentrated restriction enzyme digestion buffer compatible with both enzymes and 2.0 µL of each restriction enzyme (X and Y in Fig. 1). Mix and incubate 2 h at 37°C.
8. Prepare a standard preparative PAGE gel in 1X TBE buffer and pre-electrophorese for 30 min at 200 V. Add 7.5 µL of a standard 10X DNA loading dye to the sample and load on the preparative PAGE gel. Run the gel at 200 V until the terminal fragments resolve from the fully digested core PCRM target fragment.
9. Stain the gel in 1X TBE plus 0.5 µg/mL ethidium bromide (EthBr) for 30 min in the dark. Visualize the EthBr-stained DNA by placing the gel on a long wave UV transilluminator. Cover the transilluminator surface with plastic wrap to prevent contamination. Using an EtOH-swabbed safety razor or scalpel, cut out the band containing the fully digested PCRM fragment.
10. Rinse nuclease free dialysis tubing in distilled water and cut a piece that is four times longer than the excised band (ca. 8 cm). Close one end of the bag with a clip and place in the trough of the submarine gel tank containing 1X TAE buffer. Flame the tip of a Pasteur pipet to smooth any sharp edges that may rip the bag, then use it to rinse the bag with buffer. Fill the bag completely and drop in the gel slice. Carefully holding the bag to retain the fragment, pour some of the buffer out, and position the gel along the length of the bag. Holding the slice at the side of the bag decant almost all of the buffer out of the bag; save the last drop. Do not

squeeze all of the liquid out. Attach the second dialysis clip and position the bag in the electorphoresis tank such that the DNA will migrate out of the gel slice into the residual buffer in the bag. Be sure the buffer level is high enough to just cover the bag. Electrophorese at 100–150 V for 20 min (*see* Note 4).

11. Check elution by examining the bag briefly on a short wave UV transilluminator. All of the DNA as detected by EthBr staining should be visible as a line along the far edge of the bag. When all of the stain has left the gel slice, unclip one end and carefully remove the liquid to a 1.5-mL microfuge tube. Measure the volume collected (<400 µL) using an adjustable micropipetor. Add 1/10 vol of 3*M* NaOAc pH 7 and 2.5 vol of absolute EtOH and chill at –20°C for 30 min to overnight. Collect the precipitate by microcentrifugation 15 min at 4°C and discard the supernatant. Respin and remove the remaining fluid. Dissolve the pellet in 17 µL of 0.1X TE.
12. Check 2 µL on a miniagarose gel. Approximate the DNA concentration of the fragment by comparing it to the intensity of a known amount of size standard DNA, such as *Hind*III cut λ DNA (*see* Note 5).

3.2. Cloning of PCRM Product

1. Digest 2–5 µg of the same plasmid DNA used in the PCRM reaction with the same buffer and restriction enzymes as those used above for digestion of the target DNA PCR product.
2. Isolate the vector fragment as described for the target DNA PCR product by adding 0.1 vol of loading dye, running a preparative PAGE gel, electroeluting DNA fragments, and checking the concentration of vector fragment DNA obtained.
3. Combine 2 µL of 10X ligase buffer, 200 ng of double-digested, electroeluted PCR fragment DNA, an equimolar amount of double-digested electroeluted vector DNA, and distilled water to a total of 16 µL. Add 2 µL of 10 m*M* ATP and 2 µL of T4 DNA ligase. Mix gently; incubate 4 h at room temperature or overnight (ca. 16 h) at 16°C.
4. As a small scale test transformation, add 1 µL of ligation mix to 20 µL of competent *E. coli* cells. Keep on ice 20 min, then heat shock 2 min at 37°C. Add 200 µL LB; keep at 37°C for 20 min. Plate 100 µL onto LB plates containing the appropriate antibiotic and incubate at 37°C overnight.
5. Scale up the transformation accordingly (*see* Notes 6 and 7).

The transformants for the desired phenotype should be screened and sequenced accordingly (*see* Note 8).

4. Notes

1. Plasmid DNA purification can be done by all conventional methods. Since the misincorporation rate of *Taq* DNA polymerase is sensitive to reaction conditions, such as variable pH and the presence of divalent metal ions, it is preferable to take extra steps to purify the plasmid in order to obtain reproducible mutagenesis conditions. Conventional ethidium bromide-CsCl gradients or column chromatographic methods (e.g., Qiagen columns, Chatsworth, CA; Dusseldorf, Germany) can achieve these conditions. Generally, it is not necessary to linearize the plasmid before PCR mutagenesis. However, since linearization occasionally increases PCR product yields; a quick test with equal amounts of linearized and uncut plasmid should initially be performed.
2. Using the conventional wisdom discussed in the references for further reading, oligonucleotide primers of approx 20 nucleotides in length should be designed such that they do not contain repeats capable of intramolecular folding, intermolecular self-annealing, or annealing with the second primer. The primers should contain approx 50% G+C. Primers designed to flank the cloning site by approx 50 nucleotides facilitate clear resolution of the fully digested restriction fragment from the undigested PCR product during preparative gel purification. The universal and reverse sequencing primers of the M13/pUC series cloning vectors are useful for PCR mutagenesis of target genes cloned in the multiple cloning sites. Alternatively, restriction sites can be added by site-directed mutagenesis such that the target region is perfectly localized. Enzymes that leave noncompatible ends facilitate forced cloning of the PCRM product, thus avoiding false positive mutants owing to incorrect insert orientation or vector religation.

   We routinely purifiy primers using commercially available cartridges (Applied Biosystems, Foster City, CA). The primers may also be purified by HPLC or a standard preparative polyacrylamide-urea electrophoresis procedure involving subsequent crush and soak elution and anion exchange chromatography steps.
3. Preparation of competent *E. coli*: A 50-mL culture of the desired *E. coli* strain is incubated at 37°C in LB to approx 0.45 OD_{550}. The culture is then incubated on ice 10–15 min and cells from 35 mL of the culture are collected by centrifugation in a sterile capped tube in a SS34 Sorvall rotor at 4000 rpm for 4 min at 4°C. The pellet is resuspended in 2 mL of competent cell solution 1, an additional 14 mL is added, and the tube is incubated on ice for 15 min. The cells are pelleted as above and resuspended in 1.6 mL of competent cell solution 2. The tube is kept on

ice 15 min after which the cells are ready for transformation or storage in aliquots at –70°C.

- LB: Dissolve 5 g of NaCl, 5 g of yeast extract, 10 g of Bacto-peptone, and 1 g of glucose in 1 L H_2O. Add 10 mL of 1.0*M* Tris-HCl, pH 7.5.
- Competent cell solution 1: Dissolve 0.39 g of MES in 100 mL H_2O and adjust the pH to 6.2. Dissolve 2.42 g of $RbCl_2$, 0.22 g $CaCl_2$, and 1.98 g $MnCl_2 \cdot 4H_2O$ in 50 mL H_2O. Mix the MES and metal ion solutions and carefully adjust the pH to 5.8 with diluted acetic acid (do not back titrate with hydroxide). Adjust the volume to 200 mL with H_2O, filter sterilize, and store at 4°C.
- Competent cell solution 2: Dissolve 0.42 g of MOPS free acid, 1.66 g of $CaCl_2$, and 0.24 g of $RbCl_2$ in 100 mL H_2O. Add 30 mL of glycerol. Adjust the pH to 5.8. Add H_2O to a final volume of 200 mL, filter sterilize, and store at 4°C.

4. Electroelution of DNA in dialysis bags has generally proven adequate; however the protocol is simplified by using a commercially available kit. A variety of kits are available for simple rapid isolation of DNA fragments from agarose gels (e.g., Qiaex resin from Qiagen). When these kits are used in combination with agarose gels that can resolve DNA fragments from 8 to 1000 bp (e.g., NuSieve GTG agarose, FMC, Rockland, ME), electroelution of the digested PCRM product and plasmid vector from polyacrylamide gels can be eliminated.
5. 200 ng of *Hind*III cut λ DNA will have the following amount of DNA in each fragment (be sure to heat the sample 2 min at 42°C before loading to dissociate the 11 bp cos site between the 23 and 4.4-kb bands): 23.1 kb, 95 ng; 9.4 kb, 39 ng; 6.6 kb, 27 ng; 4.4 kb, 18 ng; 2.3 kb, 9.5 ng; 2.0 kb, 8.2 ng; 0.56 kb, 2.3 ng.
6. The method described for preparation of competent cells has worked very well. Be sure to use an *E. coli* strain that transforms with high frequency. If the mutant screening method of choice requires use of a strain that transforms poorly, first amplify the mutagenized library by transforming a high frequency strain. Then pool the transformants by soaking the colonies off of the plate. This can be done by adding LB, sealing the plate with parafilm, and gently shaking the plate on an orbital shaking platform. Transfer the cell suspension to a centrifuge tube, harvest the cells, and isolate the plasmid DNA. Use the amplified library to transform the strain for mutant isolation. Alternatively, electroporation could be tried with the poorly transforming strain.
7. Given good competent *E. coli* cells, poor transformation frequencies are most frequently a result of using too much T4 DNA ligase in the

ligaion. The excess ligase results in formation of large DNA concatamers that transform poorly. Thus, suppress the tendency to use more ligase if transformation frequencies are low.

8. Enhanced mutagenic conditions for PCRM have been reported *(20,22, 24,25)*. Leung et al. *(22)* increased the overall frequency of nucleotide misincorporation to 2% (fivefold) by using a dGTP/dATP ratio of five and adding 0.5 m*M* Mn^{2+} to the reaction. Although concern has been raised that the reported increases in error rate lie within the statistical expectations for a random sampling *(2)*, these conditions have resulted in isolation of mutant genes containing more nucleotide changes on average than when standard conditions have been employed (*24,25*; Table 1). Ikeda et al. *(20)* employed PCRM with dITP to enhance the misincorporation rate of *Taq*. They observed a three- to tenfold increase in phenotypic mutants without increasing the number of nucleotide changes per mutant. Thus, if heavily mutagenized genes are desired, the conditions of Leung et al. *(22)* should be employed. To obtain mutants generally containing single amino acid changes, standard PCR conditions or the addition of up to 200 μ*M* dITP should be employed. Elevated pH (8.2) and high concencentrations of $MgCl_2$ relative to dNTP (10–20-fold) have also been found to decrease the fidelity of *Taq* DNA polymerase *(5)*. However, if the $MgCl_2$:dNTP ratio is too high an undesirable increase in frameshift mutations is observed. A 2.5-fold molar excess of $MgCl_2$ over dNTP is best for maximizing substitution and minimizing frameshift mutations (Fig. 1 in ref. *5*). High dNTP concentrations have been reported to increase the misincorporation frequency of DNA polymerases including *Taq (2)*. Lower temperature and longer time for the elongation step have been suggested to increase the readthrough of misincorporated nucleotides *(2)*. Although reactions containing low initial template concentrations are prone to accumulate mutations, it seems likely that the number of multiple mutations would increase.

Further Reading

Erlich, H. A. (1989) *PCR Technology: Principles and Applications for DNA Amplification.* Stockton, London, New York.

Innis, M. A., Gelfand, D. H., Sninsky, J. J., and White, T. J. (1990) *PCR Protocols: A Guide to Methods and Applications.* Academic, London, New York.

Sambrook, J., Fritsch, E. F., and Maniatis, T. (1989) *Molecular Cloning: A Laboratory Manual.* Cold Spring Harbor Laboratory, Cold Spring Harbor, NY.

References

1. Saikai, R. K., Gelfand, D. H., Stoffel, S., Scharf, S. J., Higuchi, R., Horn, G. T., Mullis, K. B., and Erlich, H. A. (1988) Primer-directed enzymatic amplification of DNA with a thermostable DNA polymerase. *Science* **239,** 487–491.

2. Bloch, W. (1991) A biochemical perspective of the polymerase chain reaction. *Biochemistry* **30,** 2735–2747.
3. Bej, A. K., Mahbubani, M. H., and Atlas, R. M. (1991) Amplification of nucleic acids by polymerase chain reaction (PCR) and other methods and their applications. *Crit. Rev. Biochem. Mol. Biol.* **26,** 301–334.
4. Tindall, K. R. and Kunkel, T. A. (1988) Fidelity of DNA synthesis by the *Thermus aquaticus* DNA polymerase. *Biochemistry* **27,** 6008–6013.
5. Eckert, K. A. and Kunkel, T. A. (1990) High fidelity DNA synthesis by the *Thermus aquaticus* DNA polymerase. *Nucleic Acids Res.* **18,** 3739–3744.
6. Reiss, J., Krawczak, M., Schloesser, M., Wagner, M., and Cooper, D. N. (1990) The effect of replication errors on the mismatch analysis of PCR-amplified DNA. *Nucleic Acids Res.* **18,** 973–978.
7. Kadonaga, J. T. and Knowles, J. R. (1985) A simple and efficient method for chemical mutagenesis of DNA. *Nucleic Acids Res.* **13,** 1733–1745.
8. Myers, R. M., Lerman, L. S., and Maniatis, T. A. (1985) A general method for saturation mutagenesis of cloned DNA fragments. *Science* **229,** 242–247.
9. Shortle, D. and Nathans, D. (1978) Local mutagenesis: a method for generating viral mutants with base substitutions in preselected regions of the viral genome *Proc. Natl. Acad. Sci. USA* **75,** 2170–2174.
10. Botstein, D. and Shortle, D. (1985) Strategies and application of *in vitro* mutagenesis. *Science* **229,** 1193–1201.
11. Hermes, J. D., Parekh, S. M., Backlow, S. C., Koster, H., and Knowles, J. R., (1989) A reliable method for random mutagenesis: the generation of mutant libraries using spiked oligodeoxyribonucleotide primers. *Gene* **84,** 143–151.
12. Reidhaar-Olson, J. and Sauer, R. (1988) Combinatorial cassette mutagenesis as a probe of the information content of protein sequences. *Science* **241,** 53–57.
13. Smith, M. (1985) In vitro mutagenesis. *Annu. Rev. Genet.* **19,** 423–462.
14. Wells, J. A., Vasser, M., and Powers, D. B. (1985) Cassette mutagenesis: an efficient method for generation of mutiple mutations at defined sites. *Gene* **34,** 315–323.
15. Lhetovaara, P. M., Koivula, A. K., Bamford, J., and Knowles, J. C. K. (1988) A new method for random mutagenesis of complete genes: enzymatic generation of mutant libraries in vitro. *Prot. Eng.* **2**, 63–68.
16. Liao, X. and Wise, J. A. (1990) A simple high-efficiency method for random mutagenesis of cloned genes using forced nucleotide misincorporation. *Gene* **88,** 107–111.
17. Shiraishi, H. and Shimura, Y. (1988) A rapid and efficient method for targeted random mutagenesis. *Gene* **64,** 313–319.
18. Shortle, D., Grisafi, P., Benkovic, S. J., and Botstein, D. (1982) Gap misrepair mutagenesis: efficient site-directed induction of transition, transversion, and frameshift mutations in vitro. *Proc. Natl. Acad. Sci. USA* **79,** 1588–1592.
19. Zakour, R. A. and Loeb, L. A. (1982) Site-specific mutagenesis by error-directed DNA synthesis. *Nature* **295,** 708–710.
20. Ikeda, M., Hamano, K., and Shibata, T. (1992) Epitope mapping of anti-recA protein IgGs by region specified polymerase chain reaction mutagenesis. *J. Biol. Chem.* **267,** 6291–6296.

21. Lerner, C. G., Kobayashi, T., and Inouye, M. (1990) Isolation of subtilisin prosequence mutations that affect formation of active protease by localized random polymerase chain reaction mutagenesis. *J. Biol. Chem.* **265,** 20,085–20,086.
22. Kobayashi, T. and Inouye, M. (1992) Functional analysis of the intramolecular chaperone, mutational hot spots in the subtiliisin pro-peptide and a second-site suppressor mutation within the subtilisin molecule. *J. Mol. Biol.* **226,** 931–933.
23. Leung, D. W., Chen, E., Goeddel, D. V. (1989) A method for random mutagenesis of a defined DNA segment using a modified polymerase chain reaction. *Technique* **1,** 11–15.
24. Liebig, H.-D., Skern, T., Luderer, M., Sommergruber, W., Blaas, D., and Kuechler, E. (1991) Proteinase trapping: screening for viral proteinase mutants by a complementation. *Proc. Natl. Acad. Sci. USA* **88,** 5979–5983.
25. Gram, H., Marconi, L.-A., Barbas, C. F., Collet, T. A., Lerner, R. A., and Kang, A. S. (1992) In vitro selection and affinity maturation of antibodies from a naive combinatorial immunoglobulin library. *Proc. Natl. Acad. Sci. USA* **89,** 3576–3580.
26. Rice, G. C., Goeddel, D. V., Cachianes, G., Woronicz, J., Chen, E. Y., Williams, S. R., and Leung, D. W. (1992) Random PCR mutagenesis screening of secreted proteins by direct expression in mammalian cells. *Proc. Natl. Acad. Sci. USA* **89,** 5467–5471.
27. Zhou, Y., Zhang, X., and Ebright, R. H. (1991) Random mutagenesis of gene-sized DNA molecules by use of PCR with Taq DNA polymerase. *Nucleic Acids Res.* **19,** 6052.
28. Hanahan, D. (1983) Studies on transformation of *Escherichia coli* with plasmids. *J. Mol. Biol.* **166,** 557–580.
29. Perbal, B. (1988) *A Practical Guide to Molecular Cloning*, Wiley, New York, Chichester, pp. 411–413.

PART III

Genomic Structure

CHAPTER 11

The Simultaneous Isolation of RNA and DNA from Tissues and Cultured Cells

Frank Merante, Sandeep Raha, Juta K. Reed, and Gerald Proteau

1. Introduction

Many techniques are currently available that allow the isolation of DNA *(1–7)* or RNA *(8–23)*, but such methods allow only the purification of one type of nucleic acid at the expense of the other. Frequently, when cellular material is limiting, it is desirable to isolate both RNA and DNA from the same source. Such is the case for biopsy specimens, primary cell lines, or manipulated embryonic stem cells.

Although several procedures have been published that address the need to simultaneously purify both RNA and DNA from the same source *(24–31)*, most methods are simply a modification of the original procedure of Chirgwin et al. *(8)*. Such procedures utilize strong chaotropic agents, such as guanidinium thiocyanate and cesium trifluoroacetate *(25,27)*, to simultaneously disrupt cellular membranes and inactivate potent intracellular RNAses *(26,29,30)*. The limitations of such techniques are the need for ultracentrifugation *(26–28,30)* and long processing times (ranging 16–44 h).

Methods for isolating both RNA and DNA that circumvent the ultracentrifugation step take advantage of the fact that phenol *(1,32)* can act as an efficient deproteinization agent quickly disrupting cellular integrity and denaturing proteins *(24,31)*. The method presented here takes advantage of the qualities offered by phenol extraction

From: *Methods in Molecular Biology, Vol. 31: Protocols for Gene Analysis*
Edited by: A. J. Harwood

when it is coupled with a suitable extraction buffer and a means for selectively separating high mol wt DNA from RNA *(31)*.

The method utilizes an initial phenol extraction coupled with two phenol:chloroform extractions to simultaneously remove proteins and lipids from nucleic acid containing solutions. In addition, the constituents of the aqueous extraction buffer are optimized to increase nucleic acid recovery as discussed by Wallace *(33)*. For example, the pH of the buffer (pH 7.5), the presence of detergent (0.2% SDS), and relatively low salt concentration (100 m*M* LiCl) allow the efficient partitioning of nucleic acids into the aqueous phase and the dissociation of proteins. In addition, the presence of 10 m*M* EDTA discourages the formation of protein aggregates *(33)* and chelates Mg^{2+}, thereby inhibiting the action of magnesium dependent nucleases *(34)*.

This method differs from that presented by Krieg et al. *(24)* in that the lysis and extraction procedure is gentle enough to allow the selective removal of high mol wt DNA by spooling onto a hooked glass rod *(2,34,35)* following ethanol precipitation. This avoids additional LiCl precipitation steps following the recovery of total nucleic acids. Finally, the procedure can be scaled up or down to accommodate various sample sizes, hence allowing the processing of multiple samples at one time. The approximate time required for the isolation of total cellular RNA and DNA is 2 h. Using this method nucleic acids have been isolated from PC12 cells and analyzed by Southern and Northern blotting techniques *(31)*.

2. Materials

Molecular biology grade reagents should be utilized whenever possible. Manipulations were performed in disposable, sterile polypropylene tubes whenever possible, otherwise glassware that had been previously baked at 280°C for at least 3 h was used.

2.1. Nucleic Extraction from Nonadherent Tissue Culture Cells

1. PBS: 0.137*M* NaCl, 2.68 m*M* KCl, 7.98 m*M* Na_2HPO_4, 1.47 m*M* KH_2PO_4, pH 7.2.
2. DEPC-treated water: Diethylpyrocarbonate (DEPC)-treated water is prepared by adding 1 mL DEPC to 1 L of double-distilled water (0.1% DEPC v/v) and stirring overnight. The DEPC is inactivated by autoclaving at 20 psi for 20 min (*see* Note 1).

3. STEL buffer: 0.2% SDS, 10 m*M* Tris-HCl, pH 7.5, 10 m*M* EDTA, and 100 m*M* LiCl). The buffer is prepared in DEPC-treated water by adding the Tris-HCl, EDTA, and LiCl components first, autoclaving, and then adding an appropriate vol of 10% SDS. The 10% SDS stock solution is prepared by dissolving 10 g SDS in DEPC-treated water and incubating at 65°C for 2 h prior to use.
4. Phenol: Phenol is equilibrated as described previously *(34)*. Ultrapure, redistilled phenol, containing 0.1% hydroxyquinoline (as an antioxidant), is initially extracted with 0.5*M* Tris-HCl, pH 8.0 and then repeatedly extracted with 0.1*M* Tris-HCl, pH 8.0 until the pH of the aqueous phase is 8.0. Then equilibrate with STEL extraction buffer twice prior to use. This can be stored at 4°C for at least 2 mo.
5. Phenol:chloroform mixture: A 1:1 mixture was made by adding an equal volume of chloroform to STEL equilibrated phenol. Can be stored at 4°C for at least 2 mo. Phenol should be handled with gloves in a fume hood.
6. 5*M* LiCl: Prepare in DEPC-treated water and autoclave.
7. TE: 10 m*M* Tris-HCl, pH 8.0, 1 m*M* EDTA (pH 8.0). Prepare in DEPC-treated water and autoclave.
8. RNA guard, such as RNasin (Promega; Madison, WI).

2.2. Variations for Adherent Cell Cultures and Tissues

9. Trypsin: A 0.125% solution in PBS. For short term store at 4°C; for long term freeze.

3. Methods

3.1. Nucleic Extraction from Nonadherent Tissue Culture Cells

In this section we detail nucleic extraction from nonadherent tissue culture cells. Section 3.2. describes variations of this protocol for adherent cell cultures and tissue.

To prevent RNase contamination from skin, disposable gloves should be worn throughout the RNA isolation procedure. In addition, it is advisable to set aside equipment solely for RNA analysis; for example, glassware, pipets, and an electrophoresis apparatus.

1. Cool a cell culture that contains approx 1×10^7 cells on ice (*see* Note 2). Transfer to 15-mL polypropylene tubes and pellet by centrifugation at 100*g* for 5 min. Wash the cells once with 10 mL of ice-cold PBS and repellet. The pelleted cells may be left on ice to allow processing of other samples.

2. Simultaneously add 5 mL of STEL equilibrated phenol and 5 mL of ice-cold STEL buffer to the pelleted cells. Gently mix the solution by inversion for 3–5 min ensuring the cellular pellet is thoroughly dissolved (*see* Notes 3 and 4).
3. Centrifuge the mixture at 10,000*g* for 5 min at 20°C to separate the phases. Transfer the aqueous (upper) phase to a new tube using a sterile polypropylene pipet and reextract twice with an equal volume of phenol:chloroform (*see* Note 5).
4. Transfer the aqueous phase to a 50-mL Falcon tube. Differentially precipitate high mol wt DNA from the RNA component by addition of 0.1 vol of ice-cold 5*M* LiCl and 2 vol of ice-cold absolute ethanol. The DNA will precipitate immediately as a threaded mass.
5. Gently compact the mass by mixing and remove the DNA by spooling onto a hooked glass rod. Remove excess ethanol from the DNA by touching onto the side of the tube. Remove excess salts by rinsing the DNA with 1 mL of ice-cold 70% ethanol while still coiled on the rod. Excess ethanol can be removed by carefully washing the DNA with 1 mL of ice-cold TE, pH 8.0 (*see* Note 6).
6. The DNA is then resolubilized by transferring the glass rod into an appropriate volume of TE, pH 8.0 and storing at 4°C.
7. The RNA is precipitated by placing the tube with the remaining solution at –70°C, or in an ethanol/dry ice bath for 30 min.
8. Collect the RNA by centrifuging at 10,000*g* for 15 min. Gently aspirate the supernatant and rinse the pellet with ice-cold 70% ethanol. Recentrifuge for 5 min and remove the supernatant. Dry the RNA pellet under vacuum and resuspend in DEPC-treated water.
9. For storage as aqueous samples add 5–10 U of RNasin (RNase inhibitor) according to manufacturer's instructions. Alternatively, the RNA can be safely stored as an ethanol/LiCl suspension (*see* Notes 7 and 8).

3.2. Variations for Adherent Cell Cultures and Tissues

3.2.1. Adherent Cells

1. Remove the culture medium from the equivalent of 1×10^7 cells by aspiration and wash the cells once with 10 mL of PBS at 37°C.
2. Add 1 mL of trypsin solution and incubate plates at 37°C until the cells have been dislodged. This should take approx 10 min. Dilute the trypsin solution by addition of ice-cold PBS.
3. Follow Section 3.1., steps 2–9.

3.2.2. Procedure for Tissue

1. Rinse approx 500 mg of tissue free of blood with ice-cold PBS. Cool and mince into 3–5-cm cubes with a sterile blade.
2. Gradually add the tissue to a mortar containing liquid nitrogen and ground to a fine powder.
3. Slowly add the powdered tissue to an evenly dispersed mixture of 5 mL of phenol and 5 mL of STEL. This is best accomplished by gradually stirring the powdered tissue into the phenol:STEL emulsion with a baked glass rod. Mix the tissue until the components are thoroughly dispersed. Continue mixing by gentle inversion for 5 min.
4. Follow Section 3.1., steps 3–9.

4. Notes

1. DEPC is a suspected carcinogen and should be handled with gloves in a fume hood. Because it acts by acylating histidine and tyrosine residues on proteins, susceptible reagents, such as Tris solutions, should not be directly treated with DEPC. Sensitive reagents should simply be made up in DEPC-treated water as outlined.
2. The integrity of the nucleic acids will be improved by maintaining harvested cells or tissues cold.
3. The success of this procedure hinges on the ability to gently disrupt cellular integrity while maintaining DNA in an intact, high mol wt form. Thus, mixing of the STEL:phenol should be performed by gentle inversion, which minimizes shearing forces on the DNA.
4. The proteinaceous interface that partitions between the aqueous (upper) and phenol phase following the initial phenol extraction (Section 3.1., step 3) can be reextracted with phenol:chloroform to improve DNA recovery.
5. Chloroform is commonly prepared as a 24:1 (v/v) mixture with isoamyl alcohol, which acts as a defoaming agent. We have found that foaming is not a problem if extractions are performed by gentle inversion or on a rotating wheel and routinely omit isoamyl alcohol from the mixture.
6. The DNA may be air dried, but will then take longer to resuspend.
7. Following the selective removal of high mol wt DNA, the remaining RNA is sufficiently free of DNA contamination such that DNA is not detected by ethidium bromide staining *(31)*. If the purified RNA is to be used for PCR procedures it is strongly recommended that a RNase-treated control be performed to ensure the absence of contaminating DNA. This recommendation extends to virtually any RNA purification procedure, particularly those involving an initial step in which the DNA is sheared.

8. Typical yields of total cellular RNA range between 60–170 μg when using approx 1.5–2. × 10^7 cells with A_{260}/A_{280} values of approx 1.86 *(31)*. These values compare favorably with those obtained using guanidinium thiocyanate CsCl centrifugation methods *(8)*.

References

1. Graham, D. E. (1978) The isolation of high molecular weight DNA from whole organisms of large tissue masses. *Anal. Biochem.* **85,** 609–613.
2. Bowtell, D. D. (1987) Rapid isolation of eukaryotic DNA. *Anal. Biochem.* **162,** 463–465.
3. Longmire, J. L., Albright, K. L., Meincke, L. J., and Hildebrand, C. E. (1987) A rapid and simple method for the isolation of high molecular weight cellular and chromosome-specific DNA in solution without the use of organic solvents. *Nucleic Acids Res.* **15,** 859.
4. Owen, R. J. and Borman, P. (1987) A rapid biochemical method for purifying high molecular weight bacterial chromosomal DNA for restriction enzyme analysis. *Nucleic Acids Res.* **15,** 3631.
5. Reymond, C. D. (1987) A rapid method for the preparation of multiple samples of eukaryotic DNA. *Nucleic Acids Res.* **15,** 8118.
6. Miller, S. A. and Polesky, H. F. (1988) A simple salting out procedure for extracting DNA from human nucleated cells. *Nucleic Acids Res.* **16,** 1215.
7. Grimberg, J., Nawoschik, S., Belluscio, L., McKee, R., Turck, A., and Eisenberg, A. (1989) A simple and efficient non-organic procedure for the isolation of genomic DNA from blood. *Nucleic Acids Res.* **17,** 8390.
8. Chirgwin, J. M., Przybyla, A. E., MacDonald, R. J., and Rutter, W. J. (1979) Isolation of biologically active ribonucleic acid from sources enriched in ribonuclease. *Biochemistry* **18,** 5294–5299.
9. Auffray, C. and Rougeon, F. (1980) Purification of mouse immunoglobulin heavy-chain messenger RNAs from total myeloma tumour RNA. *Eur. J. Biochem.* **107,** 303–314.
10. Elion, E. A. and Warner, J. R. (1984) The major promoter element of rRNA transcription in yeast lies 2 Kb upstream. *Cell* **39,** 663–673.
11. Chomczynski, P. and Sacchi, N. (1987) Single-step method of RNA extraction by acid guanidinium thiocyanate-Phenol-Chloroform extraction. *Anal. Biochem.* **162,** 156–159.
12. Hatch, C. L. and Bonner, W. M. (1987) Direct analysis of RNA in whole cell and cytoplasmic extracts by gel electrophoresis. *Anal. Biochem.* **162,** 283–290.
13. Emmett, M. and Petrack, B. (1988) Rapid isolation of total RNA from mammalian tissues. *Anal. Biochem.* **174,** 658–661.
14. Gough, N. M. (1988) Rapid and quantitative preparation of cytoplasmic RNA from small numbers of cells. *Anal. Biochem.* **173,** 93–95.
15. Meier, R. (1988) A universal and efficient protocol for the isolation of RNA from tissues and cultured cells. *Nucleic Acids Res.* **16,** 2340.
16. Wilkinson, M. (1988) RNA isolation: a mini-prep method. *Nucleic Acids Res.* **16,** 10,933.

17. Wilkinson, M. (1988) A rapid and convenient method for isolation of nuclear, cytoplasmic and total cellular RNA. *Nucleic Acids Res.* **16,** 10934.
18. Ferre, F. and Garduno, F. (1989) Preparation of crude cell extract suitable for amplification of RNA by the polymerase chain reaction. *Nucleic Acids Res.* **17,** 2141.
19. McEntee, C. M. and Hudson, A. P. (1989) Preparation of RNA from unspheroplasted yeast cells *(Saccharomyces cerevisiae). Anal. Biochem.* **176,** 303–306.
20. Nemeth, G. G., Heydemann, A., and Bolander, M. E. (1989) Isolation and analysis of ribonucleic acids from skeletal tissues. *Anal. Biochem.* **183,** 301–304.
21. Verwoerd, T. C., Dekker, B. M. M., and Hoekema, A. (1989) A small-scale procedure for the rapid isolation of plant RNAs. *Nucleic Acids Res.* **17,** 2362.
22. Schmitt, M. E., Brown, T. A., and Trumpower, B. L. (1990) A rapid and simple method for preparation of RNA from *Saccharomyces cerevisiae. Nucleic Acids Res.* **18,** 3091.
23. Tavangar, K., Hoffman, A. R., and Kraemer, F. B. (1990) A micromethod for the isolation of total RNA from Adipose tissue. *Anal. Biochem.* **186,** 60–63.
24. Krieg, P., Amtmann, E., and Sauer, G. (1983) The simultaneous extraction of high-molecular-weight DNA and of RNA from solid tumours. *Anal. Biochem.* **134,** 288–294.
25. Mirkes, P. E. (1985) Simultaneous banding of rat embryo DNA, RNA and protein in cesium trifluroracetate gradients. *Anal. Biochem.* **148,** 376–383.
26. Meese, E. and Blin, N. (1987) Simultaneous isolation of high molecular weight RNA and DNA from limited amounts of tissues and cells. *Gene Anal. Tech.* **4,** 45–49.
27. Zarlenga, D. S. and Gamble, H. R. (1987) Simultaneous isolation of preparative amounts of RNA and DNA from Trichinella spiralisby cesium trifluoroacetate isopycnic centrifugation. *Anal. Biochem.* **162,** 569–574.
28. Chan, V. T.-W., Fleming, K. A., and McGee, J. O. D. (1988) Simultaneous extraction from clinical biopsies of high-molecular-weight DNA and RNA: comparative characterization by biotinylated and ^{32}P-labeled probes on Southern and Northern blots. *Anal. Biochem.* **168,** 16–24.
29. Karlinsey, J., Stamatoyannopoulos, G., and Enver, T. (1989) Simultaneous purification of DNA and RNA from small numbers of eukaryotic cells. *Anal. Biochem.* **180,** 303–306.
30. Coombs, L. M., Pigott, D., Proctor, A., Eydmann, M., Denner, J., and Knowles, M. A. (1990) Simultaneous isolation of DNA, RNA and antigenic protein exhibiting kinase activity from small tumour samples using guanidine isothiocyanate. *Anal. Biochem.* **188,** 338–343.
31. Raha, S., Merante, F., Proteau, G., and Reed, J. K. (1990) Simultaneous isolation of total cellular RNA and DNA from tissue culture cells using phenol and lithium chloride. *Gene Anal. Tech.* **7,** 173–177.
32. Kirby, K. S. (1957) A new method for the isolation of deoxyribonucleic acids: evidence on the nature of bonds between deoxyribonucleic acid and protein. *Biochem. J.* **66,** 495–504.

33. Wallace, D. M. (1987) Large and small scale phenol extractions, in *Methods in Enzymology,* vol. 152: *Guide to Molecular Cloning Techniques* (Berger, S. L. and Kimmel, A. R., eds.), Academic, Orlando, FL, pp. 33–41.
34. Sambrook, J., Fritsch, E. F., and Maniatis, T. (1989) *Molecular Cloning: A Laboratory Manual,* 2nd ed. Cold Spring Harbor Laboratory, Cold Spring Harbor, NY.
35. Davis, L. G., Dibner, M. D., and Battey, J. F. (1986) *Basic Methods in Molecular Biology.* Elsevier, New York.

CHAPTER 12

Physical Mapping of the Human Genome by Pulsed Field Gel Electrophoresis

Jacqueline Boultwood

1. Introduction

DNA fragments larger than 25 kb are not resolved effectively by standard gel electrophoresis. Pulsed field gel electrophoresis (PFGE), first described by Schwartz and Cantor in 1984 *(1)*, is a technique that allows for the separation of large DNA fragments of up to several mega base pairs (mbp) in length by periodically alternating the direction of the electric field applied to the gel. One important aspect of PFGE is the near linearity of the separation. This feature results in high resolution and allows for the effective separation of small DNA fragments between 10–20 kb as well as much larger fragments in the 100–1000 kb size range. The resolution of both large and small DNA fragments by PFGE is thought to result from the differing abilities of DNA molecules of different lengths to reorientate to the changed field direction. The PFGE equipment originally developed by Schwartz and Cantor *(1)* applied pulsed alternating orthogonal electric fields to the gel and was able to resolve DNA fragments of up to 2 mbp. However, the electric field generated by this early equipment was nonuniform and caused a skewing of the DNA tracks toward the edge of the gel with concomitant loss of resolution. A number of alternative electrophoretic configurations have been developed in order to overcome these problems and to increase the maximum size limit of resolvable DNA fragments. These include the periodic inversion a single electric

From: *Methods in Molecular Biology, Vol. 31: Protocols for Gene Analysis*
Edited by: A. J. Harwood

field (field-inversion gel electrophoresis, FIGE; *2* and *see* Chapter 13), rotating gels (Waltzer; *3*), transverse fields applied to vertical gels (TAFE; *4*), and contour-clamped homogeneous electric fields (CHEF; *5*).

PFGE has been used to separate and analyze intact yeast chromosomes *(1)*, but current technology is still unable to resolve the smallest human chromosome, of approx 50 mbp. PFGE has, however, been of enormous benefit in the physical mapping of large chromosomal regions of the human genome. Long range restriction maps generated by PFGE have, for example, been used for the analysis of multigene family structure and evolution *(6)* and for the detection of chromosomal translocation breakpoints in neoplasia *(7)*. Detailed physical maps of genomic regions harboring hereditary disease loci are also of considerable importance in the location and cloning of causative genes. The isolation of the cystic fibrosis gene was made possible through use of an extensive long range restriction map of the region generated by PFGE *(8)*. Similarly the Duchenne muscular dystrophy locus has been extensively characterized by PFGE and a 4-mbp map generated that contains a number of candidate genes *(9)*. Other human disease loci for which long range restriction maps have been constructed include: WAGR located on chromosome 11 (Wilms tumor; *10*), Huntington's disease assigned to chromosome 4p *(11)*, and the fragile X syndrome *(12)*. These examples illustrate the power and importance of the PFGE technique for the physical mapping of mammalian genomes.

In this chapter I will first describe the preparation of the sample DNA and of yeast chromosomal markers required for PFGE analysis. I will then describe how to run a pulsed field gel and give a protocol that has been used successfully for Southern analysis of this kind of gel.

2. Materials

2.1. Preparation of Agarose Plugs Containing High Mol Wt Human DNA

1. PBS/1 m*M* EDTA: 0.137*M* NaCl, 2.68 m*M* KCL, 7.98 m*M* Na_2HPO_4, 1.47 m*M* KH_2PO_4, pH 7.2 containing 1 m*M* EDTA.
2. 2% Agarose solution: 2% low melting temperature agarose (Sea-Plaque) in PBS/1 m*M* EDTA. Melt in a microwave oven just prior to use and cool to 42°C.
3. Block mold: A mold should contain a number of slots to make agarose blocks of the correct dimensions to fit the wells of the gel. It can be

made from perspex in the laboratory workshop (*see* Fig. 1) or bought from commercial manufacturers.
4. NDS: 10 m*M* Tris-HCl, 0.5*M* EDTA, 1% Sarkosyl, pH 9.5.
5. Proteinase K: Stock of 10 mg/mL in distilled water. Store at –20°C.
6. TE: 10 m*M* Tris-HCl, pH 7.6, 1 m*M* EDTA.
7. PMSF: phenylmethylsulfonyl fluoride (PMSF) should be made up as a 10 m*M* stock solution in isopropanol since it is inactivated in aqueous solutions. Store at –20°C.

2.2. Preparation of Agarose Plugs Containing Yeast DNA

8. Yeast wash: 0.05*M* EDTA pH 8, 0.01*M* Tris-HCl, pH 7.6.
9. 0.05*M* EDTA, pH 8.0.
10. L-buffer: 0.1*M* EDTA, pH 8, 0.01*M* Tris-HCl, pH 7.6, 0.02*M* NaCl.
11. 1% Agarose solution: 1% low melting temperature agarose (Sea Plaque) in L-buffer. Cool to 42°C just before use.
12. Lyticase: Make solution of 900 U/mL (i.e., 67 mg/mL) in 0.01*M* sodium phosphate containing 50% glycerol. Store at –20°C.
13. Y-buffer: 0.5*M* EDTA, pH 8.0, 0.01*M* Tris-HCl, pH 7.6, containing 1% β-mercaptoethanol. Add the β-mercaptoethanol just prior to use.
14. Sarkosyl: A 10% Sarkosyl stock in distilled water.

2.3. Restriction Enzyme Digestion of DNA in Agarose Plugs

15. Restriction enzymes: Use appropriate restriction enzymes and buffers according to manufacturer's instructions (*see* Note 1).
16. Spermidine: Add to restriction enzyme buffers at 5 m*M* when the salt concentration is between 5 and 100 m*M* and at 10 m*M* when greater than 100 m*M*.

2.4. Pulsed Field Gel Electrophoresis

17. 1% Agarose solution: 1% normal melting temperature agarose (Sigma, St. Louis, MO) in 0.5X TBE buffer. Melt prior to use. The percentage of each gel is dependent on the separation required.
18. 0.5X TBE buffer: 45 m*M* Tris-HCl, 45 m*M* boric acid, 0.5 m*M* EDTA, pH 8.3. Make as a 10X stock solution.
19. Ethidium bromide: A 10 mg/mL stock solution in distilled water.

2.5. Southern Blotting of Megabase DNA

20. Alkali denaturing solution: 0.5*M* NaOH, 1.5*M* NaCl.
21. Alkali transfer solution: 0.25*M* NaOH, 1.5*M* NaCl.
22. Neutralizing solution: 0.25*M* Tris-HCl, pH. 7.0.

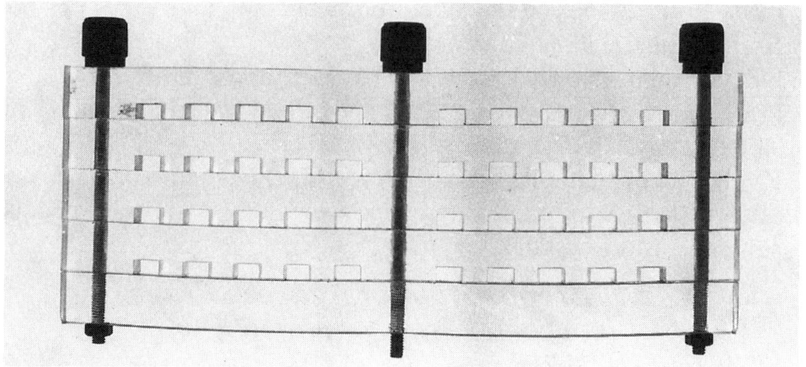

Fig. 1. A perspex block mold made in the laboratory workshop.

23. Transfer membrane: We routinely use the nylon transfer membrane Hybond-N (Amersham, Arlington Heights, IL).
24. 20X SSC stock: 3*M* NaCl, 0.3*M* sodium citrate. Dilute accordingly.
25. 50X Denhardt's: 5 g of Ficoll, 5 g of bovine serum albumin, 5 g of polyvinyl pyrrolidone in 500 mL distilled H_2O.
26. Prehybridization solution: 6X SSC, 5X Denhardts, 0.5% SDS, 100 μg/mL denatured sheared salmon sperm DNA. Prewarm to 65°C.
27. Hybridization solution: Prehybridization solution plus 0.01*M* EDTA. Before addition to the filter, add an appropriately labeled and denatured probe (*see* Notes 2 and 3) at a concentration of 1–2 ng/mL of hybridization solution.
28. X-ray film: Use film of appropriate sensitivity (e.g., XAR5, Kodak).

3. Methods

3.1. Preparation of Agarose Plugs Containing High Mol Wt Human DNA

Conventional methods of DNA extraction may lead to shearing of very large DNA molecules. Consequently, intact cells are embedded in low melting temperature agarose prior to restriction enzyme digestion and PFGE. Tissue culture cells, fractionated peripheral blood leukocytes, or whole nuclei (extracted from frozen solid tissue) are all suitable sources of human DNA for analysis by PFGE (*see* Note 4).

1. Wash the cells twice in ice-cold PBS/1 m*M* EDTA and resuspend in ice-cold PBS/1 m*M* EDTA at a concentration of approx 5×10^7 cells/mL.
2. Warm the cell suspension to 42°C and add an equal volume of the 2% agarose solution (also at 42°C) and mix gently. Immediately pipet the

cell suspension into the slots of the block mold and place it on ice for 5–10 min.

3. Push the solidified agarose plugs out of the slots using a sealed Pasteur pipet and collect the plugs in approx 50 vol of NDS.
4. Add proteinase K to 0.5 mg/mL and incubate the plugs at 50°C for 24 h. This procedure should be repeated with fresh digestion mix for a further 24 h.
5. Wash the plugs several times with fresh NDS. Transfer the plugs to approx 10 vol of TE containing 40 µg/mL of PMSF and incubate at room temperature for 30 min. Repeat this procedure with fresh TE/PMSF solution for a further 30 min.
6. Wash the plugs three times with TE over the period of 1 h to remove the PMSF. Store the plugs at 4°C in NDS.

3.2. Preparation of Agarose Plugs Containing Yeast DNA

Intact yeast chromosomes are convenient size markers for PFGE studies (*see* Fig. 2, Track 1). The method described here for the isolation of intact DNA from yeast is essentially as originally detailed by Schwartz and Cantor *(1)*.

Lambda ladders may also be used in place of, or in conjuction with, intact yeast chromosomes as DNA size markers for PFGE gels (*see* Chapter 13).

1. Collect yeast cells growing in early stationary phase by centrifugation at 3000*g* for 5 min at 4°C. Wash the cell pellet twice with yeast wash. Resuspend the cells at a concentration of 3×10^{10} cells/mL in ice-cold 0.05*M* EDTA.
2. Warm the yeast cell suspension to 42°C and add an equal volume of the 1% agarose solution (also at 42°C) to the cell suspension. Mix gently.
3. Add 75 µL of Lyticase to 10 mL and mix before the agarose sets. Pipet the cell suspension into the slots in a perspex block mold and place on ice for 5–10 min.
4. Push the solidified agarose plugs out of the slots using a sealed Pasteur pipet. Collect the plugs in approx 50 vol of Y-buffer and incubate at 37°C for 24 h. This procedure should be repeated with fresh buffer for a further 24 h.
5. Transfer the plugs to 50 vol of L-buffer containing 1 mg/mL proteinase K and 1% Sarkosyl. Incubate the plugs for 24 h at 50°C. This procedure should be repeated with fresh digestion mix for a further 24 h.
6. Wash the plugs several times with fresh L-buffer. Transfer the plugs to approx 10 vol of TE containing 40 µg/mL of PMSF and incubate at

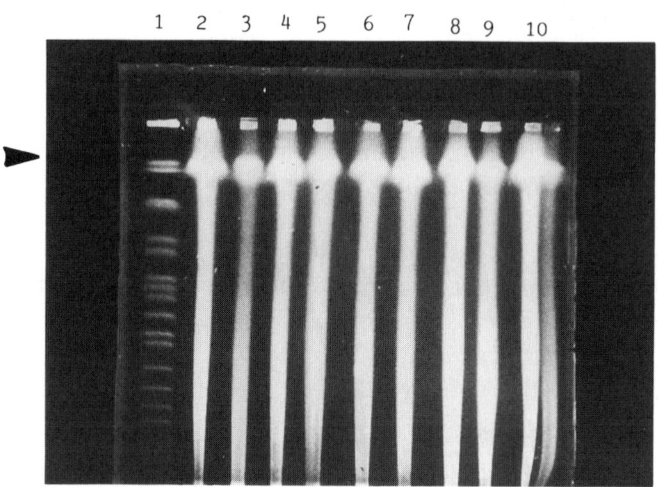

Fig. 2. Ethidium bromide stained pulsed field gel. The DNA samples (Tracks 1–10) were separated on a CHEF-DRII system (Bio-Rad, Richmond, CA) using a ramped pulse from 50–90 s over a period of 24 h. Gel was 1% agarose in 0.5X TBE and electrophoresis was at 200 V at 14°C. Track 1: Separated yeast chromosomes (*Saccharomyces cerevisiae,* strain YNN295), 15 bands are visible ranging in size from the arrow as follows: 2500, 1600, 1125, 1020, 945, 850, 800, 770, 700, 630, 580, 460, 370, 290, 245 kb respectively. Tracks 2–10: human peripheral blood leukocyte DNA digested with the following restriction enzymes; Tracks 2–4 *Bss*HII, Tracks 5–7 *Mlu*I, Tracks 8–10 *Ksp*I.

room temperature for 30 min. Repeat this procedure with fresh TE/PMSF solution for a further 30 min.

7. Wash the plugs three times with TE over the period of 1 h to remove the PMSF. Store the plugs at 4°C in 0.5*M* EDTA.

3.3. Restriction Enzyme Digestion of DNA in Agarose Plugs

1. Incubate the plugs in 50 vol of TE (pH 7.6) for 30 min at room temperature. Repeat this procedure two further times.
2. Transfer the plugs to individual Eppendorf tubes and add 10 vol of appropriate 1X restriction enzyme buffer. Equilibrate the plugs for 30 min at 4°C.
3. Replace the buffer with 100 µL of fresh 1X restriction enzyme buffer containing spermidine. Incubate for 10 min.

4. Add 20 U of enzyme for 5 μg of chromosomal DNA. The enzyme may be added in two equal portions, at the beginning and after 2 h of digestion. Incubate for 12–16 h at the specified temperature.
5. Wash the blocks for 30 min in 50 vol of TE, pH 7.6 at 4°C prior to loading into the gel.

3.4. Pulsed Field Gel Electrophoresis

A number of PFGE systems based on various electrophoretic configurations are now commercially available. PFGE systems employing contour-clamped homogeneous electric field electrophoresis (CHEF) are particularly widely used and this is the one used in the method that I describe (*see* Note 5). This protocol can readily be applied to other systems. *See* Chapter 13 for details on the use of field-inversion electrophoretic systems.

The precise details required for setting up individual CHEF systems will vary between manufacturers. Instruction manuals will normally be provided together with the PFGE apparatus purchased. Therefore, only general points concerning the running of PFGE gels are detailed in this method.

1. Prepare an agarose gel at the required concentration (typically, 1%) in 0.5X TBE buffer. Pour the molten agarose into the gel casting stand and allow to cool for 30 min at room temperature. Carefully remove the gel comb.
2. Remove the supernatant from the plugs containing digested sample DNA or markers. Push the blocks into the slots in the gel using a pipet tip.
3. Seal each slot with low melting temperature agarose at an agarose concentrate equivalent to that of the running gel. Allow the agarose to harden at room temperature for 15 min.
4. Fill the PFGE chamber with of 0.5X TBE buffer. Use a recirculating pump to allow the buffer to equilibrate to the desired temperature (typically 14°C) by passing the pump tubing through a temperature-controlled water bath.
5. Slide the gel onto the surface of the chamber and position against the gel stops. Maintain the maximum flow rate without disturbing the gel.
6. Program the switching device and constant voltage power supply. Run the gel under appropriate conditions (*see* Notes 6 and 7). At the end of the run remove the gel and place it into a 0.5 μg/mL ethidium bromide solution in electrophoresis buffer and stain for 30 min. Destain the gel in distilled water for a further 30 min and visualize the DNA on a UV transilluminator (254–360 nm). Figure 2 shows a photograph of a typical result.

3.5. Southern Blotting of Megabase DNA

DNA fragments longer than 20 kb are not efficiently transferred by standard Southern blotting; consequently DNA fragments separated by pulsed field electrophoresis must be cleaved prior to transfer onto nylon membranes.

1. UV irradiate the gel for 60 s with a 254-nm light source at an intensity of 2.5 mW/cm^2 (*see* Notes 8 and 9).
2. Incubate the gel in alkali denaturing solution for 30 min at room temperature.
3. Replace the alkali denaturing solution with alkali transfer solution and incubate for a further 15 min at room temperature.
4. Transfer the DNA onto a nylon membrane using alkali transfer solution. The Southern blot should be set up exactly as for conventional gels. Transfer for at least 24 h.
5. After transfer is complete, neutralize the membrane in 0.25*M* Tris-HCl, pH 7.0 for 5 min and then rinse in 2X SSC. Dry the membrane by blotting on 3MM Whatman paper.
6. Bake in an oven at 80°C for 20 min and UV irradiate for 2–5 min (*see* Note 10).
7. Soak the nylon membrane filter in 6X SSC for 10 min. Place the wet filter into a heat-sealable plastic bag or hybridization chamber and add approx 5 mL of prehybridization solution. Incubate the filter in prehybridization solution for 2 h at 65°C.
8. Remove the prehybridization solution from the bag or chamber and replace with an equal volume of the hybridization solution. Hybridization should be carried out for 12–16 h at 65°C.
9. Following hybridization submerge the filter in 2X SSC, 0.5% SDS at room temperature for 5 min. Transfer the filter into a chamber containing 0.5X SSC, 0.5% SDS, and incubate at 65°C for 30 min. Transfer the filter into a chamber containing 0.1X SSC, 0.5% SDS, and incubate at 65°C for 30 min.
10. Wrap the filter in cling film and place on X-ray film. Expose the film for 1–7 d at –70°C.

Figure 3 shows a typical result after autoradiography (*see* Notes 11–16).

4. Notes

1. Restriction enzymes that cleave very infrequently are used to digest human DNA for PFGE analysis and include: *Not*I, *Nru*I, *Mlu*I, *Sfi*I, *Sma*I, *Xho*I, *Sal*I, *Pvu*I, *Nar*I, *Nae*I, *Bss*HII, *Ksp*I, *Ecl*XI, *Sgr*AI, *Avi*II, *Aat*II. A number of these enzymes recognize octameric (or longer)

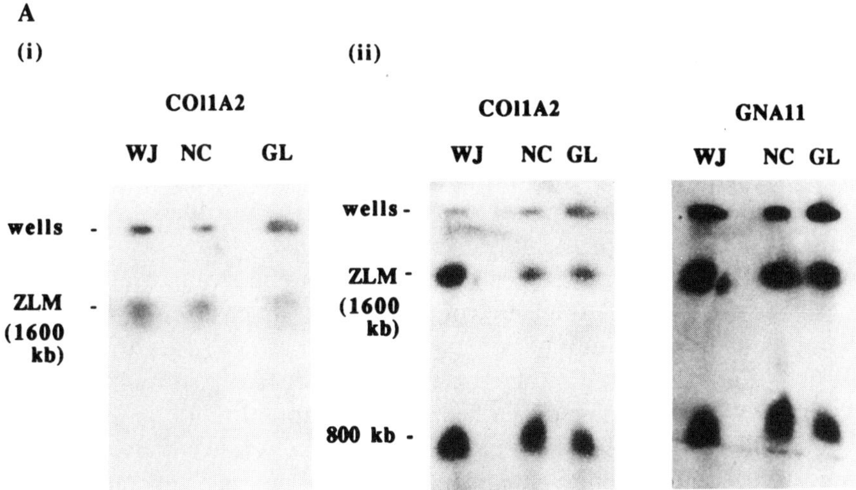

Fig. 3. Pulsed field gel electrophoresis studies on peripheral blood DNA from two leukemic patients (WJ and GL) with a chromosome 7q deletion, to look for the deletion breakpoints. DNA from a normal control (NC) was run for comparison in each case. (i): Samples were digested with *Sma*I and hybridized to the probe for the COL1A2 gene. The fragments were not resolved under the electrophoresis conditions used (they remained in the zone of limiting mobility [ZLM]). No aberrant bands were seen in the patients' samples. (ii): Samples were digested with *Bss*HII and the filter hybridized sequentially to probes for the COL1A2 and GNA11 genes. Both probes hybridized to an 800-kb fragment. No additional bands were seen in the patients' samples.

sequences and typically contain one or more CpG dinucleotides in their recognition sequence (*Sfi*I being the exception). They are rare cutters both because of the infrequency of such large recognition sequences and the underrepresentation of the CpG dinucleotide in the genome. Cleavage at these enzyme sites may be inhibited as a result of the methylation of the cytosine residue.

2. DNA probes may be radiolabeled in a number of ways. Most workers favor the method of "random primed" DNA labeling *(13)* since it gives a high specific activity in the order of 1.8×10^9 dpm/μg.
3. Labeled DNA probes should be denatured by boiling for 10 min and then cooled on ice for 3 min.
4. The requirement for cell suspensions as a DNA source for PFGE has proved restrictive in the application of this methodology to the study of solid tissues. This problem has been recently addressed in our labora-

tory by the development of a simple technique that allows for the direct analysis of solid tissue samples by PFGE *(14)*. Single frozen tissue sections are embedded without further manipulation in molten low melting temperature agarose prior to enzyme digestion and PFGE. Sufficient high mol wt DNA is yielded by this method to obtain a hybridization signal with a single copy probe. Other solid biological structures, for example human hair roots *(15)*, may also be embedded directly into agarose for subsequent analysis by PFGE.

5. The CHEF system produces high resolution and straight lanes and is able to resolve DNA fragments of up to 5 mbps. Each electrode "clamps" the voltage of its individual region of space as necessary to maintain field homogeneity. The electric field is produced from multiple electrodes that are arranged in a hexagonal contour around the horizontal gel. The hexagonal array of electrodes produces electric fields at angles of 120° or 60° depending on the placement of the gel and the polarity of the electrodes.
6. The separation of DNA fragments in agarose gels is affected by agarose concentration, buffer concentration, temperature, pulse times, voltage, and total electrophoresis run time. The agarose concentration and the pulse time determine the size range of DNA molecules separated. Agarose concentrations of 1% are suitable for separating DNA fragments up to 3 mbp in length. Gel concentrations in the 0.5–0.8% range are used to separate DNA molecules ≥3 mbp. Short pulse times between 1–10 s are used to separate DNA fragments ≥ 0.1 mbp, whereas long pulse times between 10–60 min are used to separate DNA fragments >4 mbp. Most PFGE runs require DNA separation in the 0.1–2.0-mbp size range and employ pulse times between 50–100 s.
7. The migration rate of DNA fragments through agarose gels is determined by voltage, temperature, and agrarose concentration. In general, DNA migration velocity increases with increases in voltage and temperature and decreases with increasing agarose concentration.
8. DNA must be cleaved either by acid depurination or UV irradiation to allow efficient transfer from the gel. UV irradiation is preferable to acid depurination since the latter is very temperature sensitive and thus more difficult to control.
9. If a 254-nm light source is not available, a 302-nm light may be used, but exposure times have to be lengthened approx fivefold.
10. UV at a wavelength of 312 nm is recommended. The precise time of exposure is important and will vary depending on the transilluminator used. Regular calibration using a UVX radiometer is essential.

11. The results obtained with membranes blotted from PGFE gels and hybridized to single copy DNA probes are generally less visually pleasing than those obtained from conventional agarose gels. The hybridization bands obtained from PFGE gels are rarely as sharp and discrete as those obtained with standard Southern blots and may in some instances cover a relatively wide area, thus making accurate sizing difficult. Furthermore, the hybridization results obtained from PFGE gels are generally more difficult to interpret than those obtained from conventional gels. This is primarily because the majority of enzymes used in the generation of large DNA fragments in PFGE experiments are methylation sensitive (*see* Note 1). Methylation of the cytosine residues in the enzyme recognition sequence may give rise to incomplete digestion (*see* Fig. 4, for example).
12. The effects of methylation on the rare cutter restriction enzymes makes PFGE mapping a good means of identifying islands HTFs (*Hpa*II tiny fragments) and hence coding regions. HTF islands correspond to clusters of undermethylated CpG dinucleotides and have been associated with the 5' end of some genes. HTF islands may appear on a PFGE physical map as a cluster of different rare cutter restriction sites, separated by a short distance, usually in the order of several kilobases or less *(16)*.
13. The migration of DNA fragments relative to each other can vary markedly if the DNA concentration differs between samples *(17)*. Similarly, the migration of DNA fragments relative to yeast size markers may not be the same between gels run under different conditions.
14. Two different DNA markers can appear to hybridize to the same DNA fragment, suggesting physical linkage. However, it is possible that different DNA fragments can comigrate exactly and for this reason it is important to carry out both single and double enzyme digests.
15. Unexpected DNA fragments, possibly resulting from chromosomal translocation breakpoints, should be subject to careful analysis before concluding breakpoint specificity. Rare polymorphisms or differences in the methylation states of CpG residues between individuals or tissues may give rise to additional fragments *(18)*.
16. Meiotic linkage analysis is able to position a DNA marker with respect to a neighboring marker with a resolution of between 5 c*M* and I c*M* equivalent to 5 Mbp–I Mbp of DNA. There can, however, be considerable local deviation from the relationship of 1 c*M* ≈ 1 mbp owing to nonlinearity of recombination rates along a chromosome. The physical maps of specific chromosomal regions generated by PFGE may be com-

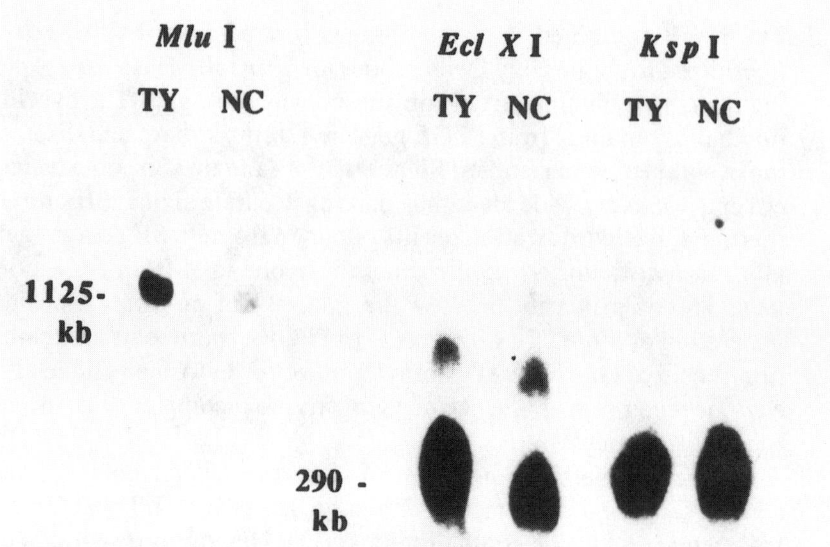

Fig. 4. Pulsed field gel electrophoresis studies on peripheral blood DNA from one leukemic patient, using a probe for GM-CSF, to look for the chromosome 5q deletion breakpoint. No additional bands were seen with the enzymes *Mlu*I, *Ecl*XI, or *Ksp*I. A fragment of 290 kb is shown with the *Ecl*XI digested DNA and a second larger fragment is also present resulting from partial digestion. TY = leukemic patient. NC = normal control.

pared with genetic linkage maps of the same area in order to examine the accuracy of the genetic map. The construction of physical maps by PFGE is achieved by the analysis of both single and double enzyme digestion results. Partial digestion resulting from methylation of cytosine residues is also of use in the formation of long range restriction maps because it allows for the mapping of several adjacent sites relative to each other.

Acknowledgments

I would like to thank the Leukemia Research Fund (UK) for funding and G. Abrahamson for generous use of Fig. 3.

References

1. Schwartz, D. C. and Cantor, C. R. (1984) Separation of yeast chromosome sized DNAs by pulsed field gradient gel electrophoresis. *Cell* **37,** 67–75.
2. Carle, G. F., Frank, M., and Olson, M. V. (1986) Electrophoretic separation of large DNA molecules by periodic inversion of the electric field. *Science* **232,** 65–68.
3. Anand, R. (1986) Pulsed field gel electrophoresis: a technique for fractionating large DNA molecules. *Trends Genet.* **2,** 278.

4. Gardiner, K. W., Laas, W., and Patterson, D. (1986) Fractionation of large mammalian DNA restriction fragments using vertical pulsed field gradient gel electrophoresis. *Somatic Cell Mol. Genet* **12,** 185–195.
5. Carle, G. F. and Olson, M. V. (1984) Separtion of chromosomal DNA molecules from yeast by orthogonal field—alternation gel electrophoresis. *Nucleic Acids Res.* **12,** 5647–5664.
6. Dunham, J., Sargent, C. A., Dawkins, R. L., and Campbell, R. D. (1989) An analysis of variation in the long-range genomic organisation of the Human Major Histocompatibility complex class II region by pulsed field gel electrophoresis. *Genomics* **5,** 787–796.
7. Borrow, J., Goddard, A. D., Sherr, D., and Soloman, E. (1990) Molecular analysis of acute promyelocytic leukemia breakpoint cluster region on chromosome 17. *Science* **249,** 1577–1580.
8. Rommens, J. M., Ianuzzi, M. C., Kerem, B., Drumm, M. L., Melmer, G., Dean, M., Rozmahel, R., Cole, J. L., Kennedy, D., Hidaka, N., et al. (1989) Identification of the cystic fibrosis gene: chromosome walking and jumping. *Science* **245,** 1059–1065.
9. Love, D., Bloomfield, J., Kenwrick, S., Yates, J., and Davies, K. (1990) Physical mapping distal to the DMD locus. *Genomics* **8,** 106–112.
10. Gessler, M. and Bruns, G. A. (1989) A physical map around the WAGR co-complex on the short arm of chromosome 11. *Genomics* **5,** 43–55.
11. Bucan, M., Zimmer, M., Whaley, W., Poustaka, A., Youngman, S., Allito, B., Ormondroyd, E., Smith, B., Pohl, T., MacDonald, M., Bates, G., et al. (1990) Physical maps of 4p16.3. The area expected to contain the Huntington Disease mutation. *Genomics* **6,** 1–15.
12. Vincent, A., Heitz, D., Petit, C., Kretz, C., Oberle, I., and Mandel, J. L. (1991) Abnormal pattern detected in Fragile–X patients by pulsed field gel electrophoresis. *Nature* **349,** 624–626.
13. Feinberg, A. P. and Vogelstein, B. (1983) A technique for radiolabelling DNA restriction endonuclease fragments to high specific activity. *Anal. Biochem.* **132,** 6–13.
14. Boultwood J., Kaklamaris, L., Gatter, K. C., and Wainscoat J. S. (1993) Pulsed field gel eletrophoresis on frozen tumour tissue sections. *J. Clin. Pathol.* **45,** 722–723.
15. Boultwood, J., Abrahamson, G. M., and Wainscoat, J. S. (1990) Structural DNA analysis from a single hair root by standard or pulsed field gel electrophoresis *Nucleic Acids Res.* **18,** 4628.
16. Burmeister, M. and Lehrach H. (1986) Long-range restriction map around the Duchenne muscular dystrophy gene. *Nature* **324,** 582–585.
17. Collins, F. S., Cole, J. L., Lockwood, W. K., and Iannuzzi, M. C. (1988) The deletion in both common types of herediatry persistence of fetal hemoglobin is approx 105 kilobases. *Blood* **70,** 1797–1803.
18. Abrahamson, G. M., Boultwood, J., and Wainscoat, J. S. (1991) Practical considerations in the analysis of chromosomal deletion breakpoints by pulsed field gel electrophoresis. *Brit. J. Haem.* **77,** 129–130.

CHAPTER 13

Field Inversion Gel Electrophoresis

Christoph Heller

1. Introduction

The field inversion gel electrophoresis (FIGE) is a special pulsed field gel electrophoresis technique that is based on the periodic inversion of a uniform electric field in one dimension *(1)*. A net migration is achieved since the product of the duration and amplitude of the "forward" pulse is larger than that of the pulse in the "backward" direction.

For FIGE, no special equipment is needed; a conventional power supply and gel box can be used. A rather simple and therefore cheap relay box, controlled by a computer or a timer, is sufficient to achieve good separation. Several designs of such relay boxes have been described *(2–8)* and there is also a selection of different devices available on the market (*see* Note 1). The homogeneous electric field means that the DNA migrates in a straight lane, making the lane-to-lane comparison easy and reliable. FIGE has been successfully used to separate linear DNA molecules with low *(9–11)*, medium *(8,12,13)*, high *(1,7)*, and very high *(14)* mol wt, covering a range from several hundred bases up to at least 6 Mbp. Beside these advantages, the FIGE technique gives the additional possibility of either separation over a wide range of mol wts at low resolution or to select a small "window" with a rather high resolution. This effect has not been described for pulsed field techniques with nonparallel fields.

These advantages are accompanied by the severe problem of band inversion, where DNA molecules of higher mol wt can travel faster than smaller ones. Using FIGE, the mobility of the DNA molecules

From: *Methods in Molecular Biology, Vol. 31: Protocols for Gene Analysis*
Edited by: A. J. Harwood Copyright ©1994 Humana Press Inc., Totowa, NJ

as a function of mol wt decreases with increasing DNA size. Eventually, however, it reaches a minimum that under certain conditions may be nearly zero. Surprisingly DNA of sizes greater than this minimium point exhibit increased mobility. This is termed "band inversion." Owing to this effect, the measuring of the size of certain DNA bands can sometimes be very difficult. By choosing appropriate conditions it might be possible to shift the separation range so that the DNA size of interest is not affected by the band inversion. Since this is not always possible, several attempts have been made to circumvent the problem. One solution is the continuous or stepwise changing of the pulse time during the run ("ramping"). Both linear and exponential pulse time ramps are possible *(15)*. More complicated pulse time regimes have also be used successfully. A second solution is the so-called "zero integrated field electrophoresis" (ZIFE). This method uses two different electric fields in the forward and backward direction in addition to two different pulse times *(16)*. Theoretical considerations and experimental data have shown that when the integral of the electrical field over one pulse cycle is nearly zero, which is achieved if the product of pulse time and electrical field in the forward direction is close to that in the backward direction, the effect of band inversion can be minimized. This method, however, suffers from very long run times.

In this chapter I describe the basic FIGE technique. FIGE can be used to separate DNA from any source; however, as an example and to complement the chapter on pulse field gel electrophoresis, I describe the preparation and separation of DNA from the bacterium *Thermus aquaticus*.

2. Materials

2.1. Preparation of Agarose Blocks

When preparing intact chromosomal DNA, it is important to work very carefully in order not to contaminate the sample with nucleases. Therefore, use only sterile buffers and solutions as well as sterile plastic and glassware wherever possible.

1. Block-formers: Molds for forming blocks that will fit into the wells of the agarose gel. These are available from a number of suppliers (e.g., Pharmacia-LKB, Piscataway, NJ cat. no. 80-1102-55).

2. Growth medium:
   a. For Lambda phage: 2X YT (1.6% w/w tryptone, 1% w/w yeast extract, 0.5% w/w NaCl), supplemented with 10 m*M* $MgSO_4$.
   b. For bacteria: Choose the appropriate growth medium.
3. RNase A and DNase: stock solutions of 4 mg/mL. Store at 4°C.
4. Low melting point (LMP) agarose: with a gelling temperature for a 2% solution of between 26 and 30°C. Electrophoresis grade agarose is sufficient for most purposes. If restriction enzymes are inhibited, try higher qualities.
5. ESP buffer: 0.5*M* EDTA, pH 9–9.5, 10 m*M* EGTA, 1% *N*-Lauroylsarcosine sodium salt (Sarkosyl), 2 mg/mL proteinase K. Before use, incubate for 30 min at 65°C to remove any nuclease contamination.
6. Storage buffer: 0.5*M* EDTA, 10 m*M* EGTA, pH 9.
7. Washing buffer: 10 m*M* Tris-HCl, pH 7.6, 1*M* NaCl.
8. Lysis buffer: 6 m*M* Tris-HCl, pH 7.6, 1*M* NaCl, 100 m*M* EDTA (pH 7.5), 0.5% Brij-58, 0.2% Na-deoxycholate, 0.5% *N*-lauroylsarcosine, 1 mg/mL lysozyme, 20 µg/mL RNase.
9. TE buffer: 10 m*M* Tris-HCl, pH 8, 1 m*M* EDTA.

2.2. Electrophoresis

1. Standard horizontal or vertical agarose gel box of about 15–25 cm gel length and a standard electrophoresis power supply. A peristaltic pump is required to recirculate the buffer.
2. Relay box to invert the electric field; able to deliver pulses at least as short as 1 s and a computer to control the relay box (*see* Note 1).
3. Electrophoresis buffer, 0.5X TBE: 45 m*M* Tris, 45 m*M* boric acid, 1 m*M* EDTA, pH 8.2. If the electric current is too high, then the EDTA concentration should be lowered to 0.2 m*M* (modified TBE), to minimize heating up of the gel (*see also* Note 2).
4. Agarose: 1% agarose gel. Use agarose with a low electroendosmosis (<0.15).

3. Methods

3.1. Preparation of Agarose Blocks

For pulsed field gel electrophoresis, specialized methods are required to obtain high molecular DNA of sufficient integrity. By embedding whole cells into agarose, the DNA can be prepared with minimal shearing *(17)*. The size of the DNA means that it is trapped in the agarose, whereas smaller molecules, including enzymes, can easily enter and exit the block by diffusion. This method has been successfully used for a number of different cell types. An overview of the general procedure for preparing intact chromosomal DNA is given in Fig. 1.

In this chapter, the preparation of lambda DNA and of bacterial chromosomes is described in particular. (For preparation of yeast chromosomes and of human DNA, *see* Chapter 12.)

3.1.1. Lambda Concatemers as Size Markers

1. Plate out lambda phage on an appropriate host bacterial lawn (for example, MC 1061), pick a fresh plaque, and transfer it into 200 mL growth medium using a Pasteur pipet. This transfers enough uninfected host cells to the culture for lambda growth. Grow overnight at 37°C. Add 2 mL chloroform and incubate for another 30 min.
2. Spin at 4000–5000*g* for 20 min and incubate the supernatant with 10 µg/mL RNase and 10 µg/mL DNase at room temperature for 1 h. Precipitate the phage by carefully dissolving 8% (w/w) PEG 2000 and 2% NaCl in the lysate and incubate at 4°C overnight.
3. Resuspend the pellet in 3 mL of TE and warm to 50°C.
4. Prepare a 1% LMP Agarose in TE and cool to 50°C. Mix the phage suspension with 1 vol of agarose and dispense carefully into precooled block-formers, avoiding air bubbles.
5. After solidification, transfer the blocks with a glass hook into a fivefold volume of ESP buffer and incubate, preferably with gentle agitation, at 50°C for 48 h. Wash the blocks in TE and transfer them into the storage buffer.

This procedure produces multimers up to about 20 times the size of the lambda genome and that are stable for at least 1 yr, when stored at 4°C.

3.1.2. Bacterial Chromosomes

There are no general protocols for preparing intact chromosomal DNA of bacteria, since the properties differ widely between different taxonomic groups. I therefore recommend that pilot experiments are made in advance, to check the optimum number of cells to be embedded and to find out the best way of embedding the cells (*see* Note 3). The following protocol has worked well for a number of different bacterial species, including *E. coli* and *T. aquaticus*.

1. Prepare a 100-mL culture and grow to a density of approx 0.3–0.5 OD_{600} (*see* Note 4). Harvest the bacteria by centrifugation at 4000–5000*g* for

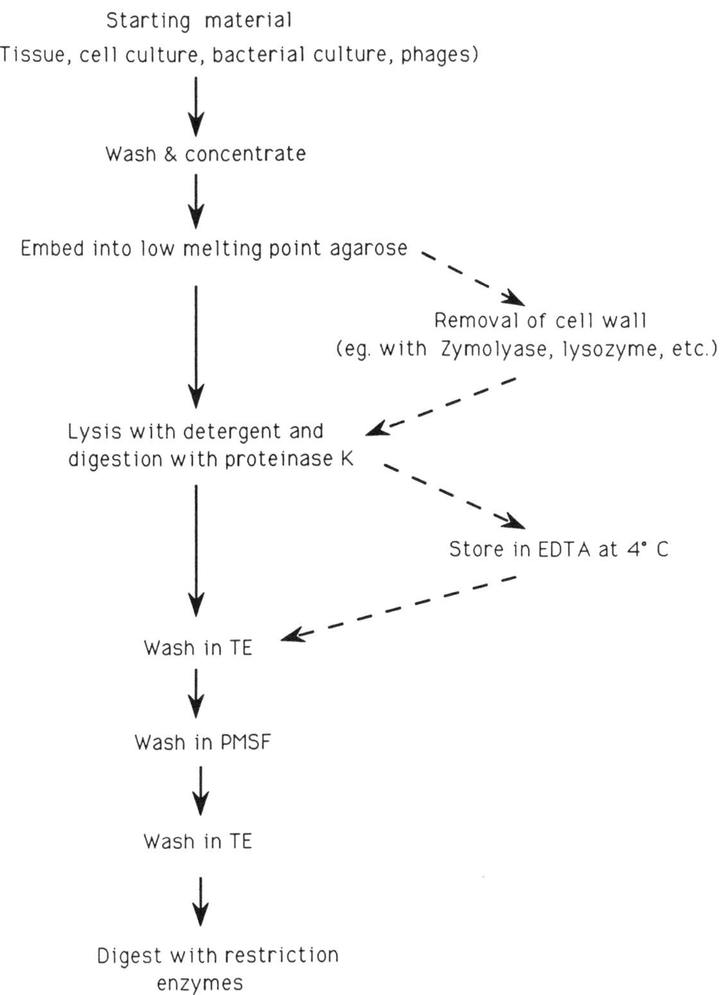

Fig. 1. Flow chart of the general procedure for preparing intact chromosomal DNA in agarose blocks.

10 min. Wash once with half the volume washing buffer and resuspend in 10 mL of washing buffer (1/10 growth volume).

2. Prepare a 1% LMP agarose solution in TE and cool to 50°C. Mix the bacterial suspension with 1 vol of agarose and dispense carefully into the precooled block-formers, avoiding air bubbles. Transfer the blocks into lysis buffer. Incubate at 37°C overnight, then transfer the blocks into ESP buffer and incubate at 50°C for a further 48 h.

3. Store in storage buffer or wash with PMSF and digest with restriction endonucleases (*see* Chapter 12).

3.2. Electrophoresis

The separation of double-stranded DNA in FIGE depends on several parameters: the forward and backward pulse times (t_f and t_b), the t_f:t_b ratio (R_t), the electric field (E), the temperature (T), buffer composition, and gel concentration. The separation range is easiest to control by alteration of the pulse times while keeping the other parameters constant (*see* Note 5).

To separate DNA of unknown size, the best approach is to initially separate over a wide size range by using a pulse time ramp to obtain an overview of the fragment size distribution. Pulse time ramps reduce the band inversion effect and separate over a wider range of sizes than constant pulses, but give a lower resolution. A ramp is achieved by commencing the pulse runs at a low t_f, for example 1 s, and then gradually increasing the pulse time to a higher value (*see* Note 6). The ratio of t_f to t_b remains constant throughout the run.

For most samples it will then be necessary to achieve better resolution in some regions of the gel, for example to resolve doublets. For this purpose, FIGE with a ramp over a narrower pulse time range or with constant pulse time is helpful. Constant pulses separate in a rather narrow mol wt range, with a high resolution, but also produce strong band inversion effects. By carefully choosing the pulse time, the separation window can be selected. The possibility to "zoom" into a particular size range seems to be a unique feature of FIGE and is not given by PFG with nonparallel fields. Figure 2 is presented as a guide to select appropriate pulse time ramps (Fig. 2A) or constant pulse times (Fig. 2B) for the separation ranges 20–1000 kbp, where R_t is 3.

The strategy described above has been successfully used to estimate the size of the *T. aquaticus* genome *(8)*, by separating several restriction enzyme digests using exponential pulse time ramps (*see* Note 6) as shown in Fig. 3.

1. Pour a 1% agarose gel to about 5 mm thick with the wells slightly larger than the sample blocks. Do not include ethidium bromide in the gel or the running buffer (*see* Note 7).

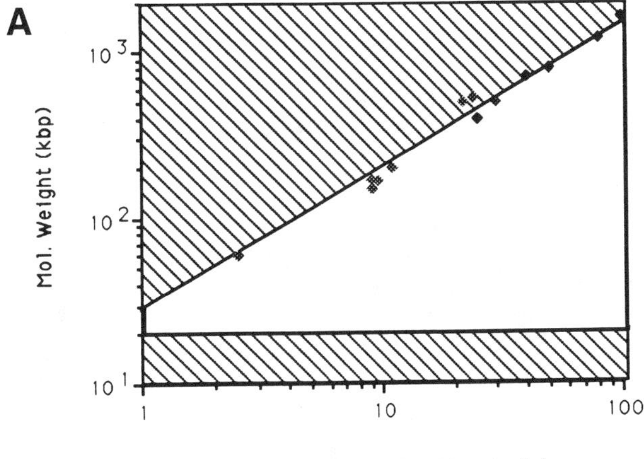

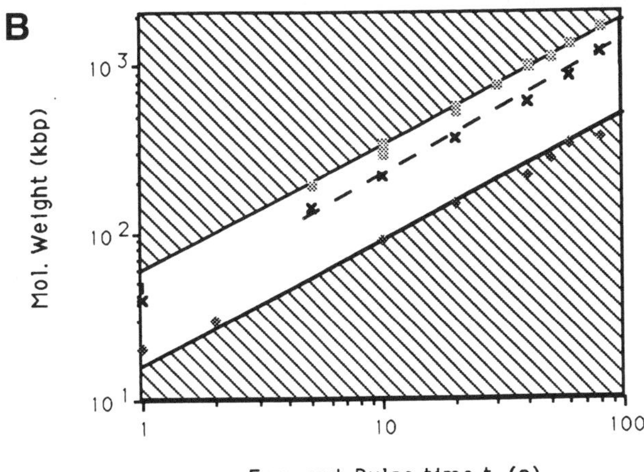

Fig. 2. (**A**) Separation range of FIGE with an exponential pulse time ramp starting with a forward pulse time of 1 s. Conditions: $E = 5.3$ V/cm, $R_t = 3$, 1% agarose in 0.5X TBE at 18°C. (**B**) Separation range of FIGE with constant pulse time. Conditions as above. The straight lines are the borders of the lower and upper compression zones, respectively, estimated by a series of experiments (marks). The dotted line shows the true upper separation limit, because of band inversion. To use the graphs, select the mol wt range you want to separate on the vertical axis, proceed to the right until it fits into the separation window, and read the pulse time t_f (or end forward pulse time, t_{fe}) on the horizontal axis.

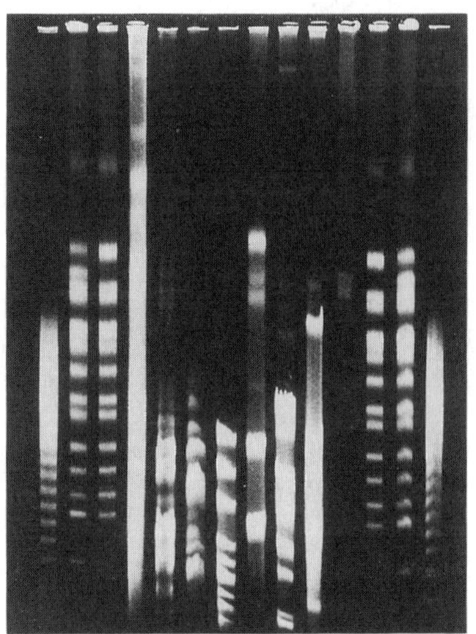

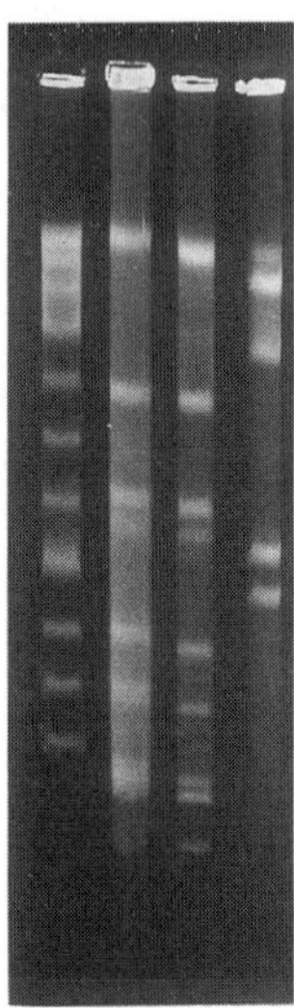

Fig. 3. (**A**) Example of the separation of different restriction fragments of the *T. aquaticus* (YT-1) genome with a pulse time ramp. Conditions: $E = 5.3$ V/cm, $R_t = 3$, $t_f = 1–120$ s, $t_b = 0.3–40$ s, exponential sawtooth pulse time ramp, duration 36 h, $T = 18°C$, 1% agarose in 0.5X TBE. The lanes are: 1. Lambda oligomers; 2. Yeast chromosomes YP 148; 3. Yeast chromosomes YNN 295, restriction digests; 4. *Asn*I; 5. *Eco*RI (partial); 6. *Hpa*I (partial); 7. *Nde*I; 8. *Spe*I; 9. *Ssp*I; 10. *Spe*I/*Asn*I (double digest); and 11. Intact chromosomal DNA.

The photo has been kindly provided by Elmar Maier. (**B**) Separation of restriction digests of the *T. aquaticus* genome with higher resolution in the lower mol wt range. Conditions: $E = 5.3$ V/cm, $R_t = 3$, $t_f = 1–30$ s, $t_b = 0.3–13.3$ s, exponential sawtooth pulse time ramp, duration 17 h, $T = 18°C$, 1% agarose in 0.5X TBE. The lanes are: 1. Lambda oligomers, 2. and 3. *Ssp*I digests of different preparations, 4. *Spe*I digest. The gel in Fig. 3A was used to get an overview over the whole size range of restriction fragments. The narrower ramp (Fig. 3B) was then used to "zoom" into the range of 50–350 kbp. The two lower bands of the *Taq–Spe*I digest that ran as one band in Fig. 3A are now clearly separated.

2. Insert the blocks, or part of the blocks, into the wells. Place them in contact with the side facing the direction of the electrophoresis. Seal the well with 1% agarose.
3. Place the gel into the gel box and fill with buffer to approx 1 mm above the gel surface. Adjust the pump flow to approx 10 mL/min (*see* Note 8). Run the DNA into the gel with a constant field of 5 V/cm for about

15 min and then apply the pulse sequence. The gel running temperature should be maintained to between 15 and 20°C (*see* Notes 2 and 5). The velocity of the DNA molecules depends on the chosen conditions. Good results have been achieved with the following set of parameters: $E = 5$ V/cm, $T = 18°C$, with a 1% agarose gel in 0.5X TBE, $R_t = 3$ (*see* Notes 9 and 10). Under these conditions lambda DNA (48.5 kbp) has a velocity of 0.45 cm/h *(8)*. I therefore recommend a run time of 18–20 h for a gel of 10 cm length.

4. Stop the run, stain in water containing 0.5–1.0 µg/mL ethidium–bromide for 30 min, destain in water for 30 min, and photograph. The gel may be further processed, for example, by Southern analysis.

As the user gets more confident with the technique he or she may prefer to use a different set of parameters (e.g., different voltage or temperature). For example, by lowering the ratio R_t, an even higher resolution can be achieved, but also the band inversion effect is enhanced *(7)*. With a few experiments the new windows of separation for a particular set of parameters can be estimated.

4. Notes

1. FIGE devices are commercially available, but it is also possible to build one. The simplest way of building a FIGE device is to use two bipolar magnetic relays (or one double relay). Such relays can be controlled by a computer via a simple amplifying circuit, using a standard npn transistor (BC series). Since mechanical relays have only a limited lifetime and are also not suitable for fast switching, it is preferrable to build a device without mechanical parts using optocouplers, steady state relays, or transistors. Detailed instructions for relay boxes are given in refs. 2–8. Devices using 4 optocouplers (ODC 5A; Opto 22, Huntington Beach, CA) connected in a similar way as a bridge-rectifier have been successful and can be directly used with most computers without an amplifying circuit. A relay box such as this *(7,8)* can easily be assembled by nonexperienced users, however, care must be taken, that all 4 optocouplers are never activated at the same time.
2. In many cases, FIGE gels can be run at room temperature without cooling, or in a cold room. If higher voltages are to be used, a cooling device is needed. Either the buffer can be recirculated through the cooler or cooling coils can be immersed into the buffer chambers. Always comply with the safety regulations.
3. It is important to embed the prokaryotes into the agarose without lysis, and this is best achieved by using an isotonic medium. Some bacteria

are sensitive to freezing and in such cases should be prepared freshly. It may be necessary to change the lysis reagents for bacteria with different cell membrane and cell wall properties.

4. By addition of chloramphenicol (180 µg/mL) at about 1 h before collection, initiation of replication is blocked. This eliminates heterogeneity owing to the presence of regions of DNA undergoing synthesis. It must be noted that if the bacterial strain contains plasmids, these will be amplified, making the interpretation of bands more difficult.
5. For changing the separation range, I generally recommend a change in the pulse time. It is also possible to change other parameters to achieve similar results. Increasing voltage or temperature shifts the separation to higher mol wts. The agarose concentration has negligible effects *(7,8)*. A change in forward to backward pulse time ratio (R_t) does not only shift the separation range, but also changes the mobility to mol wt function *(7)* and therefore should be used carefully.
6. I strongly recommend the use of repetitive pulse time ramps ("sawtooth ramps") repeating, for example, every hour. They have the advantage that an experiment can easily be interrupted (or prolonged) without losing a part of the separation window.

   I also recommend the use of exponential pulse time ramps rather than linear ones. An exponential pulse time ramp means that the increase in pulse time itself is increased during the run, and therefore emphasizes shorter pulse times. This results in a nearly linear separation in the lower mol wt range and only weak band inversion. In contrast, linear pulse time ramps still result in quite a substantial band inversion *(15)*. To avoid this effect special methods, such as ZIFE or random pulse time ramps, must be used.
7. Do not use ethidium bromide in the electrophoresis buffer or in the gel, since this reduces the mobility of the DNA fragments and shifts the separation range towards smaller molecules. Always stain and destain the gel after the run.
8. The pump rate of 10 mL/min is sufficient for small gels (12 × 12 × 0.5 cm). For larger gels it might be necessary to enhance the pump rate to maintain a constant pH and conductivity in both buffer chambers.
9. Very large molecules (>2 Mbp) can be "trapped" in the agarose block and are not able to enter the gel under the conditions mentioned above. To avoid this effect, the voltage has to be lowered substantially (to 1–2 V/cm), leading to very long run times. It has been reported that additional short pulses ("spikes") are able to "detrap" long DNA molecules *(14,18)*.
10. For systematic studies and better reproduction of experiments, it is advantageous to measure the electric field (in V/cm). It is best to mea-

sure this by dipping the two probes of a voltage meter (e.g., two platinum wires) into the buffer at the ends of the gel (or even into the gel) and dividing the reading by the distance between the probes. This is necessary because in most horizontal gel boxes only a fraction of the power supply voltage is applied to the gel; the rest of the voltage drop is owing to the resistance of the buffer. Always comply with safety regulations.

11. Only a few studies have been published about the behavior of circular DNA under pulsed field conditions, but they have shown that the movement of open as well as closed circular DNA differs dramatically from that of linear DNA of the same size. The mobility of DNA rings is largely unaffected by pulse time but they can be trapped in the gel when using higher electric fields. The different behavior of linear and circular DNA under pulsed field conditions can be used to separate these species with two dimensional pulse field gel electrophoresis devices under certain pulse regimes.

References

1. Carle, G. F., Frank, M., and Olson, M. V. (1986) Electrophoretic separations of large DNA molecules by periodic inversion of the electric field. *Science* **232,** 65–68.
2. Hennekes, H. and Kühn, S. (1989) Control of pulsed field gel electrophoresis at short switching intervals by a microcomputer. *Anal. Biochem.* **183,** 80–83.
3. Roy, G., Wallenburg, J. C., and Chartrand, P. (1988) Inexpensive and simple set-up for field inversion gel electrophoresis. *Nucleic Acids Res.* **16,** 768.
4. Larson, J. J., Nicholson, A. W., and Siegel, A. (1987) Field inversion of large DNA fragments using an inexpensive unit. *BioTechniques* **5,** 228–231.
5. Porter, C. D., Porter, A. G. F., and Archard, L. C. (1988) BBC microcomputer controlled field inversion gel electrophoresis. *Comp. Appl. Biol. Sci.* **4,** 271–273.
6. Denko, N., Giaccia, A., Peters, B., and Stamato, T. D. (1989) An asymmetric field inversion gel electrophoresis method for the separation of large DNA molecules. *Anal. Biochem.* **178,** 172–176.
7. Heller, C. and Pohl, F. M. (1989) A systematic study of field inversion gel electrophoresis. *Nucleic Acids Res.* **17,** 5989–6003.
8. Heller, C. (1990) Untersuchungen zur DNA-Bewegung im Agarose Gel. Ph.D. Thesis, University of Constance. Hartung-Gorre Verlag, Konstanz, ISBN 3-89191-360-5.
9. Lai, E., Davi, N. A., and Hood, L. (1989) Effect of electric field switching on the electrophoretic mobility of single-stranded DNA molecules in polyacrylamide gels. *Electrophoresis* **10,** 65–67.
10. Ulanovski, L. V., Drouin, G., and Gilbert, W. (1990) DNA trapping electrophoresis. *Nature* **343,** 190–192.
11. Birren, B. W., Simon, M. I., and Lai, E. (1990) The basis of high resolution separation of small DNAs by asymmetric-voltage field inversion electrophoresis and its application to DNA sequencing gels. *Nucleic Acids Res.* **18,** 1481–1487.

12. Bostock, C. J. (1988) Parameters of field inversion gel electrophoresis for the analysis of pox virus genomes. *Nucleic Acids Res.* **16,** 4239–4252.
13. Birren, B. W., Lai, E., Hood, L., and Simon, M. I. (1989) Pulsed field gel electrophoresis techniques for separating 1- to 50-kilobase DNA fragments. *Anal. Biochem.* **177,** 282–286.
14. Turmel, C., Brassard, E., Slater, G. W., and Noolandi, J. (1990) Molecular detrapping and band narrowing with high frequency modulation of pulsed field electrophoresis. *Nucleic Acids Res.* **18,** 569–575.
15. Heller, C. and Pohl, F. M. (1990) Field inversion gel electrophoresis with different pulse time ramps. *Nucleic Acids Res.* **18,** 6299–6304.
16. Turmel, C., Brassard, E., Forsyth, R., Hood, K., Slater, G. W., and Noolandi, J. (1990) High-resolution zero integrated field electrophoresis of DNA, in *Electrophoresis of Large DNA* (Lai, E. and Birren, B. W., eds.), Cold Spring Harbor Laboratory, Cold Spring Harbor, NY, pp. 101–131.
17. Schwartz, D. C. and Cantor, C. R. (1984) Separation of yeast chromosomes–sized DNAs by pulsed field gradient gel electrophoresis. *Cell* **37,** 67–75.
18. Zhang, T. Y., Smith, C. L., and Cantor, C. R. (1991) Secondary pulsed field gel electrophoresis: a new method for faster separation of larger DNA molecules. *Nucleic Acids Res.* **19,** 1291–1296.

Chapter 14

Enhanced Chemiluminescent Detection of Horseradish Peroxidase Labeled Probes

Ian Durrant

1. Introduction

Enhanced chemiluminescence (ECL) is a generic detection system that has been applied to all standard, membrane-based molecular biology techniques *(1)*. The technology is based on the well-characterized horseradish peroxidase (HRP) catalyzed oxidation of luminol in the presence of peroxide *(2)*. In this reaction a cyclic scheme, involving the enzyme, leads to the formation of luminol radicals *(3)*. These radicals decay through an excited intermediate compound (3-aminophthalate), and as this compound returns to the ground state light is emitted. The enhanced reaction offers a major improvement over the basic scheme in that the light output is increased over 1000-fold, whereas the background is reduced *(4)*. The light emitted has a maximum wavelength of 428 nm and can be captured with high efficiency by blue-light sensitive X-ray film. Consequently, the final result is very similar in appearance to the results obtained with traditional procedures for the detection of radioactive labels; a hard copy is obtained, multiple exposures can be taken, and the results can be analyzed by densitometry and easily reproduced for presentations and publications.

ECL can be used in any application provided that an HRP group is available at the end point of the process, including the detection of

From: *Methods in Molecular Biology, Vol. 31: Protocols for Gene Analysis*
Edited by: A. J. Harwood Copyright ©1994 Humana Press Inc., Totowa, NJ

hapten (for example fluorescein *[5]*) labeled probes with an antibody-HRP conjugate, the detection of proteins via similarly labeled antibodies on Western blots *(6)* or in immunoassays *(7)*, and the detection of nucleic acid probes directly labeled with HRP *(8)* (*see* Note 18). This chapter deals specifically with the labeling of long nucleic acid probes. Oligonucleotide labeling with HRP is described in Chapter 15.

The labeling method used to introduce HRP directly onto the nucleic acid probe is based on the process first described by Renz at EMBL *(9)*. Initially, HRP is made relatively positively charged by crosslinking it to polyethyleneimine (PEI) via the action of parabenzoquinone (PBQ; Fig. 1) to form the labeling reagent consisting of PEI bearing either 1, 2, or 3 HRP molecules. The negatively charged phosphate backbone of single-stranded nucleic acid interacts electrostatically with the positively charged HRP complexes leading to a labeling frequency of approx 1 complex every 50 bases. The labeled probe is stabilized by crosslinking the labeling reagent to the nucleic acid by the action of the short range crosslinking agent glutaraldehyde. Labeled probes are stable and the presence of the large labeling complexes does not affect the hybridization properties of the probe *(10)*.

Hybridization is performed in a specially prepared and optimized buffer. This buffer is based on 6*M* urea as a denaturant in which the T_m of a typical probe is reduced to around 67°C, in a similar manner to buffers based on 50% formamide *(11)*. Hybridization rate is known to be maximal at a temperature of T_m –25°C *(12)* leading to a hybridization temperature optimum of 42°C. This temperature is also a maximum temperature for this system owing to the proteinaceous nature of the label that will lose activity if subjected to extended incubations at elevated temperatures. In addition to urea, the buffer contains a blocking agent (to prevent the HRP portion of the probe binding nonspecifically to the membrane) and a new rate enhancement system, which has yielded an approx 10-fold increase in sensitivity compared to earlier versions of the buffer.

The system is capable of routinely detecting 0.5 pg of a single copy gene in human genomic DNA digest Southern blotted onto nylon membrane (Fig. 2). For systems with higher target concentrations,

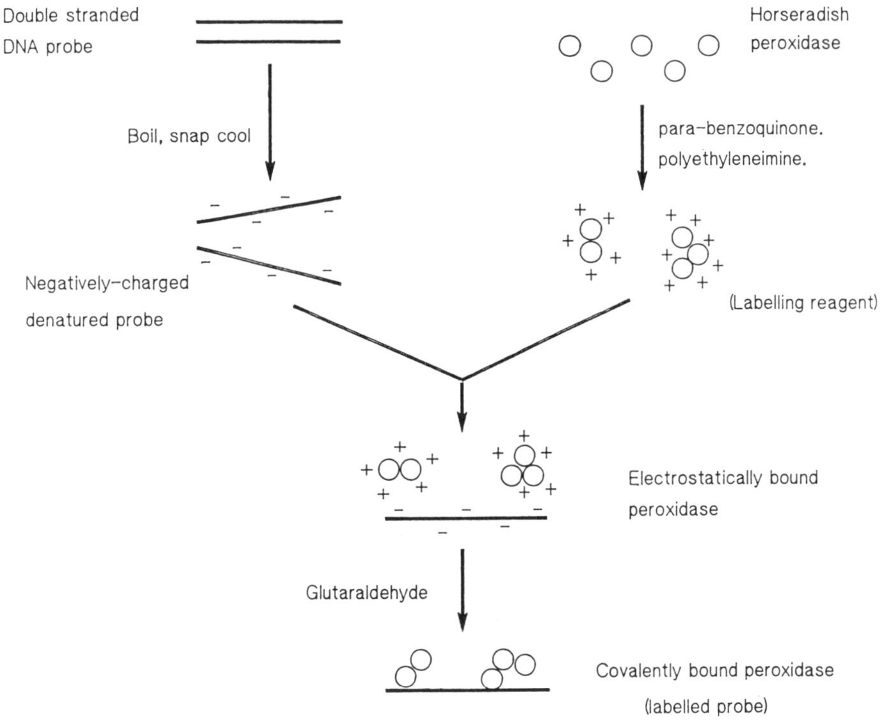

Fig. 1. Outline scheme for the production of enzyme labeled probes.

for example plaque and colony screening, the high sensitivity of the ECL reaction combined with the rate enhancement system of the hybridization buffer enable results to be achieved in a single working day (Fig. 3).

2. Materials

All the reagents necessary for labeling nucleic acid probes, the hybridization buffer and the ECL detection reagents are available from Amersham International (Amersham, UK; ECL-Direct systems RPN 3000, RPN 3001). The system can be used in conjunction with both nylon and nitrocellulose membranes. Both Hybond™-N$^+$ (nylon) and Hybond-ECL (nitrocellulose; Amersham) have been found to work well in this system.

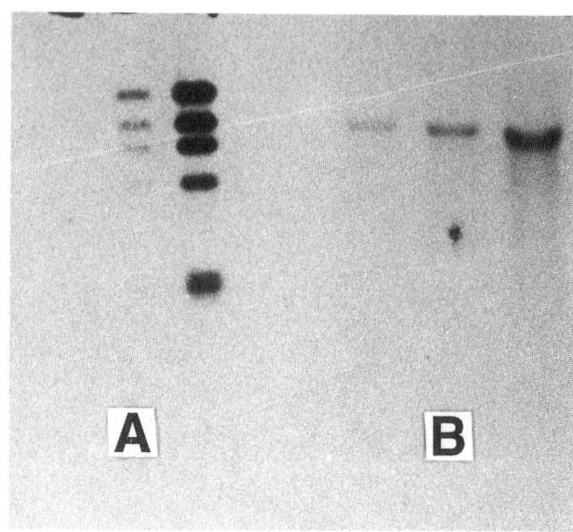

Fig. 2. The new ECL-direct system. (**A**) Loadings of 1, 10, and 100 pg of lambda DNA digested with *Hind*III Southern blotted onto Hybond-N+. (**B**) Loadings of 1, 2, and 5 μg human genomic DNA digest blotted onto Hybond-N+. Probed with (A) Lambda/*Hind*III or (B) 1.5-kb N-*ras* specific sequence, at 10 ng/mL for 17 h at 42°C; 60-min exposure.

2.1. Probe Labeling

1. Labeling reagent: This comprises HRP crosslinked to PEI. Its production is described in refs. *9* and *10*. It can be purchased ready made (Amersham) and the use of this compound is described in the following protocol.
2. Glutaraldehyde solution: 1.5% (v/v) glutaraldehyde solution in deionized water. Although in the pure form glutaraldehyde is classified as harmful, in the ECL kit it is present at a concentration of only 1.5% (v/v) at which level it is not classified as harmful. However, like all chemicals, it should be handled within the principles of good laboratory practice.

2.2. Hybridization

3. Hybridization buffer: The buffer can be obtained ready made (Amersham). In addition to 6*M* urea it contains a detergent, a pH stabilizing compound, and a novel rate enhancement system. (The new rate enhancer is directly responsible for a significant increase in sensitivity over earlier versions of the ECL direct system. The new hybridization buffer

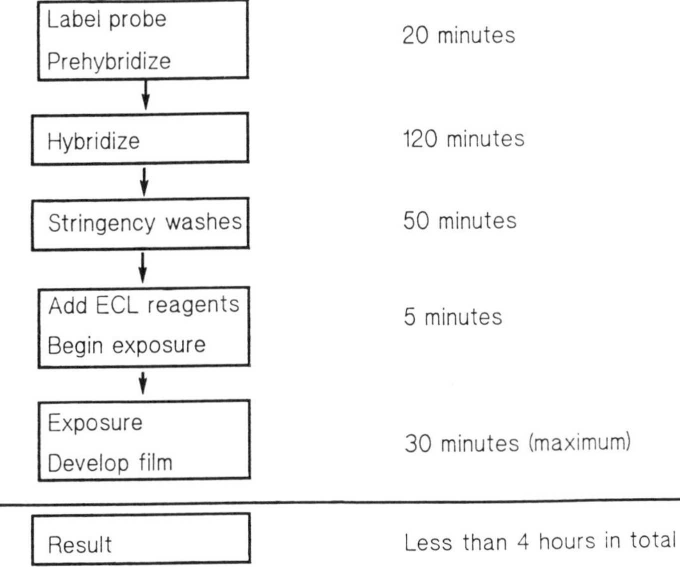

Fig. 3. Outline procedure for rapid hybridizations.

ensures a higher level of target coverage and promotes an increase in the stability of the enzyme label during the hybridization process.)

4. Sodium chloride.
5. Blocking agent: 5 g/100 mL of hybridization buffer. The blocking agent supplied (Amersham) has been specifically selected for use with this system.
6. 20X SSC stock solution: 3*M* NaCl, 0.3*M* tri-sodium citrate.
7. Primary wash solution: 6*M* Urea, 0.4% SDS, 0.5X SSC.
8. Secondary wash solution: 2X SSC.

2.3. Detection

9. Detection Solution 1: This contains specially purified luminol and the ECL enhancer compound in a borate buffer.
10. Detection Solution 2: This contains a peracid salt in borate buffer.
11. SaranWrap™: A product of the Dow Chemical Company. It is recommended at various stages of the protocol and is particularly suitable for molecular biology applications since it does not absorb UV or blue light. SaranWrap is often available from local suppliers.

12. X-ray film: A film optimized for the light detection is available (Hyperfilm™-ECL, Amersham). However, other X-ray films are suitable, e.g., Hyperfilm-MP, Kodak-AR. It is also possible to detect the light output on a charge-coupled device (CCD) camera *(13)*.

3. Methods

The following protocol is written for the hybridization of a 20 × 20 cm membrane. Obviously, for different size pieces of membrane the amounts of probe and various solutions should be varied accordingly.

3.1. Probe Labeling

1. Dilute 1 µg of the DNA to be labeled to a concentration of 10 ng/µL in deionized water, ensuring that the final concentration of monovalent cations in the DNA solution is less than 10 m*M* (*see* Note 1).
2. Denature double-stranded DNA in a vigorously boiling water bath for 5 min (*see* Note 2). Immediately place on ice and leave to cool for 5 min. Briefly spin the DNA solution in a microcentrifuge to settle all the liquid to the bottom of the tube (*see* Notes 3–5).
3. Add 100 µL of labeling reagent and mix thoroughly by pipeting the solution up and down.
4. Add 100 µL of glutaraldehyde solution. Mix gently by pipeting up and down. Incubate at 37°C for 10 min.
5. Labeled probes can be stored on ice for 10–15 min but for longer term storage, up to at least 6 mo, add glycerol to a final concentration of 50% (v/v) and store at –20°C. Do not boil the probe before use in hybridizations, even after long term storage.

3.2. Hybridization

1. Prepare 100 mL of hybridization buffer in advance (*see* Note 6). Add NaCl to give a final concentration of 0.5*M* (*see* Note 7). Add blocking agent to a final concentration of 5% (w/v). Immediately after addition of the block swirl the buffer to avoid coagulation of the blocking agent and then place the buffer, at room temperature, on a magnetic stirrer or roller mixer, for 1 h. Warm the buffer to 42°C, with occasional mixing, for 30–60 min. The buffer can be used immediately or stored in aliquots at –20°C for up to 3 mo.
2. Prehybridize the blot in the hybridization buffer for 1 h (*see* Note 8).
3. Add 300 µL of labeled probe to the buffer used in the previous step. This gives a final concentration of 10 ng/mL (*see* Notes 9 and 10).

4. Hybridize at 42°C overnight with constant agitation.
5. Place blot into a clean container and cover with 2 mL/cm^2 of primary wash buffer, prewarmed to 42°C. Incubate for 20 min at 42°C in a shaking water bath (*see* Notes 11 and 12).
6. Replace wash buffer with fresh wash solution and incubate for a further 20 min at 42°C.
7. Place blot into a fresh container and cover with 2 mL/cm^2 of secondary wash buffer. Incubate with agitation at room temperature for 5 min.
8. Replace wash with fresh solution and repeat the 5 min incubation. Blots can be stored in the final wash solution for up to 30 min before proceeding to the ECL detection stage (*see* Note 13).

3.3. ECL Detection

1. Mix equal volumes of the detection reagents 1 and 2 to give 50 mL. This is sufficient working reagent to cover the membrane at 0.125 mL/cm^2 (*see* Notes 14 and 15).
2. Remove the blots from the secondary wash buffer, allow to drain and place DNA side up onto a clean surface (for example, a piece of SaranWrap). Cover each blot with detection reagent and allow to stand for 1 min only.
3. Drain off excess detection reagent and wrap blots in SaranWrap. Place in a film cassette DNA side up and, in a darkroom, add a piece of X-ray film. Close the cassette and time the exposure for 1 min only.
4. Remove the film and replace with a fresh piece of X-ray film; begin timing the second exposure. Develop the first film and on the basis of the signal strength decide how long the second exposure should last. Typically, a 10–30 min exposure would be required for high target applications, with an exposure of up to 2 h for the most sensitive levels of detection.

4. Notes

1. The probe labeling reaction is most efficient in low levels of monovalent cations. High levels (>10 m*M*) will inhibit the initial electrostatic interactions of the labeling process.
2. It is vital that the probe is fully denatured before labeling to ensure maximum labeling efficiency. Experiments have shown that heating blocks may not be adequate to perform this task.
3. The reaction will label single-stranded DNA and RNA without the need to denature the probe first.

4. The reaction can be performed on as little as 100 ng of material and as much as 2 μg. The probe length can vary from 300 bases to 50 kb.
5. In general, cloned probes should be excised and separated from the vector sequences by one of the commonly used methods for insert preparation *(14)*, since vector sequences may cause some background. It is possible to label DNA fragments within low melting point agarose gel slices provided that the concentration of agarose in the gel does not exceed 0.7% (w/v).
6. Hybridizations can be performed in boxes, bags, or hybridization ovens. The volume of buffer recommended can be altered with experience, in particular when using large blots in bags or ovens or when hybridizing a number of membranes in the same chamber. The only criterion is that the individual membranes should move freely in the buffer.
7. The salt concentration can be used to control the stringency of the hybridization. 0.5*M* NaCl is strongly recommended since it provides optimal results with a wide variety of probes and achieves the maximum rate of hybridization. However, in certain cases variation of the salt concentration, in the range 0.05–1.0*M*, may yield better results.
8. For maximum sensitivity, the prehybridization step is routinely performed for 1 h, although as little as 15 min may be used in many cases.
9. When adding the probe to the hybridization solution avoid adding it directly onto the membrane as the very high localized probe concentration can cause nonspecific binding. Add the probe away from the membrane or remove some of the hybridization solution and premix it with the probe before adding the entire mixture back to the hybridization chamber.
10. The basic system uses a probe concentration of 10 ng/mL and the hybridization is performed overnight. For hybridization to membranes containing a high level of target, for example colony and plaque screening, it is possible to reduce the probe concentration to as low as 2 ng/mL or to reduce the hybridization time to as little as 2 h.
11. The stringent wash is the primary wash that contains 6*M* urea as a denaturant. The wash temperature is generally set at 42°C to avoid HRP denaturation with the stringency being controlled by the SSC concentration. For many probes a buffer containing 0.5X SSC has been found to be suitable.
12. An alternative procedure is to perform the stringency washes at 55°C in the absence of urea but using 0.1–0.5X SSC, 0.4% SDS depending on the level of stringency required. These washes should only be performed

for 2 × 10 min rather than 2 × 20 min, to avoid denaturation of the enzyme label.

13. The secondary wash removes the SDS from the membranes that may interfere with the ECL detection reaction. Blots can be held in this buffer for a short period of time but for longer term storage (overnight) the blots should be wrapped, wet, in SaranWrap and stored at 4°C or immersed in 50% glycerol and stored at –20°C. Briefly rinse stored filters in fresh 2X SSC before proceeding to the ECL detection reaction.
14. The ECL detection reagents are supplied as two ready-to-use solutions, one of which contains the luminol and enhancer and the other of which contains a peracid salt (enzyme substrate). The reagents are stored separately since a low level of chemical degradation may occur on long term storage of the mixture. However, the working ECL reagent can be premixed and stored for at least 48 h at 4°C without any loss in light output.
15. Care should be taken to avoid chemical contamination of the substrates. If possible, avoid wearing gloves that are powdered since the powder can inhibit the ECL reaction and thus cause blank patches to appear on the final result.
16. Rehybridization. The light output from the ECL reaction peaks at approx 10 min and then decays with a half life of 2 h owing to enzyme inactivation. For most applications the required sensitivity can be achieved in a maximum of 1 h. The level of light falls below the threshhold of detection on X-ray film after 5–6 h. It is possible to rehybridize a membrane without removing the previous probe since during the subsequent ECL detection the HRP label on the first probe, which is inactivated, will not be detected *(8)*. The avoidance of stripping protocols preserves the target, saves time, and allows as many as 10 reprobings on nylon membranes without difficulty. In some applications where the target levels are high and the result has been achieved very quickly it is possible that the first probe will retain some enzyme activity and in these cases it would either be necessary to remove it or to leave the blots for some time in the ECL reagents before performing the second hybridization.
17. Rewashing. It is possible to envisage the situation where a result has been obtained but it is felt that the washes were not stringent enough. Provided that the ECL reaction has not proceeded for significantly longer than 30 min it is possible to remove the blots from the cassette and perform fresh stringency washes and ECL detection. Since some of the

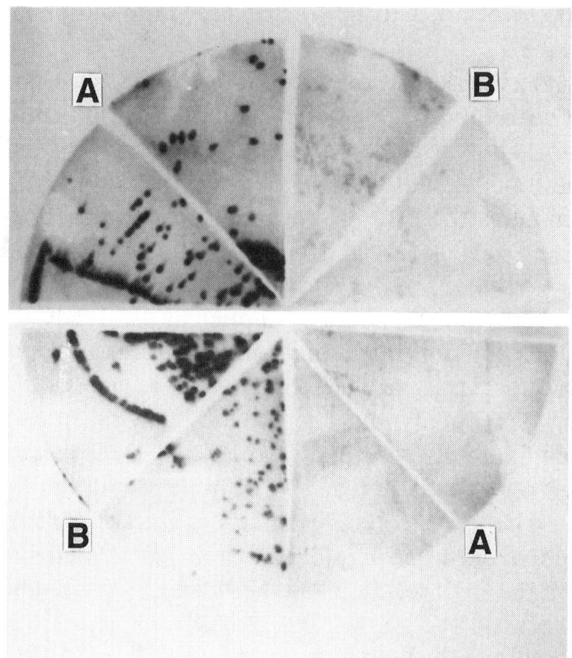

Fig. 4. Cosmid screening. Overnight growth of Cosmid PTL5 (top half) and Cosmid Lorist 4 (bottom half) on Hybond-N^+. Hybridized with **(A)** PTL5 and **(B)** Lorist 4 specific HRP labeled probes for 17 h at 42°C; 10-min exposure.

light capacity of the system will have been used, the second exposure will generally need to be longer than the first to achieve a similar sensitivity.

18. Applications
    a. Colony and plaque screening. The ECL system can be used to screen all the commonly used colonies and plaque systems including plasmid and cosmid colonies (Fig. 4) and M13 and lambda plaques *(8)*. An important point is the removal of bacterial debris, particularly from colony lifts, where nonspecific binding of HRP to the debris may occur. Hybond N^+ disks are ideal for this task. The colony lift is placed on a pad of 0.5*M* NaOH, which lyses and fixes the DNA simultaneously in a maximum of 5 min. The debris is removed by vigorously washing the filter in 5X SSC (for example by holding the membrane in forceps and agitating in the buffer) or by rubbing the membrane with a gloved finger. Specificity can also be aided by

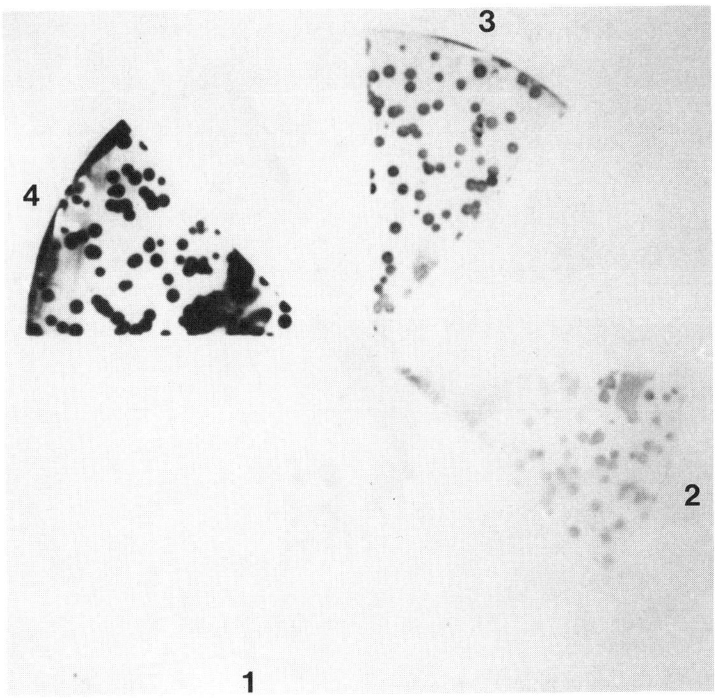

Fig. 5. Probe concentration. M13mp8 plaques containing N-*ras* specific insert hybridized at 42°C with (1) 1 ng/mL; (2) 2 ng/mL; (3) 5 ng/mL; (4) 10 ng/mL of homologous probe for 2 h; 10-min exposure.

avoiding overgrowth of the colony and plaque targets. The combined lysis/fixation procedure is applicable to all the colony and plaque systems tested. For most of these systems the maximum hybridization time required is only 4 h and may be as low as 1 h. The probe concentration should initially be kept at 10 ng/mL but with experience of the system may be reduced to as little as 2 ng/mL depending on the level of target (Fig. 5).

b. Northern blotting. For Hybond-N$^+$ membrane the formaldehyde/MOPS denaturing gel system *(15)* is recommended in the production of Northern blots. The ECL buffer, including the blocking agent, has not been found to contain any measurable levels of RNase activity and provided that a reasonable level of target is present in the RNA

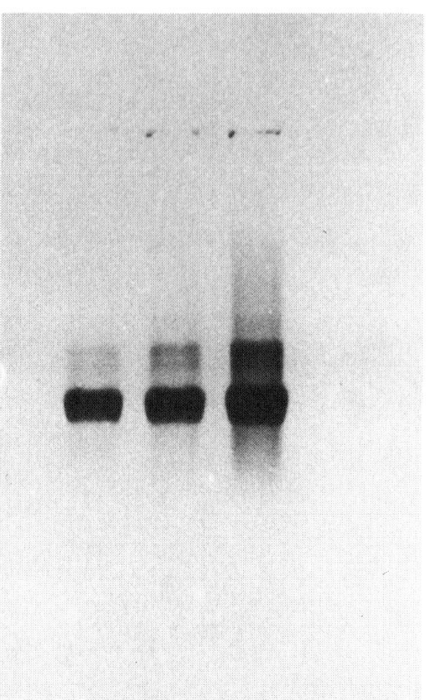

Fig. 6. Northern blot analysis. Poly-A$^+$ RNA derived from Hela cells was blotted onto Hybond-N$^+$ and hybridized overnight at 42°C with 20 ng/mL of an actin probe; loadings of 250 ng, 500 ng, and 1 µg; 30-min exposure.

sample loaded on the gel, the ECL system is readily applicable to the detection of Northern blots (Fig. 6).

c. Southern blotting. Southern blots are used for a variety of applications encompassing a range of sensitivity levels. The ECL-direct system is tested by detecting 0.5 pg of a human, single copy gene on Hybond-N$^+$. This represents the level of target available in a 1 µg loading of a genomic DNA digest when hybridized with a 1.5-kb probe (*see* Fig. 2); a 2 µg loading of human genomic DNA is equivalent to 1 attomole of target. At the other extreme, in terms of target levels, is the detection of PCR products in which the direct labeled probes are very successful *(16)*. Intermediate target levels are found in the detection of fingerprints or the screening of YAC digests (Fig. 7).

Acknowledgements

I would like to thank C. Bellanne of CEPH for the YAC screening result and my colleagues at Amersham, in particular Tim Stone and

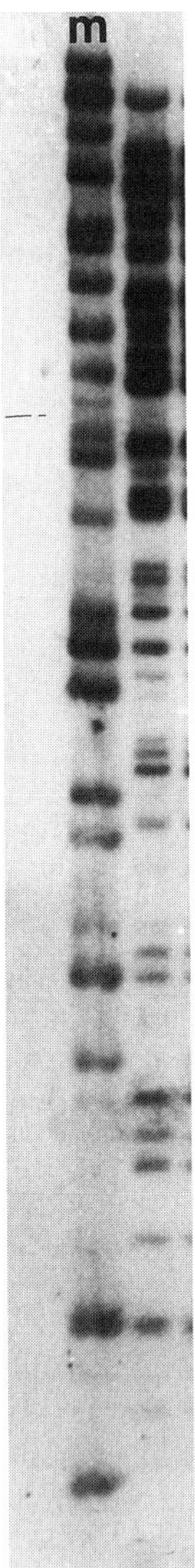

Fig. 7. Screening YAC vectors. *Pvu*II digest of YAC vector containing human genomic sequences blotted onto Hybond-N$^+$. Probed with 10 ng/mL of HRP labeled total human genomic DNA. M = marker track; 20 ng of ΦX174/*Nru*I digest and 90 ng lambda/*Hind*III digest, probed with 0.35 ng/mL ΦX174 and 0.75 ng/mL lambda (both HRP labeled). All three probes were hybridized simultaneously for 17 h at 42°C; 30-min exposure.

Louise Benge, for their contributions to this work. This chapter is dedicated to the memory of Louise Benge.

References

1. Cunningham, M. W. (1991) Nucleic acid detection with light. *Life Sci.* **6,** 1–5.
2. Thorpe, G. H. G. and Kricka, L. J. (1986) Enhanced chemiluminescent reactions catalyzed by horseradish peroxidase. *Meth. Enzymol.* **133,** 331–353.
3. Thorpe, G. H. G. and Kricka, L. J. (1987) Enhanced chemiluminescent assays for horseradish peroxidase: characteristics and applications, in *Bioluminescence and Chemiluminescence: New Perspectives* (Scholmerich, J., Andreesen, R., Kapp, A., Ernst, M., and Woods, W. G., eds.), Wiley, New York, pp. 199–208.
4. Thorpe, G. H. G., Kricka, L. J., Moseley, S. B., and Whitehead, T. P. (1985) Phenols as enhancers of the chemiluminescent horseradish peroxidase-luminol-hydrogen peroxide reaction: application in luminescence-monitored enzyme immunoassays. *Clin. Chem.* **31,** 1335–1341.
5. Cunningham, M., Harvey, B., Benge, L., and Wheeler, C. (1991) ECL random prime system—protocol variations. *Highlights* **2,** 9–10.
6. Wisdom, B. and Winter, P. (1990) Detection of galactosyltransferase by immunoblotting using enhanced chemiluminescence. *Life Sci.* **3,** 13–14.
7. Thorpe, G. H. and Kricka, L. J. (1989) Incorporation of enhanced chemiluminescent reactions into fully automated enzyme immunoassays. *J. Bioluminescence Chemiluminescence* **3,** 97–100.
8. Durrant, I., Benge, L. C. A., Sturrock, C., Devenish, A. T., Howe, R., Roe, S., Moore, M., Scozzafava, G., Proudfoot, L. M. F., Richardson, T. C., and McFarthing, K. G. (1990) The application of enhanced chemiluminescence to membrane-based nucleic acid detection. *Biotechniques* **8,** 564–570.
9. Renz, M. and Kurz, C. (1984) A colorimetric method for DNA hybridization. *Nucleic Acids Res.* **12,** 3435–3444.
10. Pollard-Knight, D., Read, C. A., Downes, M. J., Howard, L. A., Leadbetter, M. R., Pheby, S. A., McNaughton, E., Syms, A., and Brady, M. A. W. (1990) Non-radioactive nucleic acid detection by enhanced chemiluminescence using probes directly labeled with horseradish peroxidase. *Analyt. Biochem.* **185,** 84–89.
11. Hutton, J. R. (1977) Renaturation kinetics and thermal stability of DNA in aqueous solutions of formamide and urea. *Nucleic Acids Res.* **4,** 3537–3555.
12. Britten, R. J. and Davidson, E. H. (1985) *Hybridization Strategy in Nucleic Acid Hybridization: A Practical Approach* (Hames, B. D. and Higgins, S. J., eds.), IRL, Oxford, pp. 3–15.
13. Boniszewski, Z. A. M., Comley, J. S., Hughes, B., and Read, C. A. (1990) The use of charge-coupled devices in the quantitative evaluation of images, on photographic film or membranes, obtained following electrophoretic separation of DNA fragments. *Electrophoresis* **11,** 432–440.
14. Sambrook, J., Fritsch, E. F., and Maniatis, T. (1989) Recovery and purification of DNA fractionated on agarose gels, in *Molecular Cloning: A Laboratory Manual*, 2nd ed., Cold Spring Harbor Laboratory, Cold Spring Harbor, NY, pp. 6.22–6.35.

15. Sambrook, J., Fritsch, E. F., and Maniatis, T. (1989) Electrophoresis of RNA through gels containing formaldehyde, in *Molecular Cloning: A Laboratory Manual,* 2nd ed., Cold Spring Harbor Laboratory, Cold Spring Harbor, NY, pp. 7.43–7.50.
16. Sorg, R., Enczmann, J., Sorg, U., Kogler, G., Schneider, E. M., and Wernet, P. (1990) Specific non-radioactive detection of PCR-amplified HIV sequences with enhanced chemiluminescence labeling (ECL). *Life Sci.* **2,** 3,4.

CHAPTER 15

Nonradioactive Oligonucleotide Probe Labeling

Ian Durrant and Sue Fowler

1. Introduction

Oligonucleotide probes are increasingly being used in various applications, such as in the detection of PCR products; for *in situ* hybridization, and in colony and plaque screening *(1)*. Oligonucleotide probes are typified by being short, defined sequences of around 15–50 bases in length; although most frequently 20–30 bases long. If the base composition of the probe is known it is possible to quantify and predict the associated hybridization properties *(2)*. Oligonucleotide probes are capable of discriminating between perfectly matched and mismatched target sequences, where the mismatch can be as little as one base in twenty. With the recent advances in automated oligonucleotide synthesis machines, oligonucleotides are easily produced in large amounts. The hybridization process using oligonucleotide probes is relatively rapid owing to the high molar concentrations of probe generally used.

Traditionally, oligonucleotide probes have been labeled with radioactivity. These labeling reactions tend to operate on a relatively small scale and licences are required for the handling and disposal of radioactive materials. Consequently, a role for nonradioactive labeling methods has emerged.

Recent advances in nucleotide chemistry have enabled researchers to produce oligonucleotides with defined chemical modifications. Oligonucleotides can be synthesized easily bearing a reactive thiol *(3)* or amino group *(4)* at the 5'-end. The amino group has been used

From: *Methods in Molecular Biology, Vol 31: Protocols for Gene Analysis*
Edited by: A. J. Harwood Copyright ©1994 Humana Press Inc., Totowa, NJ

to label oligonucleotides with biotin, fluorescein, or other haptens. This method allows the introduction of only one reporter group, leading to low sensitivity, and involves a number of time-consuming purification procedures since the overall efficiency of the incorporation may be as low as 30–50%. Recent improvements have led to the availability of biotin phosphoramidite compounds that are used to add biotin directly to the oligonucleotide during synthesis. Some of these compounds can be used in a repetitive manner on the synthesizer to form a tail of biotin residues leading to a potential increase in sensitivity *(5)*. The extremely high coupling efficiency of these compounds (>94%) allows the labeled probe to be used with a minimum of purification. These hapten-based approaches rely on the detection of the hapten with enzyme-linked immunological molecules.

An alternative is to link the enzyme directly to the amino or thiol modified oligonucleotide probe. The amino link route has been demonstrated with alkaline phosphatase *(6)*. The chemistry is complex and requires a number of purification processes. The linking of horseradish peroxidase (HRP) to thiol-linked oligonucleotide probes is, however, much simpler *(7)*. The enzyme is first derivatized with a thiol reactive group. This derivatization does not interfere with the enzyme function and the product can be stabilized by freeze-drying in conveniently sized aliquots. The reaction of a fivefold molar excess of derivatized enzyme with thiol-linked oligonucleotide is simple and rapid and achieves such a high labeling efficiency that no further probe purification is required. The remaining free HRP does not interfere with the subsequent hybridization and detection processes.

The above methods rely on the synthesis of a specially modified probe sequence. For occasions where this is not convenient, the alternative is to label using an enzyme catalyzed reaction at the 3'-end of a nascent oligonucleotide *(8)* (in an analogous manner to that traditionally done with radiolabels). In theory, any hapten modified nucleotide can be used in conjunction with terminal deoxynucleotidyl transferase (Tdt) to produce a 3'-end tail on the probe sequence. In practice, the reaction must be carefully modified to take into account altered kinetics of the addition reaction in the presence of a hapten modified nucleotide. Ideally, the tailing reaction must be controlled to produce a short tail of 4–10 nucleotides in length. This gives the maximum sensitivity without compromising the stringency of the probe.

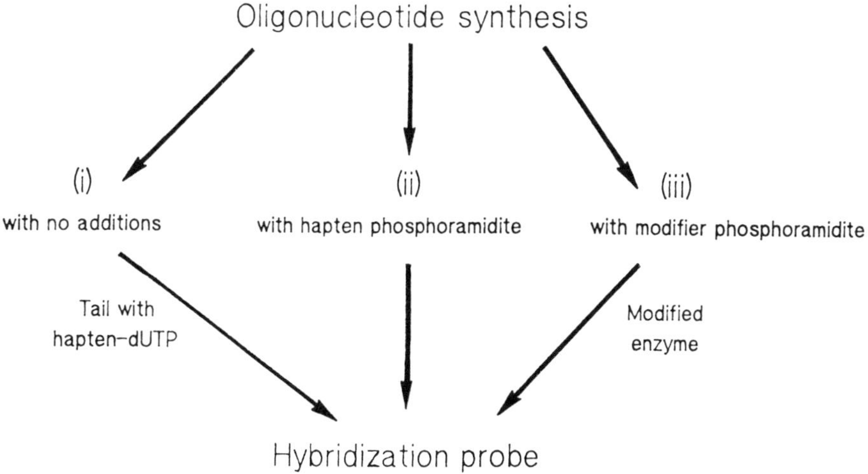

Fig. 1. Outline of oligonucleotide labeling methods.

Thus, there are three basic approaches available for nonradioactive oligonucleotide labeling (Fig. 1). Each of these methods is specifically illustrated by the following systems:

1. Tailing with fluorescein dUTP at the 3'-end of an unmodified oligonucleotide using terminal deoxynucleotidyl transferase *(8)*.
2. Directly labeling a thiol-modified oligonucleotide with the enzyme horseradish peroxidase *(7)*.
3. Addition of biotin directly to the oligonucleotide while being synthesized on an automated machine using a biotin phosphoramidite *(5)*.

The hybridization properties of probes labeled by these methods will also be discussed. A convenient, rapid, and sensitive detection system will be used to show examples of the types of application for which these probes can be used.

2. Materials

2.1. 3'-End Labeling

1. Fluorescein-11-dUTP.
2. Cacodylate buffer: 10X stock solution containing sodium cacodylate pH 7.4, dithiothreitol, and cobalt chloride. *Note: Cacodylate is an arsenic based compound and is classified as highly toxic. Avoid contact with skin and do not swallow. It is a possible carcinogen and there are dangers of cumulative effects. This compound, like all others, should*

be used within the principles of good laboratory practice. Wear appropriate protection, such as overalls, gloves, and safety glasses.

3. Terminal transferase: 2U/μL stock solution.

The above materials are available commercially in an optimized format (Amersham, Arlington Heights, IL RPN2131).

2.2. Biotin Labeling

4. Biotin phosphoramidite: available commercially (e.g., Amersham, RPN 2012).
5. Oligonucleotide synthesizer: Various commercial systems are available (e.g., Applied Biosystems, Foster City, CA).

2.3. Enzyme Labeling

6. Thiol modifier: available commercially (e.g., Amersham, RPN 2112).
7. Oligonucleotide synthesizer: as above.
8. Derivatized HRP: Horseradish peroxidase derivatized with a bifunctional crosslinking agent that leaves a thiol reactive group attached to the enzyme. This material is available in a freeze-dried form (Amersham, RPN2111).

2.4. Hybridization

9. 20X SSC stock: 3*M* NaCl, 0.3*M* tri-sodium citrate.
10. Blocking agent: Optimized blocking agent available commercially (Amersham; in RPN2111 and RPN2130).
11. *N*-lauroyl sarcosine, sodium salt.
12. SDS stock: 20%(w/v) in distilled water.
13. Hybridization buffer: 5X SSC, 0.5% (w/v) blocking agent. 0.02% (w/v) SDS, 0.1% (w/v) *N*-lauroylsarcosine.

3. Methods

3.1. 3'-End Labeling

1. Add the labeling reaction components (*see* Notes 1–3) in the following order: 100 pmol of oligonucleotide, 10 μL of fluorescein-dUTP, 16 μL of cacodylate buffer, sterile water to a volume of 144 μL. Then add 16 μL of terminal transferase and incubate for 90 min at 37°C.
2. Store the labeled probe on ice for immediate use or store at –20°C for long term storage of up to 6 mo.

3.2. Enzyme Labeling

1. Prepare thiol-modified oligonucleotide as detailed by the thiol modifier supplier and in compliance with the operating instructions of the oligonucleotide synthesis machine (*see* Note 4).

2. Purify thiol-modified oligonucleotide by a recommended method. Optimal results will be achieved by HPLC (*see* Note 5). Thiol-labeled oligonucleotides can be stored at –20°C for several months.
3. Desalt 5 µg of oligonucleotide (*see* Note 6) in a volume of 50 µL on a Sephadex G25 spin column, or equivalent, equilibrated in water *(9)*.
4. Immediately add the desalted oligonucleotide to a tube containing approx 200 µg freeze-dried, derivatized HRP (*see* Note 7). Mix gently in the pipet tip to redissolve the HRP. Incubate at room temperature for 1 h (*see* Note 8).
5. Store labeled probe at a concentration of 1 µg/100 µL in 50% glycerol for up to 6 mo.

3.3. Biotin Labeling

1. Use biotin phosphoramidite as detailed by the supplier (*see* Note 9) to perform a standard oligonucleotide synthesis under normal operating conditions (*see* Note 10), specifying biotin at the desired positions; usually at the 5'-end, in any number (*see* Note 11).
2. Purify the labeled oligonucleotide, preferably by HPLC (*see* Note 12) and store labeled sequences at –20°C for at least 6 mo.

3.4. Hybridization

1. The probes generated from all three labeling systems can be used with the same simple hybridization buffer (*see* Note 13).
2. Prehybridize blots in the hybridization buffer for 15 min at the hybridization temperature to be used.
3. Hybridize for 1–2 h at the desired temperature, typically 42°C, at a probe concentration of 10 ng/mL (*see* Note 14).
4. Perform post-hybridization washes at the required stringency, dependent on the melting temperature of the oligonucleotide probe (*see* Note 15). Stringency is controlled by a combination of salt concentration and temperature, as traditionally performed with radioactively labeled probes. Even for probes directly linked to HRP the temperature can be raised to 60°C for the short stringency wash incubations (*see* Note 16). A typical wash protocol would be: (a) Wash in 2–5X SSC, 0.1% SDS for 2 × 5 min at room temperature. (b) Wash in 1–3X SSC, 0.1% SDS for 2 × 20 min at the desired stringency temperature.

3.5. Detection

After hybridization the various labels used on the probes have to be visualized. There are a number of methods available, principally involving HRP and alkaline phosphatase, for the production of colored precipitates or light as a final endpoint *(10)*.

1. A particularly simple, rapid, and sensitive endpoint that can be applied to all three oligonucleotide labeling systems is the light producing reaction, enhanced chemiluminescence (ECL), catalyzed by HRP *(11)* (*see* Chapter 14 and Note 17). Blots hybridized with probes directly labeled with HRP are placed in the ECL substrate immediately after the stringency washes. After 1 min incubation the blots are exposed to X-ray film for the desired period (typically 10–30 min) (*see* Note 18) to achieve a maximum sensitivity of 25 amol.
2. The two indirect systems require incubation of the blots in immunological reagents (e.g., antifluorescein-HRP conjugate or streptavidin-HRP complex for fluorescein and biotin, respectively). The blots are first blocked (using the blocking agent as in the hybridization buffer) for 30–60 min and then incubated in the appropriate dilution of antibody for a similar length of time. After four brief washes to remove unbound antibody the blots are processed with the ECL reagents as for directly labeled probes. Exposure times are again around 10–30 min for most applications (*see* Note 18 and Figs. 2–6) with an absolute sensitivity of 10–20 amol for both biotin- and fluorescein-labeled probes.

4. Notes

1. The labeling reaction can be performed using as little as 25 pmoles and as much as 250 pmoles with equal success. The volumes of the various components should be scaled accordingly.
2. One hundred picomoles is equivalent to 560 ng of a 17-base sequence and 1000 ng of a 30-base sequence.
3. Oligonucleotide concentration is typically calculated from the absorbance at 260 nm and assuming that a concentration of 33 µg/mL to be equivalent to 1 A_{260} U. This is only an approximation and for optimum results with this system and others, it is necessary to accurately assess the concentration of the oligonucleotide probe. The relationship between concentration and absorbance is given by the following equation *(12)*:

$$E = (A \times 15200) + (G \times 12010) + (C \times 7050) + (T \times 8400)$$

where A, G, C, and T represent the number of times each base occurs in the probe sequence and E is the molar extinction coefficient (the absorbance of a 1*M* solution).

4. Thiol modifier (Amersham, RPN2112) is dissolved in 1.6 mL anhydrous acetonitrile and placed on the synthesizer. The thiol-modifier phosphoramidite is programmed into the sequence to be placed at the 5'-end and a normal synthesis scheme followed; i.e., 15–30 s coupling time to achieve a coupling efficiency of >95%. Increasing the coupling

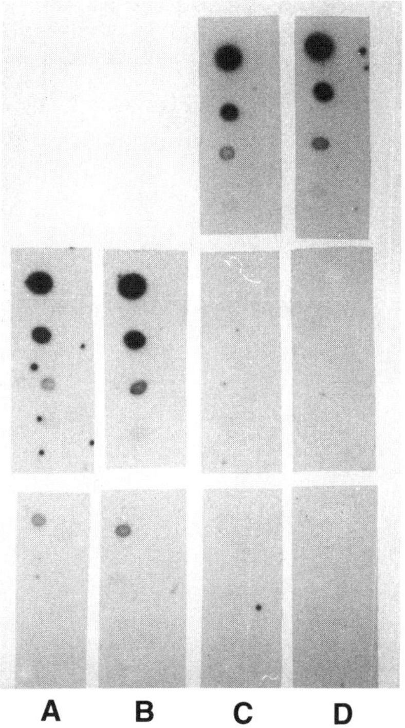

Fig. 2. Demonstration of stringency control using oligonucleotides carrying a 3'-tail of fluorescein-dUTP. Columns **(A)** and **(C):** Probes labeled with Fl-dUTP at the 3'-end and ^{32}P at the 5'-end. Columns **(B)** and **(D):** Probes labeled at the 5'-end only, with ^{32}P. Top row: 17-base probe perfectly matched to the target. Middle row: 17-base probe with one mismatch for the target. Bottom row: 17-base probe with two mismatches for the target. All filters were hybridized at 42°C for 2 h with a probe concentration of 5 ng/mL. Stringency washes were performed in 1X SSC, 0.1% SDS at room temperature (columns [A] and [B]) or at 50°C (columns [C] and [D]). Autoradiographed overnight at –70°C.

time or decreasing the volume of acetonitrile used to dissolve the compound may raise the coupling efficiency to >98%.

5. Thiol modifier suppliers will probably recommend the relevant purification procedures to be followed when dealing with thiol-linked probes. During synthesis the thiol group is protected by a trityl group that can be used as a lipophillic tag to purify the thiol oligonucleotides from failure sequences using reverse-phase HPLC. The trityl protecting group is subsequently removed using silver nitrate and dithiothreitol (DTT).

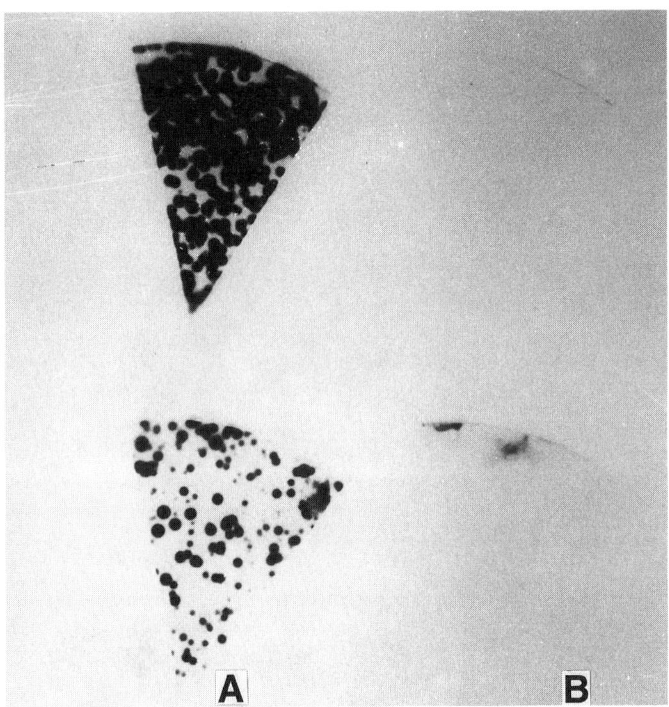

Fig. 3. Plaque screening with 3'-end labeled probes. M13 mp8 plaques containing human N-*ras* proto-oncogene sequences were lifted onto Hybond-N+. Probed with 5 ng/mL of (Column **[A]**) a 25-base antisense, fluorescein-dUTP tailed, probe; (Column **[B]**) a 25-base sense, fluorescein-dUTP tailed, probe. Hybridized at 42°C for 2 h; stringency wash of 1X SSC, 0.1% SDS at 42°C; detected with antifluorescein HRP and ECL; 10 min exposure.

6. Thiol-modified oligonucleotides are stored after synthesis and purification in DTT to stop the thiol groups dimerizing. The DTT has to be removed immediately prior to coupling to the modified HRP which is best achieved by use of a spin column. The desalted oligonucleotide should be added to the modified HRP within 5 min of preparation. Note that if using commercially available spin columns they will first have to be equilibrated with water. Failure to do this will alter the reaction buffer for the coupling step leading to significantly lower levels of labeling.
7. The use of spin columns (*see* Note 6) and the use of freeze-dried derivatized HRP ensures that a high concentration of reactants is achieved in the labeling process. It is possible to use freshly prepared

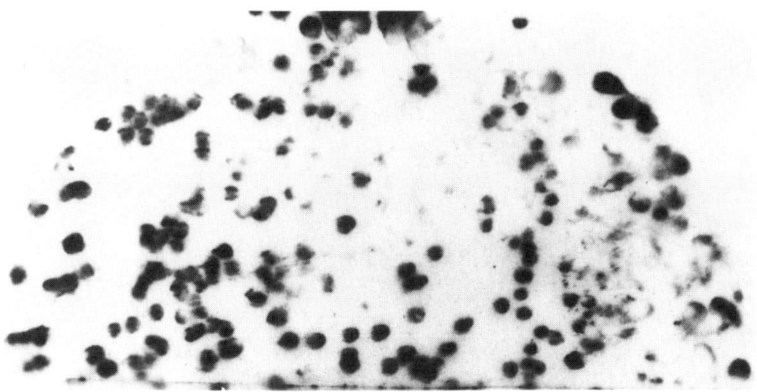

Fig. 4. Colony screening with enzyme labeled probes. *E. coli* colonies transformed with pSP65 containing human N-*ras* proto-oncogene sequences lifted onto Hybond-N^+. Hybridized with 5 ng/mL of an HRP-labeled 25-base probe at 42°C for 1 h; stringency wash of 3X SSC, 0.1% SDS at 42°C; ECL detection; 15 min exposure.

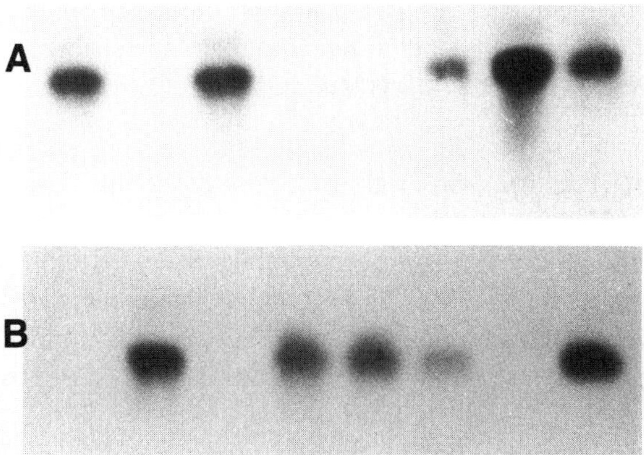

Fig. 5. PCR product detection with 5' hapten-labeled probes. Samples of human genomic DNA from normal, cystic fibrosis carrier and affected patients were subjected to PCR to amplify a 112-base fragment covering the 508 deletion locus. PCR samples were Southern blotted onto Hybond-N^+ and hybridized with (**A**) a 21-base, 1-biotin labeled, probe specific for the normal allele; (**B**) a 21-base, 1-biotin labeled, probe specific for the CF deletion allele. Hybridized at 42°C for 1 h; stringency wash of 1X SSC, 0.1% SDS at 42°C; detected with streptavidin-HRP and ECL; 1 min exposure.

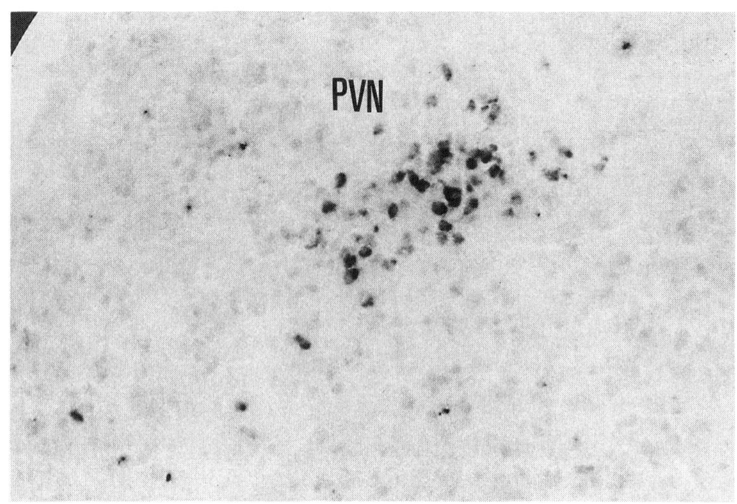

Fig. 6. *In situ* hybridization with 5' hapten-labeled probes. Cryostat sections (15 μm) were cut from the rat brain (PVN = paraventricular nucleus) and fixed with paraformaldehyde. Hybridized with 160 ng/mL of a 30-base, 10 biotin-labeled, probe specific for oxytocin mRNA for 17 h at 37°C; stringency wash of 1X SSC at 55°C; detected with streptavidin-alkaline phosphatase (Amersham) and NBT/BCIP color reagents: overnight color development.

derivatized HRP in solution or alternative methods of desalting the oligonucleotide. However, the volume may increase with either of these options and the reaction may not reach maximum efficiency.

8. The labeling reaction couples >85% of the oligonucleotide to the enzyme in a 1-h incubation at room temperature. However, if more convenient, it is possible to leave the coupling reaction to proceed overnight at 4°C without any difference in the final result.
9. The biotin phosphoramidite as supplied, in a solid form, is stable at room temperature for up to 1 mo. For longer term storage, store at –20°C. Only make up in acetonitrile on the day to be first used and, for maximum coupling efficiency, use the entire contents within 3–4 d of placing on the machine.
10. The 0.12 g bottle of biotin phosphoramidite (Amersham, RPN2012) will give 10 biotinylation reactions when used in conjunction with a standard small scale synthesis (0.2 μmol) program. The coupling of each biotin group exceeds 94% and will frequently exceed 98% when the synthesizer is set up well.

11. If it is not possible to use the entire contents of the bottle in the specified period, it is possible to remove the bottle from the machine and store the biotin phosphoramidite in acetonitrile at –20°C for at least 1 wk.
12. Biotin labeled oligonucleotides should always be synthesized with the final trityl group left on. This is done to avoid a possible chemical reaction involving elimination of the final biotin during deprotection of the bases of the probe in a hot ammonia solution. The trityl group can then be removed separately either before or after reverse phase HPLC purification. In many cases, owing to the very high coupling efficiency achieved, it may not prove necessary to purify the probes after deprotection and trityl removal.
13. The oligonucleotide hybridization buffer can be prepared in advance and stored at –20°C for at least 3 mo. The blocking agent dissolves relatively slowly. The contents of the buffer should be mixed and heated to 50–60°C with constant stirring for approx 30–60 min.
14. The temperature of hybridization is to a large extent dependent on the length and base composition of the probe and independent of the type of method used for labeling. For many probes a temperature of 42°C will be found to be adequate although this will have to be determined empirically for each system, particularly for probes greater than 20 bases. The probe concentration recommended will work well in most applications of nonradioactive oligonucleotide probes. In some cases it may prove necessary to alter the probe concentration but this should only be done in the range of 5–20 ng/mL.
15. The melting temperature of an oligonucleotide probe approximates to the following formula *(13)*:

$$T_m = (4 \times \text{number of G + C bases}) + (2 \times \text{number of A + T bases})$$

    Stringency washes are generally performed at 3–5°C below the melting temperature if it is necessary to distinguish between perfectly matched and mismatched probe:target hybrids. It is not possible to give exact conditions for all probes but it should be noted that in general the stringency washes previously used with radioactive labels are equally effective with nonradioactive systems as described here. No significant effect on melting temperature has been observed on the introduction of any of the three labels discussed.
16. With HRP labeled probes care must be taken in the stringency washes to avoid denaturation of the enzyme label. Normal stringency washes would be performed for 2 × 15 min at temperatures below 42°C. At higher temperatures, up to 60°C maximum, the washes should be reduced to 2 × 5 min.

17. The ECL detection system is based on the HRP catalyzed oxidation of luminol to produce blue light. The enhancement system stimulates the peak light output and sustains the light output for a number of hours *(14)*. The light can be collected on X-ray film producing a hard copy image similar to those traditionally obtained with radioactive labels. The system is equally applicable to directly labeled probes and the detection of haptens mediated by HRP labeled antibodies *(15)*.
18. All three systems can be used in a variety of applications as illustrated (Figs. 2–6). The direct labeled probes are particularly suitable for high throughput systems where bulk labeling of the probe may be an advantage. The fluorescein labeled probes offer slightly higher sensitivity and a more flexible approach. The biotin labeled probes are useful in *in situ* hybridization and have a separate role in capture-based techniques such as PCR based sequencing and genomic walking.

Acknowledgments

The authors would like to thank their colleagues at Amersham, in particular Patricia Chadwick, and Sarah Augood (MRC Group, Babraham) for their contributions to this chapter.

References

1. Keller, G. H. and Manak, M. M. (1989) Hybridization formats and detection procedures, in *DNA Probes.* Stockton, New York, pp. 149–214.
2. Thein, S. L. and Wallace, R. B. (1986) The use of synthetic oligonucleotides as specific hybridization probes in the diagnosis of genetic disorders, in *Human Genetic Disease: A Practical Approach* (Davies, K. E., ed.), IRL, Oxford, pp. 33–50.
3. Connolly, B. A. and Rider, P. (1985) Chemical synthesis of oligonucleotides containing a free sulphydryl group and subsequent attachment of thiol specific probes. *Nucleic Acids Res.* **13,** 4485–4502.
4. Connolly, B. A. (1987) The synthesis of oligonucleotides containing a primary amino group at the 5 terminus. *Nucleic Acids Res.* **15,** 3131–3139.
5. Misiura, K., Durrant, I., Evans, M. R., and Gait, M. J. (1990) Biotinyl and phosphotyrosinyl phosphoramidite derivatives useful in the incorporation of multiple reporter groups on synthetic oligonucleotides. *Nucleic Acids Res.* **18,** 4345–4354.
6. Murakami, A., Tada, J., Yamagata, K., and Takano, J. (1989) Highly sensitive detection of DNA using enzyme-linked DNA probe. 1. Colourimetric and fluorometric detection. *Nucleic Acids Res.* **17,** 5587–5595.
7. Fowler, S. J., Harding, E. R., and Evans, M. R. (1990) Labeling of oligonucleotides with horseradish peroxidase and detection using enhanced chemiluminescence. *Technique* **2,** 261–267.

8. Durrant, I. (1991) Stringency control with 3'-tailed oligonucleotide probes. *Highlights* **2,** 6,7.
9. Sambrook, J., Fritsch, E. F., and Maniatis, T. (1982) Spun column procedures, in *Molecular Cloning: A Laboratory Manual*, Cold Spring Harbor Laboratory, Cold Spring Harbor, NY, pp. 466–467.
10. Pollard-Knight, D. (1990) Current methods in nonradioactive nucleic acid labeling and detection. *Technique* **2,** 113–132.
11. Durrant, I. (1990) Light based detection of biomolecules. *Nature (London)* **346,** 297–298.
12. Wallace, R. B. and Miyada, C. G. (1990) Oligonucleotide probes. *Meth. Enzymol.* **152,** 438.
13. Wallace, R. B., Shaffer, J., Murphy, R. F., Bonner, J., Hirose, T., and Itakura, K. (1979) Hybridization of synthetic oligodeoxyribonucleotides to ΦX174: the effect of single base pair mismatch. *Nucleic Acids Res.* **6,** 3543–3556.
14. Thorpe, G. H. G. and Kricka, L. J. (1987) Enhanced chemiluminescent assays for horseradish peroxidase: characteristics and applications, in *Bioluminescence and Chemiluminescence: New Perspectives* (Scholmerich, J., Andreesen, R., Kapp, A., Ernst, M., and Woods, W. G., eds.), Wiley, New York, pp. 199–208.
15. Cunningham, M. (1991) Nucleic acid detection with light. *Life Science* **6,** 2–5.

CHAPTER 16

Analysis of DNA Restriction Enzyme Digests by Two-Dimensional Electrophoresis in Agarose Gels

Edward L. Kuff and Judy A. Mietz

1. Introduction

Two-dimensional (2D) electrophoretic fractionation of digested DNA has been useful in analyzing and comparing the genomic structure of viruses and prokaryotic organisms *(1–4)*. For eukaryotes, it has been used to examine the genomic distribution of multicopy gene families *(5)*, to determine the methylation status of individual members of such families *(6–8)*, and to analyze the physical organization of complex genetic loci *(9,10)*. The basic principle is straightforward (Fig. 1): High-mol-wt DNA is digested in solution with a restriction enzyme of choice and the digest electrophoresed in the first dimension. The sample lane is cut as a longitudinal strip from the gel, equilibrated with appropriate buffer, and incubated with a second restriction enzyme. The strip is then embedded in a second gel and electrophoresed in a direction perpendicular to the first. Although simple in concept, the procedure involves many manipulations: thus the chances of accident are high in the beginning. Close attention to detail and a stoic attitude toward sudden failure are useful in persevering to a good series of high resolution gels. We describe first the basic 2D procedure as practiced in our laboratory *(7,8)* and additionally comment on the literature with regard to reported technical descriptions and applications (*see* Notes).

From: *Methods in Molecular Biology, Vol. 31: Protocols for Gene Analysis*
Edited by: A. J. Harwood

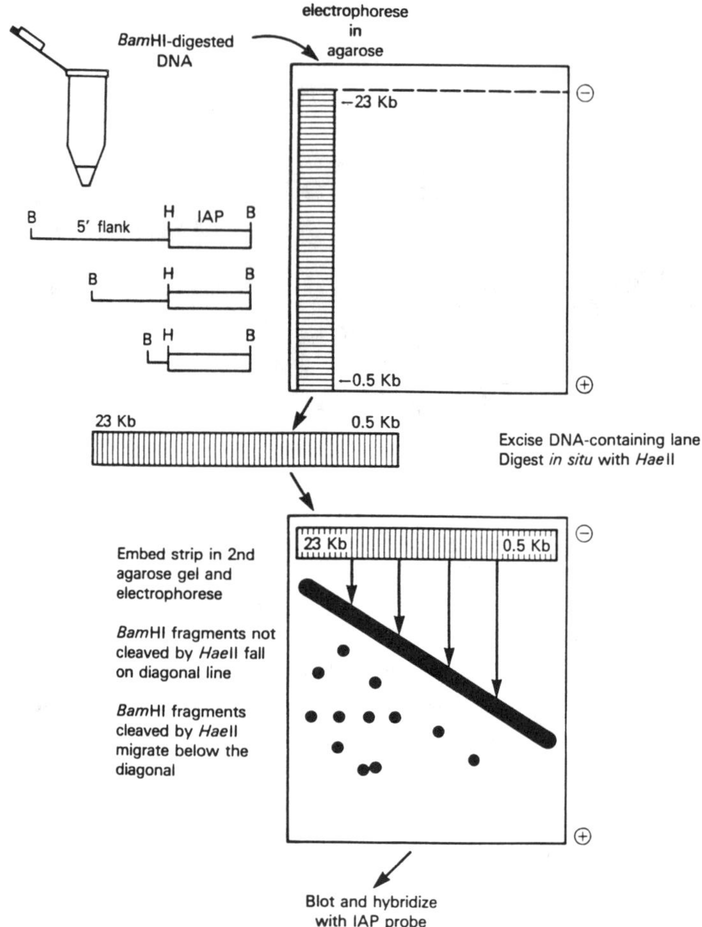

Fig. 1. The basic procedure for 2D electrophoretic analysis, as applied to a multicopy gene family. DNA fragments generated by restriction enzyme 1, in this case *Bam*HI, are applied to the first-dimension gel. The set of junction fragments generated by the enzyme from members of the gene family is represented to the middle left. A strip is excised from the sample lane of the first-dimension gel. The buffer is exchanged and the fragments digested *in situ* with restriction enzyme 2, in this case *Hae*II. The gel strip is then embedded transversely in the second-dimension gel and electrophoresed as before. After staining, the gel is hybridized with a probe specific for the gene family. A major diagonal is found, as shown in the figure, that contains fragments not cut by the second enzyme. Fragments cut by both enzymes run ahead of the diagonal in the second gel and appear as discrete hybridizing spots.

2. Materials

Glass redistilled water and highest purity chemicals are used for all solutions. Gels are prepared from electrophoresis grade ultrapure agarose (EEO –0.12). All reactions are carried out in heat sterilized Eppendorf or glass tubes.

1. Electrophoresis equipment: We have found commercially available horizontal gel electrophoresis equipment to be satisfactory (*see* Note 1). A sheet of graph paper is precisely aligned and taped face up beneath the gel bed of the box used for first-dimension electrophoresis. Sample wells at least 8 mm wide and 1 mm thick should be used during the first-dimension electrophoresis. Power is provided by a 500 V dc supply regulated by a pulse controller (Models PS500x and PC750, respectively, from Hoefer Scientific Instruments, San Francisco, CA).
2. Restriction enzymes and buffers: Use as recommended by the manufacturers.
3. Phenol:chloroform: 50:50:1 phenol:chloroform:isoamyl alcohol.
4. Chloroform: 50:1 chloroform:isoamyl alcohol mix.
5. 10*M* lithium chloride.
6. Ethanol: both absolute and 70% solutions.
7. TE: 10 m*M* Tris-HCl, 1 m*M* EDTA, pH 7.5.
8. Digestion tubes: siliconized glass or plastic tubes 0.8 cm internal diameter and 2–3 cm longer than the gel strip itself.
9. BSA: A 100X stock solution of 10 mg/mL of pentex grade bovine serum albumin. Store at –20°C.
10. 10X loading buffer: 25% Ficoll 400, 1% SDS, 0.25% bromophenol blue and 0.25% xylene cyanol.
11. 10X TBE: 1*M* Tris base, 1*M* boric acid, 0.02*M* sodium EDTA, pH 8.3. Dilute 1/20 for use.
12. Molecular weight markers: 0.5 µg/µL solution of a combined *Hind*III/*Eco*RI digest of λ phage DNA.

3. Methods

3.1. First-Dimension Electrophoresis

1. Digest 25–50 µg of DNA (*see* Note 2) for 3–5 h at 37°C in a 250–500 µL reaction and a restriction enzyme at a concentration of 2 U/µg of DNA.
2. Remove a 10-µL sample of the digest, heat for 10 min at 65°C to dissociate sticky ends, and analyze on an agarose minigel (*see* Note 3). If necessary add more enzyme and incubate further.

3. Deproteinize the enzyme digests by extraction with phenol:chloroform followed by chloroform. Add 4 vol of 10*M* lithium chloride to the aqueous phase and place overnight at 5°C to precipitate RNA (*see* Note 4). Centrifuge at 5°C and 15,000*g* for 30 min. Collect the clear supernatant, precipitate the DNA by addition of 2 vol of absolute ethanol, and wash the pellet with 70% ethanol. Take up the pellet in TE buffer to an expected concentration of 0.5–1 µg/µL. Dissolve thoroughly by vortexing or pipeting and store at 5°C.
4. Prepare a gel solution of between 0.7–1.0% agarose in 0.5X TBE in a tared flask by boiling for 3 min in a microwave oven. Readjust to the final volume by weighing and cool to 50°C before pouring. Place the gel in an appropriate comb. When the gel is set, fill the electrophoresis apparatus with running buffer to a depth of 0.5–0.6 cm above the gel. Then remove the comb.
5. Mix 10–12 µg of digested DNA in 25 µL TE with 3 µL of 10X loading buffer. Heat for 10 min at 65°C in a stoppered tube and spin briefly to collect condensate.
6. Remove the comb and pipet samples into the wells. Take care to avoid pulling the sample to the surface when the pipet tip is lifted since this can lead to a surface streaming phenomenon that draws sample from the well (*see* Note 5). Other wells should be filled with size markers for staining or hybridization; with a 1 µg sample of each digest to monitor by staining, or with blank loading mix. It is useful to leave a blank well between each main sample well.
7. Commence the electrophoretic run at room temperature with a field strength of 6–7 V/cm gel length. When the two dyes have resolved in the gel (after about 15 min) reduce the power supply to provide 4 V/cm for 13 h (for a 15 cm gel) or 20 h (for a 24 cm gel). During this time set the power supply pulse controller to periodically reverse the current flow maintaining a constant forward-to-reverse time ratio of 3:1. "Ramp" the pulse regime starting with an initial cycle of 30:10 ms that is lengthened by a factor of 0.1/h (*see* Note 6).
8. On completion of the run remove the buffer to below the level of the gel bed. Cut away the part of the gel containing the markers and monitor lane and stain with ethidium bromide. Immobilize the remaining portion of the gel by pressing gently along its length with a plastic bar and, beginning in the well, cut a 6-mm wide strip down the center of the nearest sample lane using two parallel razor blades fixed to a plastic spacer (*see* Note 7). If the blades are clean, wet, and presented to the gel at an angle of about 30°, they can be drawn smoothly and slowly through the gel following a reference line on the graph paper positioned

below the gel. Mark each gel slice as it is cut by punching one or more small holes at the ends with a Pasteur pipet.

9. Transfer the gel strip to a wide glass baking dish containing 500 mL of the second restriction digest buffer.
10. Stain the regions of gel from between the strips with ethidium bromide and reassemble on a UV transilluminator to ensure that the cuts were well aligned.

3.2. Second-Dimension Electrophoresis

1. Equilibrate the gel slices for a total of 3 h against three changes of buffer. Transfer to digest tubes by immersing the tube and floating the strip into the filled cavity. Cap the tube at one end, remove from the fluid, and drain off the buffer.
2. Add 1.0–1.5 mL, depending on length of slice, of fresh buffer solution that contains 100 µg/mL of BSA and place the tube vertically in an ice bath for 15 min. Add between 250–300 U of the second restriction enzyme (*see* Note 8) and seal the tube. Mix the contents by rocking (*see* Note 9).
3. Rock the tube slowly overnight at 5°C such that the fluid phase moves back and forth along the cut edges of the gel strip. Transfer the tube to a stationary 37°C water bath for 3–4 h.
4. Drain off the enzyme solution and float out the gel strip into 500 mL of 0.5X TBE.
5. Pour the second dimension gel with the same composition as the first and to a depth of 0.6 cm. Form a trough of 0.5–0.6 cm width at the top of the gel by placement of a transverse plastic bar when the gel is poured (*see* Note 10). The trough should be designed to fit the gel slice from the first dimension.
6. Transferring the gel strip to the second dimension gel requires particular care. Drain the buffer from the gel strip(s) by careful decanting or aspiration, leaving just enough fluid to slide the strips about easily. Against a dark background, maneuver the gel strip onto a plastic strip long enough to fit into the gel trough and position the strip along the edge of the plastic so that the origin of the first-dimension run, identified by the hole(s), points toward the operator's side of the electrophoresis apparatus. Lower the plastic strip into the trough and lay it vertically against the back gel surface to which it should adhere. The gel strip should face forward in the trough. Pipet 1–2 mL of molten 0.5% agarose in 0.5X TBE into the open space between the gel strip and the front of the trough. Gently advance the gel strip to the front edge of the trough where it should remain attached as the gel sets. Avoid trapping bubbles between

the gel strip and running gel since this will interfere with the DNA mobility. The plastic strip may be withdrawn after a minute. The remainder of the trough may then be filled with molten agarose.

7. Inject 1 or 2 μL of mol wt markers for staining and/or hybridization directly into the ends of the embedded gel strip using a yellow pipet tip.
8. Run the second-dimension gel under the same voltage gradient, time, and pulse intervals as the first.
9. After electrophoresis the gel may be stained or, more usefully, Southern blotted and hybridized with an appropriate probe (*see* Notes 11 and 12).

Figure 2 shows a 2D analysis of the proviral IAP element in the mouse genome. There are approx 1000 copies of the IAP element per haploid mouse genome *(11)*. About 350 individual spots, representing junction fragments between provirus and 5' flanking DNA, can be resolved on the original of the figure. Taking into account the fact that a number of junction fragments do not contain a second enzyme site (resulting in the heavy diagonal), we estimate that the procedure is yielding close to the maximum resolution. With the method as described here, we have been able to detect insertion of a single novel IAP provirus in the DNA of a subcloned mouse myeloid cell line (unpublished observation). For other applications of 2D electrophoresis, *see* Note 13.

4. Notes

1. We use three sizes of horizontal lucite boxes (manufactured by the Aquebogue Machine and Repair Shop, Inc., Long Island, NY). All have gel bed widths of 13.3 cm; bed lengths are 15 cm (Model 800), 22 cm (Model 850), and 42 cm (model 842). Model 800 is used for both dimensions of small 2D gels (13 × 13 cm). Model 842 can accommodate three samples in the second dimension by embedding the strips at intervals of 13 cm. Model 850 is used to provide longer gel strips (20 cm) for larger 2D gels; the second dimension is run in an apparatus with a removable gel bed 20 × 24 cm in size (Model HRH, International Biotechnologies, Inc., New Haven, CT). This apparatus is equipped for cooling during electrophoresis but we have not found this useful to do. We use an 8-place comb with wells of 12 × 1.5 mm for the 13-cm wide gels.
2. DNA may be prepared from whole cells, tissue homogenates or, preferably, from isolated nuclei to reduce RNA contamination. We use a standard DNA extraction scheme involving digestion with proteinase K at 37°C in the presence of sodium dodecyl sulfate (SDS), deproteinization

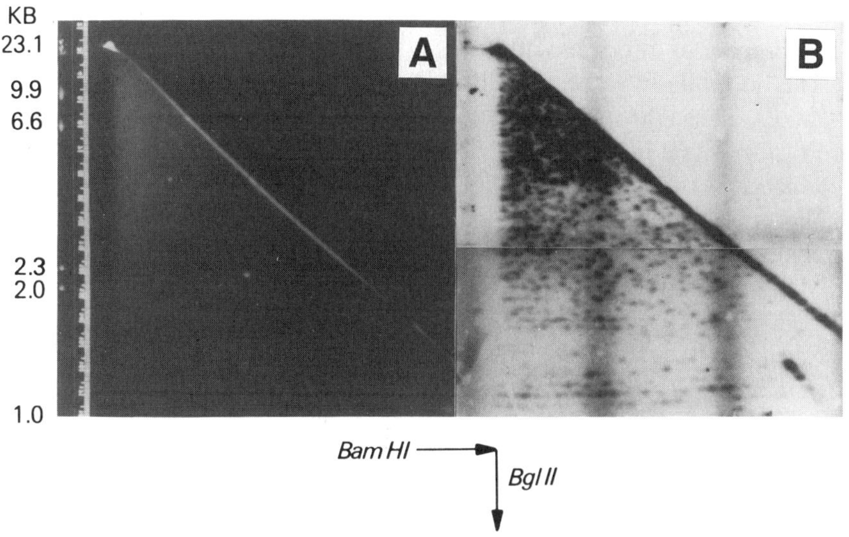

Fig. 2. Two-dimensional electrophoretic analysis of mouse genomic DNA. Twelve mg of *Bam*HI-digested DNA was run in a 13 × 24 cm first-dimension gel of 0.9% agarose in 0.5X TBE. A 0.5 × 20 cm strip, representing about 6 μg of digest, was cut longitudinally from the sample lane and incubated with *Bgl*II. The gel strip was embedded transversely in a 20 × 24 cm second dimension gel and electrophoresed as before. **(A)** Ethidium-stained second dimension gel showing λ phage *Hind*III size markers along the left edge. The major diagonal is formed by *Bam*HI fragments that were not cut by *Bgl*II, whereas the minor diagonals, horizontals, and prominent spots represent fragments derived from families of repetitive genomic sequences. **(B)** Autoradiograph of the gel after hybridization with a nick-translated probe (specific activity of 2×10^8 cpm/μg) specific for members of the multicopy family of IAP (intracisternal A-particle) proviruses *(8,12)*. The upper and lower sections of panel B were derived from separate autoradiographic exposures.

with phenol:chloroform:isoamyl alcohol (50:50:1) and chloroform: isoamyl alcohol (50:1), and either dialysis or ethanol precipitation to obtain the final DNA preparation. The detailed procedure given by Brilliant et al. *(11)* has provided consistently good high-mol-wt material.

3. Many restriction enzymes produce visible staining bands from repetitive DNA elements in the genome, and the investigator will become familiar with the appearance of favorable digests.
4. DNA preparations from whole cells or homogenates can be nearly half RNA, which must be largely removed to (a) permit a reliable estimate of DNA concentration; and (b) avoid overloading in the first-dimen-

sion wells. Transfer RNA is not removed by this procedure but does not appear to interfere with loading.

5. This is unlikely to occur with the more usual glycerol-based loading mixes. These, however, give more prominent edge effects.
6. Pulse reversal is not essential but we found it useful for increasing the separation of fragments in the 2–10-kb size range.
7. We cut a narrow strip to avoid more slowly migrating DNA on the edges of the lane.
8. Many restriction enzymes are active when diffused into gels formed of electrophoretic grade (low electroendosmotic properties and low sulfate content) agarose *(3)*. The number includes *AluI, AvaI, BamHI, BglII, BstEII, EcoRI, HaeII, HaeIII, HhaI, HindIII, MspI, SalI,* and *TaqI,* as well as a variety of rare cutters used for producing very large fragments from DNA *(5)*.
9. Care is taken that the strip lies flat along the inner lower surface of the tube; air spaces seem to collect material that can cause dragging in the second-dimension gel.
10. An alternative is to form the trough by slicing a preformed gel about 2 cm from the starting edge and sliding the severed piece backward the required distance.
11. For hybridization, DNA fragments may be transferred to nitrocellulose or nylon membranes by standard blotting techniques. Alternatively, we find that hybridization in gels with end-labeled oligonucleotides *(8)* or extensively nick-translated larger fragments, such as in Fig. 2B, provides increased sensitivity and improved resolution compared to the blotted preparations. For this the gels are denatured and then dried. After rehydration, hybridization is carried out directly in the gel *(8)*.
12. Two-dimensional electrophoresis in agarose gels can provide high resolution of restriction fragments between 1–10 kb (*see* Fig. 2). Resolution in the upper mol wt range can be improved by running the gels for a longer time, although, of course, smaller fragments are lost when this is done.
13. Reported applications:
    a. Multicopy gene families: Excellent resolution of several other multicopy gene families has been reported, e.g., the group of 6 endogenous mouse mammary tumor proviruses *(6)*, the mouse *H-2* class I gene family with about 30 crosshybridizing members, and the VL30 and IAP type II families of defective proviruses, each with 150–200 members *(5)*. The latter authors have described their 2D procedure in detail, including a specially designed electrophoretic apparatus.
    b. Viral and prokayrotic genomes: Two-dimensional electrophoresis has been used to resolve and compare restriction digests of less complex

genomes, such as λ phage (1), *E. coli (4),* and the myxobacterium *Myxococcus xanthus (2).* Yi et al. (4) developed patterns containing 300–400 individual spots and were able to detect a single copy of incorporated λ phage in the bacterial restriction pattern. Hybridization probes were not required since the restriction fragments were end-labeled for detection in the final gel (*see also* ref. *3*). Detailed comparison between two similar but nonidentical genomes was greatly enhanced by differential autoradiography of patterns developed from ^{32}P and ^{35}S labeled digests run together *(4).*

c. Complex genetic loci: The organization of complex genetic loci is amenable to 2D electrophoretic analysis. Woolf et al. *(9)* used first-dimension field inversion gel electrophoresis of large restriction enzyme (*Sal*I) fragments combined with *Hind*III digestion of the excised sample lane and a conventional second-dimension run. The data permitted them to map the genomic organization of the multicomponent murine T-cell receptor gamma locus. In an analysis of the immunoglobulin heavy chain locus *(10),* first-dimension pulse field electrophoresis of DNA cleaved to large fragments with the enzyme *Sfi*I was followed by *Eco*RI digestion of the sample strip and standard electrophoresis in the second dimension. Woolf et al. *(9)* describe in detail a technique for concentrating the bands in the first-dimension gel strip that greatly enhances resolution in the final patterns.
d. Detection of DNA methylation: The methylation status of selected CpG sites in the *Myxococcus xanthus* genome *(2)* and within individual members of mammalian multicopy gene families *(6–8)* has been assessed by 2D electrophoresis using methylation-sensitive restriction enzymes in the second dimension.
e. Related procedures: A variety of additional techniques for 2D electrophoretic separation of DNA fragments (use of polyacrylamide gels, denaturing gel systems, specialized apparatus including vertical rigs) have been described *(13–15).* Several investigators have used two-dimensional techniques in which the first dimension products are physically separated by slicing the sample lane into multiple sections, and the DNA in each fraction digested with a second enzyme either *in situ* or after extraction and applied individually to separate lanes in the second gel *(14,16,17).*

References

1. Rosenvold, E. C. and Honigman, A. (1977) Mapping of AvaI and XmaI cleavage sites in bacteriophage DNA including a new technique of DNA digestion in agarose gels. *Gene* **2,** 273–288.

2. Yee, T. and Inouye, M. (1982) Two-dimensional DNA electrophoresis applied to the study of DNA methylation and the analysis of genome size in *Myxococcus xanthus*. *J. Mol. Biol.* **154,** 181–196.
3. Boehm, T. L. J. and Drahovsky, D. (1984) Two-dimensional restriction mapping by digestion with restriction enzymes in agarose and polyacrylamide gels. *J. Biochem. Biophys. Meth.* **9,** 153–161.
4. Yi, M., Lo-chung, A., Ichikawa, N., and Ts'o, P. O. P. (1990) Enhanced resolution of DNA restriction fragments: a procedure by two-dimensional electrophoresis and double labeling. *Proc. Natl. Acad. Sci. USA* **87,** 3919–3923.
5. Sheppard, R. D., Montagutelli, X., Jean, W. C., Tsai, J., Rose, A., Guénet, J. L., Cole, M. D., and Silver, L. M. (1991) Two-dimensional gel analysis of complex DNA families: methodology and apparatus. *Mammalian Genome* **1,** 104–111.
6. Fanning, T. G., Hu, W., and Cardliff, R. D. (1985) Analysis of tissue-specific methylation patterns of mouse mammary tumor virus DNA by two-dimensional Southern blotting. *J. Virol.* **54,** 726–730.
7. Mietz, J. and Kuff, E. L. (1990) Tissue and strain-specific patterns of endogenous proviral hypomethylation analyzed by two-dimensional gel electrophoresis. *Proc. Natl. Acad. Sci. USA* **87,** 2269–2273.
8. Mietz, J., Fewell, J. W., and Kuff, E. L. (1992) Selective activation of a discrete family of endogenous proviral elements in normal BALB/c lymphocytes. *Mol. Cell. Biol.* **12,** 220–228.
9. Wolf, T., Lai, E., Kronenberg, M., and Hood, L. (1988) Mapping genomic organization by field inversion and two-dimensional gel electrophoresis: application to the murine T-cell receptor-gene family. *Nucleic Acids Res.* **16,** 3863–3875.
10. Walter, M. A. and Cox, D. W. (1989) A method for two-dimensional DNA electrophoresis (2D-DE): application to the immunoglobulin heavy chain variable region. *Genomics* **5,** 157–159.
11. Kuff, E. L. and Lueders, K. K. (1988) Intracisternal A-particle gene family: structural and functional aspects. *Adv. Cancer Res.* **51,** 183–276.
12. Brilliant, M. H., Gondo, Y., and Eicher, E. M. (1991) Direct molecular identification of the mouse pink-eyed unstable mutation genome scanning. *Science* **252,** 566–569.
13. Fischer, S. G. and Lerman, L. S. (1979) Two-dimensional electrophoretic separation of restriction enzyme fragments of DNA. *Meth. Enzymol.* **68,** 183–192.
14. Parker, R. C. and Seed, B. (1980) Two-dimensional gel electrophoresis "SeaPlaque" agarose dimension. *Meth. Enzymol.* **65,** 358–363.
15. De Wachter, R. and Fiers, W. (1982) Two-dimensional gel electrophoresis of nucleic acids, in *Gel Electrophoresis of Nucleic Acids: A Practical Approach.* IRL, Oxford, pp. 77–116.
16. Potter, S. S. and Newbold, J. E. (1976) Multidimensional restriction enzyme analysis of complex genomes. *Analytical Biochem.* **71,** 452–458.
17. Smith, S. S. and Thomas, C. A. (1981) The two-dimensional restriction analysis of Drosophila DNAs: males and females. *Gene* **13,** 395–408.

CHAPTER 17

Inverse Polymerase Chain Reaction

Daniel L. Hartl and Howard Ochman

1. Introduction

The inverse polymerase chain reaction (IPCR) was the first extension of the conventional polymerase chain reaction to allow the amplification of unknown nucleotide sequences without recourse to conventional cloning. In the conventional polymerase chain reaction (PCR), synthetic oligonucleotides complementary to the ends of a known sequence are used to amplify the sequence *(1,2)*. The primers are oriented with their 3' ends facing each other, and the elongation of one primer creates a template for annealing the other primer. Repeated rounds of primer annealing, polymerization, and denaturation result in a geometric increase in the number of copies of the target sequence. However, regions outside the boundaries of the known sequence are inaccessible to direct amplification by PCR. Since DNA synthesis oriented toward a flanking region is not complemented by synthesis from the other direction, there is at most a linear increase in the number of copies of the flanking sequence.

Nevertheless, the applications of a procedure that could selectively amplify flanking sequences are numerous. They include the determination of genomic insertion sites of transposable elements, in which case the sequence of the transposable element can be used to choose oligonucleotide primer sequences; as well as the determination of the upstream and downstream genomic regions flanking cDNA coding sequences, in which case the 5' and 3' sequences of the cDNA can be used to choose the primers. Another class of applications includes the determination of the nucleotide sequences at the ends of cloned

From: *Methods in Molecular Biology, Vol. 31: Protocols for Gene Analysis*
Edited by: A. J. Harwood

DNA fragments, particularly DNA fragments as large as those in yeast artificial chromosomes *(3,4)* or bacteriophage P1 *(5,6)*; in these applications the primer sequences can be chosen based on the nucleotide sequence of the vector flanking the cloning site.

The possibility of amplifying flanking sequences was sufficiently interesting that several groups undertook the challenge, and successful protocols were developed independently and nearly simultaneously by three research groups *(7–9)*. All three methods are based on the same idea, which is outlined in Fig. 1. Panels A and B exemplify the two main types of application. The dark stippling denotes regions of known sequence, and the light stippling represents regions of unknown sequence. In panel A the known sequence (for example, a coding region) is continuous, and the problem is to amplify either the left (L) or right (R) flanking sequence in genomic DNA. In panel B the known sequence (for example, in a cloning vector) is interrupted, and the problem is to amplify the junction with the unknown sequence on either the left (L) or the right (R). For convenience of further discussion, we will call the known sequence the "core" sequence and the unknown sequence the "target" sequence.

The first step in the method is to identify restriction enzymes that will cleave within the core and target sequences in order to liberate DNA fragments containing core sequence juxtaposed with target sequence. The cutting sites of the restriction enzymes are denoted by the arrows in Fig. 1, and normally each junction requires a different enzyme. For each junction, a set of candidate restriction enzymes can be created by choosing enzymes that will cleave within the core sequence 50–1,000 bp from the junction. Final choice among the candidates is based either on trial and error or, more reliably, on a Southern blot *(10)* using the core sequence as a probe. Restriction enzymes with four-base cleavage sites are generally to be preferred, since they are more likely to cleave at an acceptable point within the target sequence.

After cleavage, the next step in the procedure is circularization, in which the ends of the fragment liberated by the restriction enzyme are ligated together. The circular molecule produced in this way has two core-target junctions, one of which was present in the original molecule, and a second one produced by the ligation. The ligation step is followed by a PCR using oligonucleotide primers that anneal to the core sequences near the core–target junctions. The "inverse"

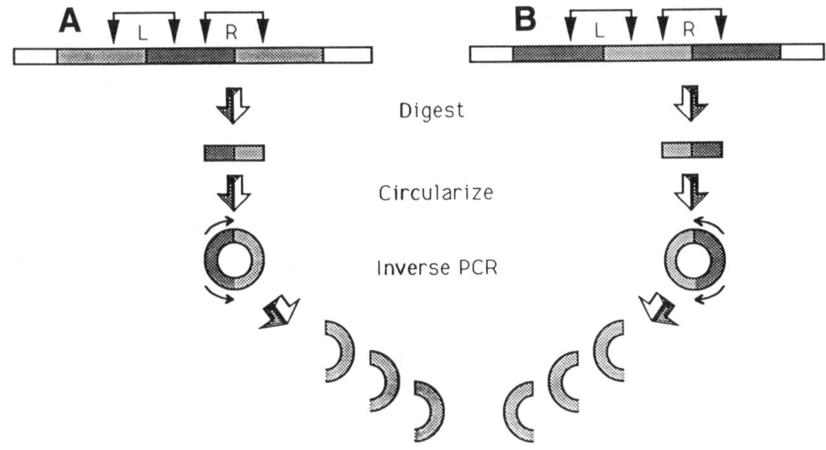

Fig. 1. Major applications of IPCR. Panels **(A)** and **(B)** illustrate known DNA sequences by dark shading and unknown flanking sequences of interest by light shading. In panel A the known sequence is contained within the unknown sequences; in panel B. the unknown sequence is contained within known sequences. The brackets indicate the possible positions of restriction sites that allow IPCR of either the left (L) or right (R) flanking sequence. After digestion with the appropriate restriction enzyme, molecules are circularized, bringing two positions in the known sequence into juxtaposition with the unknown sequence. The known sequences then serve as oligonucleotide priming sites (curved arrows) to amplify the unknown sequence.

part of IPCR comes from the fact that the 3' ends of the primers are oriented outward, toward the target sequence (curved arrows in Fig. 1). Because of the circular configuration of the substrate, the product of each polymerization can serve as template for annealing of the opposite primer. Hence, repeated rounds of annealing, polymerization, and denaturation result in a geometric increase in the number of copies of the target sequence.

IPCR has proven useful in recovering genomic sequences flanking viral and transposable element insertions *(7–9)*, in generating probes corresponding to the ends of cloned inserts for use in chromosome walking *(11–14)*, and in the direct cloning of unknown cDNA sequences from total RNA *(15)*. Since the time that IPCR was developed, a number of other methods have been proposed that can either serve as an alternative to IPCR or that are more suited to certain specialized applications. For example, one technique for human genomic

DNA uses vector sequence for one primer and the Alu consensus sequence for the other primer in order to amplify the insert ends of YAC or cosmid clones *(16)*. The most relevant of these techniques are discussed briefly (*see* Notes).

2. Materials

2.1. Preparation of Restriction Fragments

1. Restriction enzymes and buffers: These are generally supplied together. Digests are carried out as indicated by the manufacturer and differ according to the particular enzyme.
2. $T_{0.1}E$: 10 m*M* Tris-HCl, 0.1 m*M* EDTA, pH 8.0.

2.2. Ligation to Form Circles

3. 10X Ligation buffer: 500 m*M* Tris-HCl, pH 7.4, 100 m*M* $MgCl_2$, 100 mM dithiothreitol, 10 m*M* ATP.
4. T4 DNA ligase: Supplied at a concentration 1,000 Weiss U/µL.
5. TE: 10 m*M* Tris-HCl, pH 7.5, 1 m*M* EDTA.

2.3. PCR Amplification

6. 10X PCR buffer: 100 m*M* Tris-HCl, pH 8.4, 500 m*M* KCl, 5 m*M* $MgCl_2$, and 0.1% gelatin.
7. dNTP mix: Combine equimolar amounts of each deoxynucleotide triphosphate (dNTP) to produce a 100 m*M* stock solution (25 m*M* of each dNTP).
8. *Taq* polymerase: supplied at a concentration of 5 U/µL (*see* Note 1).
9. Primers: The length of primers varies in each application but averages 20 nucleotides (*see* Note 2). Dilute stocks to 50 pmol/µL.

3. Methods

3.1. Preparation of Restriction Fragments

The single largest variable in this step is the quality and complexity of the DNA. Best results are usually obtained with DNA that has been extracted with phenol, and with DNA extracted from simple genomes such as bacteria, yeast, nematodes, and *Drosophila* (*see* Notes 3–7).

1. Restriction enzyme digest for 2 h at 37°C according to the manufacturer's conditions. Usually 1–5 µg of genomic DNA is used for the initial digestions that are carried out in a volume of 20–50 µL. Since the DNA sample is precipitated prior to ligations, the reaction volume is not critical.

2. Phenol extract by the addition of an equal volume of equilibrated phenol. Invert the sample, spin briefly in a microfuge to separate phases, and remove the aqueous phase (*see* Note 8).
3. Chloroform:isoamyl alcohol (24:1) extract the aqueous phase from the previous extraction as described in step 2.
4. Ethanol precipitate by addition of 0.1 vol of 3*M* sodium acetate (pH 5.3) to the DNA sample, followed by 2 vol of absolute ethanol. Precipitate the DNA at –20°C for 30 min and harvest by centrifugation for 20 min at 4°C. Wash the DNA pellet with 70% ethanol and dry under vacuum.
5. Resuspend the sample in $T_{0.1}E$ at 100 ng/µL.

3.2. Ligation to Form Circles

It is advisable to carry out three simultaneous reactions with different DNA concentrations (20, 100, and 500 ng/µL) in order to increase the probability that one of the dilutions will be favorable for the formation of monomeric circles.

1. Set up 3 tubes containing 0.2, 1.0, and 5.0 µL of digest.
2. Add 10 µL of 10X ligation buffer and water to final volume of 95 µL (i.e., 93.8, 93, and 89 µL of distilled H_2O, respectively).
3. Add 0.05 Weiss U of T4 DNA ligase/µL. Incubate at 14°C for least 2 h. (Ligations generally proceed overnight.)
4. Heat inactivate the ligation mixture at 65°C for 15 min and extract with phenol and chloroform as described in the previous section. Precipitate the DNA as described above and resuspend DNA in 10 µL of TE.

3.3. PCR Amplification

1. For each ligation product, add 4 µL of the DNA from the ligation (*see* Note 9) to a tube suitable for PCR. Add 5 µL of 10X PCR buffer, 0.4 µL of the dNTP mix, 1 µL of each primer, and water to a final volume of 50 µL. Layer 50 µL of mineral oil on top of the reaction mix.
2. Heat the reaction mixture at 95°C for 10 min. This facilitates amplification by introducing nicks into the circular templates. Add 0.2 µL of *Taq* polymerase.
3. Standard amplification conditions are 30 cycles of denaturation at 95°C for 40 s, primer annealing at 56°C for 30 s, and primer extension at 72°C for 100 s (*see* Note 10).
4. Run 1/10 of each sample on a 1% agarose gel to analyze amplification products. The IPCR products should be confirmed by Southern blotting and hybridization to a DNA probe derived from core (*see* Note 11).

Once confirmed, the PCR product may then be either cloned (*see* Chapter 3) or sequenced directly (*see* Chapter 23).

3.4. Secondary Amplification of IPCR Products

If nonspecific products are amplified, it is sometimes helpful to carry out a second amplification. This can be done in two different ways.

3.4.1. Nested Primers

In this case, one of the original primers is replaced with another, which anneals within the core sequence closer to the core–target junction (i.e., still within the amplified fragment).

1. Take 1 µL of the first PCR reaction to be used as the amplification template.
2. Repeat original PCR with the alternate primers, but reduce the number of cycles to 15.

3.4.2. Gel Purification

Alternatively, the products of the initial PCR reaction can be separated in an agarose gel.

1. Run 5 µL of the first PCR reaction on a low melting point temperature agarose gel.
2. Extract a fragment of the correct size by poking with the narrow end of a Pasteur pipet. This should remove an agarose plug of 5–10 µL.
3. Dispense the agarose plug into 100 µL of distilled water and heat to 95°C for 5 min to melt the agarose.
4. Use 1 µL to repeat the PCR with either both of the original primers or a nested primer. Follow the protocol for the nested primer, as in Section 3.4.1.

4. Notes

1. When amplifying fragments from enteric bacteria, we use *Taq* polymerase purified from *Thermus aquaticus*; cloned thermostable polymerase (AmpliTaq) is used in all other applications.
2. Specificity of the PCR amplification depends heavily on the choice of oligonucleotide primers. We typically use oligonucleotide primers 20 nucleotides in length that are approx 50% in their G + C content, which yields a theoretical average melting temperature of 60°C using the empirical rule of thumb that the T_d approx equals 2× (number of As plus Ts) + 4× (number of Gs plus Cs).
3. Choice of restriction enzyme is determined by the distribution of restriction sites in the target region, established, if necessary, by Southern blots *(10)*, and by the maximum size limit of PCR amplification. Frag-

ments containing <3 Kb of target sequence are preferable, since these amplify best. Fragments of satisfactory size containing either the left or right core–target junction (L and R in Fig. 1) can sometimes be obtained in a single restriction digestion by suitable choice restriction enzymes.

4. In some cases digestion with two different restriction enzymes is required in order to generate a core–target fragment of satisfactory size; if the restriction enzymes produce incompatible ends, then it is necessary to render the ends of the fragments blunt using the Klenow polymerase or T4 DNA polymerase prior to the ligation step.
5. In particular applications of IPCR, we have had good success using *Pst*I to recover sequences flanking insertion sequence IS*1* in *Escherichia coli (7)*, and using *Cla*I or *Taq*I to recover sequences flanking insertion sequence IS*30 (17)*. Most common YAC vectors allow recovery of the insert junction nearest the centromere using *Eco*RV and the recovery of the insert junction nearest the telomere using *Hinc*II *(12)*.
6. When using restriction enzymes that have convenient four-base cleavage sites within the core sequence, the usual experience in most laboratories is that IPCR is successful in about half the cases without performing preliminary Southern blots *(10)*. The success rate probably results from the fact that four-base restriction sites are sufficiently frequent (theoretically, one in every 256 base pairs) that chance favors the target sequence having a site neither too close to the core sequence nor too far from it. Failing in an initial attempt, subsequent trials should be based on preliminary data from Southern blots to identify a suitable restriction enzyme (or a pair of enzymes).
7. It should be emphasized again that the success of IPCR depends on the complexity of the starting material, with simpler genomes yielding more reliable results. The procedure has worked well with cloned material and with *E. coli*, *Caenorhabditis,* and *Drosophila.* Genome complexity increases the chance of spurious PCR products and decreases the efficiency of the circularization reaction. Recovery of flanking sequences from more complex genomes may be improved by enrichment of DNA fragments of a suitable size class. For example, restriction fragments of the desired size (determined from preliminary Southern blots), can be extracted from agarose gels with glass powder, electroelution, or DEAE membranes. There should be a total of about 1 μg of cleaved DNA in order to conduct circularization at several DNA concentrations and PCR.
8. In cases where the digested sample can be heat treated to inactivate the restriction endonuclease (68°C for 10 min), samples may be diluted and ligation can often proceed without intermediate purification procedures (steps 2–5).

9. Some reports indicate that the efficiency of IPCR is improved by the amplification of linear rather than circular molecules *(9)*. In this case, the circularized fragments are cleaved at a unique restriction site within the core sequence. However, finding the appropriate restriction enzyme introduces additional steps and potential complications.
10. When amplification products greater than about 3 kb are expected, the denaturation time should be increased to 60 s and the extension time to 150 s.
11. Do not use the PCR primers as a probe since they will hybridize to all products generated by these primers.
12. Alternatives to IPCR: Among the alternatives to IPCR that may be considered for recovering flanking sequences from genomic DNA are the following. Experimental details can be found in the original citations and additional general discussion in Ochman et al. *(17)*.
    a. Ligation-mediated PCR employs a double-stranded cassette consisting of a 24-mer and an 11-mer complementary to the 3' end of the 24-mer. The blunt end of the cassette is ligated onto DNA fragments created by primer extension using an oligonucleotide that anneals to the core sequence. Experimental details can be found in refs. *(18–20)*.
    b. Vectoret (or "bubble") PCR has been used to recover the vector–insert junctions in YACs *(21)*. DNA from yeast cells containing a YAC is digested with restriction enzymes that cleave at a convenient site in the core sequence. A double-stranded oligonucleotide cassette ("vectorets") containing a "bubble" region of noncomplementary bases is ligated onto the end of the target sequence. Amplification proceeds from a primer specific to the YAC vector, which produces a newly synthesized strand complementary to one unpaired loop in the bubble, and polymerization in the reverse direction uses an oligonucleotide that anneals to this region in the newly synthesized strand.
    c. Oligo-cassette mediated PCR employs a double-stranded 28-mer that is ligated onto the target sequence after digestion with an appropriate restriction enzyme *(22)*. Approx 50 cycles of primer extension are carried out using a biotinylated primer complementary to the core region, and the biotinylated products are trapped with streptavadin-coated magnetic beads. The isolated product is then subjected to conventional PCR using a nested primer within the core sequence and a reverse primer for the oligo-cassette.

Acknowledgments

This work was supported by NIH grants GM40322 and HG00357 to D. L. Hartl and GM40995 to H. Ochman.

References

1. Saiki, R. K., Scharf, S. J., Faloona, F., Mullis, K. B., Horn, G. T., Erlich, H. A., et al. (1985) Enzymatic amplification of β-globin genomic sequences and restriction site analysis for diagnosis of sickle cell anemia. *Science* **230,** 1350–1354.
2. Saiki, R. K., Gelfand, D. H., Stoffel, S., Scharf, S. J., Higuchi, R. G., Horn, G. T., et al. (1988) Primer-directed enzymatic amplification of DNA with a thermostable DNA polymerase. *Science* **239,** 487–491.
3. Burke, D. T., Carle, G. F., and Olson, M. V. (1987) Cloning of large segments of exogenous DNA into yeast by means of artificial chromosome vectors. *Science* **236,** 806–812.
4. Hieter, P., Connelly, C., Shero, J., McCormick, M. K., Antonarakis, S., Pavav, W., et al. (1990) Yeast artificial chromosomes: promises kept and pending, in *Genetic and Physical Mapping,* vol. 1 (Davies, K. E. and Tilghman, S. M., eds.), Cold Spring Harbor Laboratory, Cold Spring Harbor, NY, pp. 83–120.
5. Sternberg, N. (1990) Bacteriophage P1 cloning system for the isolation, amplification, and recovery of DNA fragments as large as 100 kilobase pairs. *Proc. Natl. Acad. Sci. USA* **87,** 103–107.
6. Pierce, J. C. and Sternberg, N. L. (1993) Using the bacteriophage P1 system to clone high molecular weight (HMW) genomic DNA. *Meth. Enzymol.* (in press).
7. Ochman, H., Gerber, A. S., and Hartl, D. L. (1988) Genetic applications of an inverse polymerase chain reaction. *Genetics* **120,** 621–623.
8. Triglia, T., Peterson, M. G., and Kemp, D. J. (1988) A procedure for in vitro amplification of DNA segments that lie outside the boundaries of known sequences. *Nucleic Acids Res.* **16,** 8186.
9. Silver, J. and Keerikatte, V. (1989) Novel use of polymerase chain reaction to amplify cellular DNA adjacent to an integrated provirus. *J. Virol.* **63,** 1924–1928.
10. Southern, E. M. (1975) Detection of specific sequences among DNA fragments separated by gel electrophoresis. *J. Mol. Biol.* **98,** 503–517.
11. Garza, D., Ajioka, J. W., Carulli, J. P., Jones, R. W., Johnson, D. H., and Hartl, D. L. (1989) Physical mapping of complex genomes. *Nature* **340,** 577,578.
12. Ochman, H., Medhora, M. M., Garza, D., and Hartl, D. L. (1990) Amplification of flanking sequences by IPCR, in *PCR Protocols: A Guide to Methods and Applications* (Innis, M., Gelfand, D., Sninsky, J., and White, T., eds.), Academic, New York, pp. 219–227.
13. Silverman, G. A., Ye, R. D., Pollack, K. M., Sadler, J. E., and Korsmeyer, S. J. (1989) Use of yeast artificial chromosome clones for mapping and walking within human chromosome segment 18q21.3. *Proc. Natl. Acad. Sci. USA* **86,** 7485–7489.
14. Silverman, G. A., Jockel, J. I., Domer, P. H., Mohr, R. M., Taillon-Miller, P., and Korsmeyer, S. J. (1991) Yeast artificial chromosome cloning of a two-megabase-size contig within chromosomal band 18q21 establishes physical linkage between BCL2 and plasminogen activator inhibitor type-2. *Genomics* **9,** 219–228.
15. Huang, S., Hu, Y., Wu, C., and Holcenberg, J. (1990) A simple method for direct cloning cDNA sequence that flanks a region of known sequence from

total RAN by applying the inverse polymerase chain reaction. *Nucleic Acids Res.* **18,** 1922.

16. Breukel, C., Wijnen, J., Tops, C., Klift, H. V., Dauwerse, H., and Meera Khan, P. (1990) Vector-Alu PCR: a rapid step in mapping cosmids and YACs. *Nucleic Acids Res.* **18,** 3097.
17. Ochman, H., Ayala, F. J., and Hartl, D. L. (1993) Use of the polymerase chain reaction to amplify segments outside the boundaries of known sequences. *Meth. Enzymol.* **218,** 309–321.
18. Pfeifer, G. P., Steigerwald, S. D., Mueller, P. R., Wold, B., and Riggs, A. D. (1989) Genomic sequencing and methylation analysis by ligation mediated PCR. *Science* **246,** 810–813.
19. Mueller, P. R. and Wold, B. (1989) In vivo footprinting of a muscle specific enhancer by ligation mediated PCR. *Science* **246,** 780–786.
20. Fors, L., Saavedra, R. A., and Hood, L. (1990) Cloning of the shark Po promoter using a genomic walking technique based on the polymerase chain reaction. *Nucleic Acids Res.* **18,** 2793–2799.
21. Riley, J., Butler, R., Ogilvie, D., Finniear, R., Jenner, D., Powell, S., et al. (1990) A novel, rapid method for the isolation of terminal sequences from yeast artificial chromosome (YAC) clones. *Nucleic Acids Res.* **18,** 2887–2890.
22. Rosenthal, A. and Jones, D. S. C. (1990) Genomic walking and sequencing by oligo-cassette mediated polymerase chain reaction. *Nucleic Acids Res.* **18,** 3095,3096.

PART IV

SEQUENCE VARIATIONS

CHAPTER 18

Use of Silver Staining to Detect Nucleic Acids

Lloyd G. Mitchell, Angelika Bodenteich, and Carl R. Merril

1. Introduction

Silver stains are useful for the detection of nanogram amounts of proteins or nucleic acids in acrylamide gels or on various membranes. They have been shown to be more sensitive than organic stains in detecting proteins and DNA. They are capable of detecting as little as 0.03 ng/mm^2 of DNA *(1)*. In addition, silver stains avoid the mutagenic hazards presented by both ethidium bromide and radioactive detection methods.

The thirteenth century alchemist, Count Albert von Bollstadt, was the first to record that silver nitrate would stain organic material, including human skin. However, silver was not employed in scientific studies until 1844 when Krause used silver to stain histological tissues. Histological silver stains were further developed by the anatomists Golgi and Cajal. The first silver stains used on polyacrylamide gels were adapted from these histological stains. They were quite tedious, requiring numerous steps and solutions. They took 3 h to perform *(2)*. Continued research on the use of silver staining to detect proteins and nucleic acids separated in polyacrylamide has resulted in a number of simplified protocols. It has also extended our knowledge of the mechanisms underlying these stains.

Silver images of protein or nucleic acid patterns are produced by a difference in the oxidation-reduction potential in regions occupied

From: *Methods in Molecular Biology, Vol. 31: Protocols for Gene Analysis*
Edited by: A. J. Harwood Copyright ©1994 Humana Press Inc., Totowa, NJ

by nucleic acids or proteins compared to the surrounding gel or membrane *(3,4)*. This redox potential catalyzes the reduction of ionic to metallic silver. A positive (dark) image will be produced if the region occupied by nucleic acid or protein has a higher redox potential than the surrounding region, whereas a negative image will result if the redox potential is higher in the surrounding matrix. The redox potentials may be altered by manipulating the chemistry of the staining solutions. Experiments with nucleic acids and their components have implicated the purines as the active subunits in the silver staining reaction *(5)*.

There are three general chemistries utilized in silver staining: diamine or ammoniacal stains, nondiamine chemical reduction stains, and photochemical stains. The diamine silver stains, which were initially developed for the visualization of nerve fibers *(6)*, were the first type of silver stain employed to detect biopolymers separated on polyacrylamide gels. The diamine stains rely on the formation of silver diamine complexes in the presence of ammonium hydroxide. To produce an image, the solution containing the silver diamine complexes must be acidified, most commonly with citric acid, in the presence of formaldehyde. The resulting decrease in ammonium ions liberates silver ions from the silver diamine complexes, thereby facilitating the reduction of the silver ions with the formaldehyde to metallic silver. The concentration of citric acid in the image developing solution must be precisely controlled to reduce background deposition of silver.

The photo-development stains rely on light to reduce ionic to metallic silver. These stains are simple and quick, requiring fewer solutions, but they lack the sensitivity of the other methods *(7)*.

Most nondiamine chemical reduction stains were developed by adapting photochemical protocols *(3)*. They employ silver nitrate, which reacts with biopolymers under acidic conditions. This is followed by the selective reduction of ionic silver by formaldehyde under alkaline conditions. Sodium carbonate is often utilized in these stains to maintain alkalinity during image development since formic acid is generated by the oxidation of formaldehyde. The nondiamine chemical stains are fast and easy to perform. They also work well for gels that are under 1 mm thick. A nondiamine silver staining protocol, such as the one described in this chapter, can be performed in 40 min using solutions that are stable for months.

2. Materials

All reagents must be prepared with ultrapure deionized water with a conductivity of <1 µmho. The water should also be free of both ionic and organic contaminates. Such contaminants are often responsible for background staining. Care must also be taken to prevent proteins from skin or other potential contaminating sources from contact with the gels or the reagents, since this silver stain is also an excellent protein detector.

Silver can be used to stain native and denaturing acrylamide gels, as well as dehydratable gels using discontinuous buffer systems and varying amounts of crosslinkers *(8)*. We recommend the use of gels that are 0.4 mm thick (*see* Note 1).

The solutions are easy to prepare and all but the developer may be kept for at least 6 mo. They can also be made during the procedure.

1. Fixative: 10% ethanol (v/v) (*see* Note 2).
2. Oxidizer: 0.0034*M* potassium dichromate (1 g/L) and 0.0032*M* nitric acid (or 0.2 mL of concentrated nitric acid, 16*M*/L) (*see* Note 3). Avoid direct contact with the skin since this solution is an irritant. It is a strong oxidizer and should be kept away from reducing agents. It is also photosensitive and should be stored in an amber or dark bottle.
3. Silver solution: 2% silver nitrate (2 g/L) (*see* Note 3). Silver nitrate solution is caustic to eyes and mucous membranes, and it will stain both skin and clothing. Store in an amber or dark bottle.
4. Developer: 0.28 sodium carbonate (30 g/L) to which 0.5 mL of 37% formalin/L (*see* Note 4) is added just prior to use. This solution should be used at 4°C and it may be stored at that temperature. Since formaldehyde vapor may be a carcinogen, this solution should be made and used in a fume hood.
5. Stop solution: 5% acetic acid (v/v).

3. Method

Following the electrophoretic separation of nucleic acids, each polyacrylamide gel is placed in a separate clean glass or plastic tray. The use of clean vinyl or latex gloves, with the powder washed off, and a lab coat will reduce the risk of contaminating the gel and the staining of the investigator. All solutions should be at room temperature except the developer, which is precooled at 4°C. Subdued laboratory lighting will help to decrease background staining. Development should not be done on a light box. The gel should be handled gently to avoid crushing that may produce artifacts. Solutions should not be poured directly on the gel; instead, tilt the tray and pour solutions into a corner or use individual trays for each solution and move the gels from tray to tray. Solution volumes should be sufficient to completely immerse the gels. The trays should be gently agitated during staining.

1. Following electrophoretic separation soak the gel in fixative for 10 min (*see* Note 5). Use a volume of fixative double that needed to cover the gel (*see* Note 6). The gel may be stored indefinitely in the fixative at this point if development needs to be delayed.
2. The gel is then placed in the dichromate oxidizer soak for 3–5 min (*see* Note 7). Then decant the oxidizer. Wash the gel three times with deionized water, for 3–5 min each (*see* Notes 7 and 8).
3. Place the gel in the silver nitrate solution for 15–20 min (*see* Note 9). Briefly wash the gel (for 20–30 s) in deionized water to remove silver nitrate from the gel surface (*see* Note 10).
4. Form the image by washing the gel with precooled developer for approx 30 s with constant agitation (*see* Note 11). Observe the gel closely during this process and discard the solution as soon as it loses clarity (*see* Note 12). Replace with fresh developer and again discard as soon as it loses clarity. Repeat this process until the gel image has formed or until the background staining becomes objectionable.
5. When the DNA bands reach the desired intensity relative to the background, stop development by placing the gel in 5% acetic acid (*see* Notes 13 and 14). The DNA bands normally are dark brown or black on a faint yellow/brown background (*see* Fig. 1).

Stained gels cast on polyester backing may be dried in a microwave oven or left at room temperature to dry. Nonbacked gels can be dried on cellophane or filter paper under vacuum with mild heat. For a permanent record, the gel should be photographed on a light box since gels silver-stained with this procedure may darken over time.

4. Notes

1. Thinner gels can be stained faster, but they are more fragile and difficult to work with whereas gels thicker than 0.7 mm require longer to stain. To increase the durability of gels that are thinner than 0.7 mm, we recommend casting them on a polyester backing (such as GelBond PAG, FMC, Rockland, ME or Gel-Fix, Serva, Heidelberg, Germany). Alternatively, the gel may be cast and bonded to silanized glass (Bind-

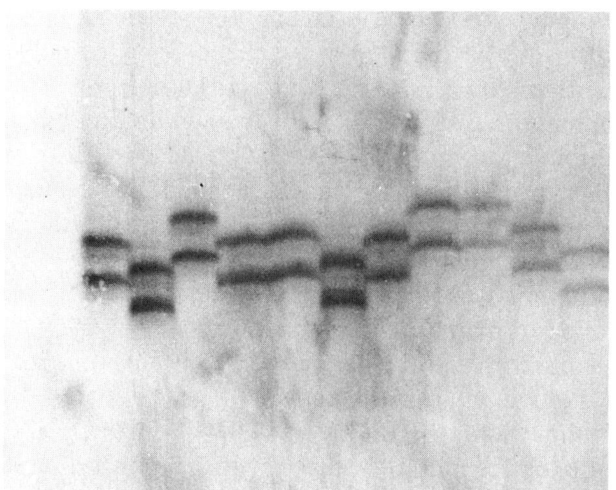

Fig. 1. A typical silver-stained gel. This gel contains a region of the mitochondrial DNA, from several unrelated individuals, that is rich in AC repeat sequences. The DNA was PCR amplified and 1 μL (8–35 ng/μL) from each reaction was denatured by adding 3 μL of a loading buffer containing formamide followed by heating at 96°C for 1 min. The DNA was electrophoresed at 55°C on 6% polyacrylamide gel containing 6*M* urea. This gel was cast on GelBond PAG (FMC, Rockland, ME). This example also illustrates some of the surface staining artifacts that may occur with this technique.

Silane, Pharmacia-LKB, Piscataway, NJ). Gels that are attached to polyester sheets offer the advantage that after electrophoresis they can be trimmed to remove the regions that do not contain DNA, thereby decreasing the amount of staining solutions required.

If gels bound to polyester sheets are utilized, the polyester sheets must be flattened and held tightly by capillary attraction against the backing glass plate of the gel apparatus prior to casting. To accomplish this, a few milliliters of distilled water or 5% glycerol is pipeted onto the backing gel plate, then the polyester sheet is placed on the plate. One must be careful to place the nonbinding surface against the backing plate. The sheet is then flattened with a rubber print roller or pipet. It is important that all bubbles and irregularities be removed so that the gel will have uniform thickness. To prevent the polyester sheet from moving while rolling, a thin film of silicon grease can be applied between the sheet and plate on lower end of the gel. The polyester sheet should be large enough to cover and include the gel, loading wells, and the spacers that are placed on top of the sheet. Once the polyester sheet is in place with the spacers, the gel is cast as usual.

2. Methanol may be substituted, but it is toxic and may be absorbed by contact through the skin or by respiration. Many fixative solutions for DNA and proteins separated on polyacrylamide gels contain acetic acid. However, this protocol does not include acetic acid in the fixative, since darker bands were obtained in its absence.
3. This solution may also be made as a 10× stock.
4. Concentrated formalin stock is generally 37%.
5. Fix for 20 min with gels thicker than 0.7 mm. This time may be reduced by using fixative preheated to between 40–50°C.
6. Besides minimizing the diffusion of DNA in the gel, fixation also removes the buffer ions and other chemicals (such as urea) that could interfere with staining.
7. Soak for 10 min for gels thicker than 0.7 mm.
8. It is important to wash the gel until the concentration of the dichromate oxidizer is reduced such that the gel is no longer visibly yellow. If the gel is still yellow or turns red in the dichromate oxidizer, an acid wash of 0.01*M* nitric acid may be used to remove the coloration.
9. Soak gels thicker than 0.7 mm for 25–30 min. If any contaminants are present at this point in the procedure, particularly chlorides, a silver precipitate may form. The presence of such a precipitate usually results in a reduction of sensitivity of the silver stain and an increase in the background staining.
10. A wash that is too long at this step will allow silver to diffuse out of the gel and will diminish the sensitivity of the staining.
11. Cooled developer is used to permit controlled image development.
12. A brownish silver carbonate precipitate forms as silver ions escape from the gel. This precipitate will deposit on the gel surface if the solutions are not changed frequently.
13. Gels that are overdeveloped or that have a mirror-like appearance owing to surface deposition of silver can be destained with the following procedure: Dissolve 37 g of anhydrous cupric sulfate and 37 g of sodium chloride in 850 mL of deionized water. With constant stirring add concentrated ammonium hydroxide until the solution becomes deep blue and there is no precipitate, then bring the volume to 1 L. A second solution containing 436 g sodium thiosulfate pentahydrate in 1 L of deionized water is also prepared. Equal parts of these two solutions are combined just prior to use. The gel is soaked in this solution until almost completely clear. The destaining is stopped by 10% acetic acid for 15 min,

followed by a minimum of 1 h washing with several changes of deionized water. The gel may then be restained by beginning the procedure at the silver nitrate step (step 3). The dichromate step is not required for restaining. This destaining procedure is also useful for the removal of silver stains from clothing.

14. Some artifacts and precipitate on the surface of the gel can be removed by rubbing the dried gel gently with a damp piece of absorbant cotton. This technique should be tried in a noncritical region. The procedure should be stopped when the surface becomes tacky. However, it can be repeated after the gel is allowed to dry.

References

1. Goldman, D. and Merril, C. R. (1982) Silver staining of DNA in polyacrylamide gels: linearity and effect of fragment size. *Electrophoresis* **3,** 24–32.
2. Merril, C. R., Switzer, R. C., and Van Keuren, M. L. (1979) Trace polypeptides in cellular extracts and human body fluids detected by two dimensional electrophoresis and a highly sensitive silver stain. *Proc. Natl. Acad. Sci. USA* **76,** 4335–4339.
3. Merril, C. R. (1987) Development and mechanisms of silver stains for electrophoresis. *Acta Histochem. Cytochem.* **19,** 655–667.
4. Merril, C. R. (1990) Silver staining of proteins and DNA. *Nature* **343,** 779–780.
5. Merril, C. R. and Pratt, M. E. (1986) A silver stain for the rapid quantitative detection of proteins nucleic acids on membranes or thin layer plates. *Anal. Biochem.* **156,** 96–110.
6. Bielschowsky, M. (1904) Die silberimpragnation der neurofibrillen. *J. Psychol. Neurol.* **3,** 169–189.
7. Merril, C. R., Harrington, M., and Alley, V. (1984) A photodevelopment silver stain for the rapid visualization of proteins separated on polyacrylamide gels. *Electrophoresis* **5,** 289–297.
8. Allen, R. C., Graves, G., and Budowle, B. (1989) Polymerase chain reaction amplification products separated on rehydratable polyacrylamide gels and stained with silver. *Biotechniques* **7(7),** 736–744.

CHAPTER 19

A Nonradioactive Method for the Detection of Single-Strand Conformational Polymorphisms (SSCP)

Peter J. Ainsworth and David I. Rodenhiser

1. Introduction

The diagnosis of disease by direct analysis is an increasingly important branch of laboratory medicine. The detection of DNA sequence anomalies in different at-risk individuals may be effected by indirect techniques (e.g., linkage analysis), or by more direct methods involving the detection of sequence deviation. Large insertions or deletions can be detected by pulsed field gel electrophoresis or by using Southern blotting/RFLP techniques, however the more subtle alterations in DNA sequence such as point mutations or small deletions/insertions are below the level of resolution of these techniques. Several protocols have been reported that now allow detection of small deviations in DNA sequence *(1,2)*. Among the most promising of these is the technique for the detection of Single-Strand Conformational Polymorphisms (SSCPs).

The first report of SSCP analysis to detect DNA sequence deviations was made by Orita et al. in 1989 *(3)*. This technique involved radiolabeling the target DNA sequences amplified by PCR, rendering the DNA to a single-stranded state, then identifying sequence-dependent shifts in the electrophoretic mobility of the single-strand DNA (ssDNA). Most single base changes in DNA fragments up to 200 base pairs in size could be detected as mobility shifts, thereby

From: *Methods in Molecular Biology, Vol. 31: Protocols for Gene Analysis*
Edited by: A. J. Harwood Copyright ©1994 Humana Press Inc., Totowa, NJ

allowing the rapid screening for mutations and avoiding tedious cloning and sequencing of multiple DNA samples. This methodology has several drawbacks. It requires the use of radioisotopes and and electrophoresis on large sequencing gels to effect separation of the single-strand conformers. Both of these limitations increase the time and expense of the technique.

We have reported earlier a nonradioactive SSCP-based method (*see* Fig. 1) for detection of a B1 variant of Tay Sachs disease *(4,11)*, as have other groups for phenylketonuria *(5)* and for p53 mutants *(6)*. In this chapter, we present a refined SSCP methodology presently in use within our laboratory for the detection of mutations associated with cystic fibrosis and neurofibromatosis. This improved method eliminates the necessity of the asymmetric PCR amplification step of the original protocols (*see* Fig. 1) and takes advantage of the semiautomated Phastgel™ system.

2. Materials

General reagents for PCR are required as for standard protocols, e.g., a 10 m*M* stock of deoxynucleotides (dNTPs), distilled water, and mineral oil. We routinely use *Taq* polymerase (Amplitaq, Cetus, Norwalk, CT) together with its stock buffer mix and $MgCl_2$ concentrate, since we have found that these give more consistent results than other available thermostable polymerases. Oligonucleotide primers are synthesized on site and resuspended as a stock at a concentration of 25 pmol/µL.

2.1. PCR-SSCP

1. Denaturant solution: 97.5% deionized formamide, 4.6*M* urea, 0.3% bromophenol blue, 0.3% xylene cyanol, 10 m*M* EDTA.
2. 2X TBE: Dissolve 24.2 g of Tris-base, 12.4 g of boric acid, and 1.49 g of EDTA in 1 L of dH_2O.
3. Polyacrylamide gel electrophoresis: precast 12.5% homogeneous polyacrylamide gels acrylamide:bisacrylamide (30:1) (available from Pharmacia, Piscataway, NJ).
4. Silver stain: The components can easily be made up in the laboratory and are used in a method, adapted from ref. 7, in the Phastgel system.

3. Methods

3.1. PCR-SSCP

1. Amplify target DNA from genomic DNA (*see* Note 1) using conventional PCR protocols. Conditions for amplification will depend on the particular PCR primers and will need to be established empirically.

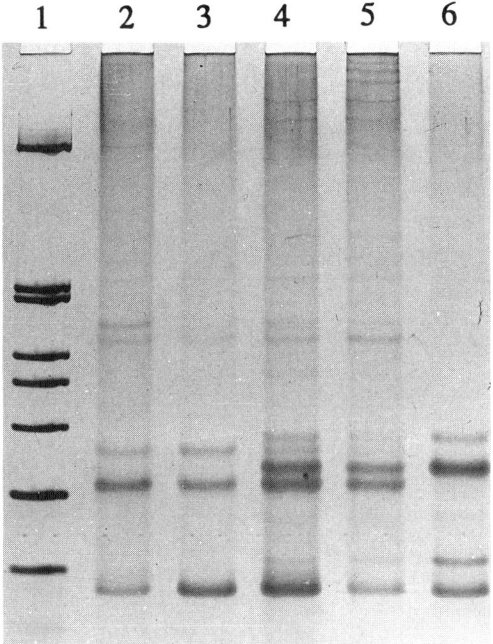

Fig. 1. Single-stranded 135-bp DNA fragments were generated by 40 cycles of asymmetric amplification from an unpurified PCR product using the 5' primer of a PCR product derived from the α subunit of Hexosamidiase A *(11)*. Electrophoretic separation was effected on a 90 × 70 × 1.5 mm 9% polyacrylamide gel (MiniProtein, BioRad), the DNA being visualized by a manual silver stain technique (BioRad). Lane 1: DNA marker (*Hinf*I digest of pBR322); lane 2: control DNA; lane 3: paternally-derived DNA that is normal for this region; lane 4: maternally-derived DNA that is heterozygous for a mutation in this region; lane 5: DNA derived from the proband that is also heterozygous; lane 6: a plasmid that contains the same mutant DNA segment.

2. Take 2 µL of the PCR product and mix with 2 µL of denaturant (*see* Note 2). Heat at 94°C for 10 min. Immediately incubate at 65°C for a further 5 min, followed by room temperature (20°C) for 15 min.
3. While denaturing the PCR products, prerun two precast 12.5% polyacrylamide Pharmacia Phastgels gels to equilibrate the buffers. Prerun one gel at room temperature (20°C) and one at 4°C, both for the 100 volt-hours (vh).
4. Separate 1 µL samples on each gel. Run the gel at 20°C for a total of 350 vh and the gel at 4°C for 600 Vh; in both cases include the prerun time in the calculation (*see* Note 3).

5. After electrophoresis, transfer the gels to the Phastgel™ development unit, fix the gels for 5 min in 20% trichloroacetic acid at 20°C, followed by 2 min in 50% ethanol/10% acetic acid at 50°C. Wash twice, first for 2 min then for 4 min, in 10% ethanol/5% acetic acid at 50°C.
6. Sensitize by soaking in 8.3% gluteraldehyde for 6 min at 50°C. Wash at the same temperature twice with 10% ethanol/5% acetic acid (for 3 and then 5 min); followed by washing twice in dH_2O for 2 min each.
7. Stain for 10 min in 0.5% silver nitrate solution at 40°C, and then wash twice in dH_2O for 30 s at 30°C. Develop for 1 min at 30°C in 150 mL of a freshly made solution of 2.5% sodium carbonate containing 75 µL of 40% formaldehyde. After a minute, replace with fresh solution and incubate for a further 3.5 min.
8. Stop in 5% acetic acid for 2 min at 50°C, and then preserve the processed gel by soaking in 10% acetic acid/5% glycerol for 3 min at 50°C.

Figure 2 shows results using the PCR-SSCP technique to screen of a 425 bp DNA fragment containing exon 11 of the gene responsible for cystic fibrosis (*8; see* Notes 4 and 5).

4. Notes

1. Any routine protocol that yields high purity DNA from whole blood can be used *(9,10)*.
2. It is not necessary to gel purify the PCR product if the PCR amplification conditions are optimized to produce a single band. Optimization may require varying Mg^{2+} ion concentration, the annealing temperature, and the oligonucleotide sequence. We find that primers of ≥24 bp with a GC to AT ratio of 50% generally work well.
3. Maintenance of a constant gel temperature throughout the run is a necessity for optimized resolution of the single-strand DNA variants. A comparison of Figs. 2A and B shows the effect of temperature on the sensitivity of the procedure; at 20°C all four mutants could be identified (Fig. 2A), whereas at 4°C only one of the mutant alleles is detected (Fig. 2B, lane 5).
4. It is apparent from Fig. 2B that the same single-stranded DNA molecule may have more than one conformational form (conformer). The basis for this is unknown, but it may be surmised that more than one pattern of secondary structure formation would be stable at a particular temperature. The fact that several different conformers of the same single-strand may form might be expected to increase the diagnostic sensitivity of the technique. Furthermore, screening both sense and antisense DNA strands simultaneously also improves the diagnostic efficiency.

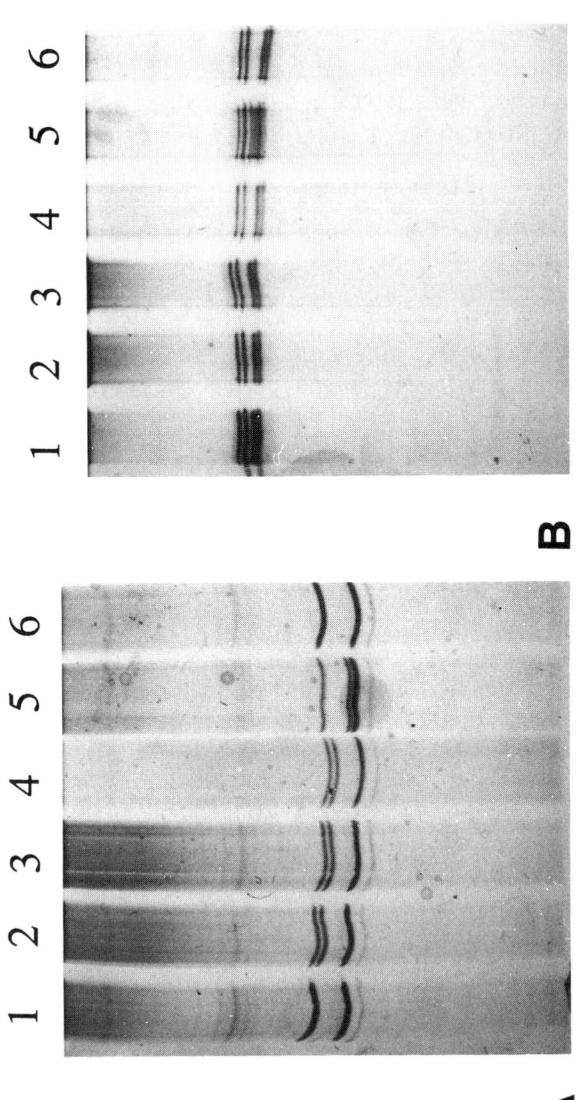

Fig. 2. PCR amplified exon 11 of the CFTR gene *(8)* was denatured as in Section 3.2. One microliter of aliquots of each sample were separated on a 12.5% polyacrylamide Phastgel™ run at either **(A)** 20°C or **(B)** 4°C. Lane 1 and 6 normal CFTR exon 11 DNA; lanes 2–4: DNA that contains a G > A mutation at position 1784; lane 5: DNA that contains a G > A mutation at position 1778.

5. The absolute percentage of mutations detected by this technique remains uncertain, although in our experience it seems to be efficient when screening DNA fragments of 450 bp or less. This appears especially true if electrophoretic separation is effected at more than one temperature. We are particularly encouraged by preliminary results in our laboratory where several SSCP variants in two 200–400 bp exons of the Neurofibromatosis type 1 gene have been demonstrated and led to the detection of several different point mutations and/or base insertions.

References

1. Cotton, R. G. H. (1989) Detection of single base changes in nucleic acids. *Biochem. J.* **263,** 1–10.
2. Rossiter, B. J. F. and Caskey, C. T. (1990) Molecular scanning methods of mutation detection. *J. Biol. Chem.* **265,** 12,753–12,756.
3. Orita, M., Suzuki, Y., Takao, S., and Hayashi, K. (1989) Rapid and sensitive detection of point mutations and DNA polymorphisms using polymerase chain reaction. *Genomics* **5,** 874–879.
4. Ainsworth, P. J., Surh, L. C., and Coulter-Mackie, M. B. (1991) Diagnostic single-strand conformational polymorphism (SSCP): a simplified non-radioisotopic method as applied to a Tay-Sachs B1 variant. *Nucleic Acids Res.* **19,** 405–406.
5. Dockhorn-Dworniczak, B., Dworniczak, B., Brommellkamp, L., Bulles, J., Horst, J., and Bocker, W. (1991) Non-isotopic detection of single-strand conformational polymorphism (PCR-SSCP): a rapid and sensitive technique in diagnosis of phenylketonuria. *Nucleic Acids Res.* **19,** 2500.
6. Mohabeer, A., Hiti, A., and Martin, J. (1991) Nonradioactive single-strand conformational polymorphism (SSCP) using the Pharmacia "PhastSystem™." *Nucleic Acids Res.* **19,** 405,406.
7. Andersson, A. and Johansson, J. (1987) Rapid purity checking of synthetic oligonucleotides with PhastSystem™. *J. Biochem. Biophys. Meth.* **14 (Suppl.),** 37.
8. Cutting, G., Kasch, L., Rosenstein, B., Zielenski, J., Tsui, L. C., Stylianis, E., and Kazazian, H. (1990) A cluster of cystic fibrosis mutations in the first nucleotide-binding fold of the cystic fibrosis conductance regulator protein. *Nature* **346,** 366–368.
9. Jeanpierre, M. (1987) A rapid method for purification of DNA from blood. *Nucleic Acids Res.* **15,** 9611,9612.
10. Miller, S. A., Dykes, D. D., and Polesky, H. F. (1988) A simple salting out method for extracting DNA from human nucleated cells. *Nucleic Acids Res.* **16,** 1213.
11. Ainsworth, P. J. and Çoulter-Mackie, M. B. (1990) A novel mutation in the gene for the α subunit of β-*N*-acetylhexosaminidase associated with the B1 variant of Tay-Sachs disease. *Amer. J. Hum. Genet.* **47 (Suppl.),** 810.

CHAPTER 20

Temperature Gradient Gel Electrophoresis (TGGE) for the Detection of Polymorphic DNA and RNA

Karsten Henco, Jutta Harders, Ulrich Wiese, and Detlev Riesner

1. Introduction

The potency of gel electrophoresis to analyze conformational transitions of nucleic acids has opened up a new field of applications. Helix coil transitions and, at a refined level, minor variations because of mutations and mismatches, may be studied experimentally by solvent-gradient and temperature-gradient gel electrophoresis. The latter method is the subject of this chapter.

In the simplest form of temperature-gradient gel electrophoresis (TGGE), a nucleic acid sample is loaded in a broad slot at the negative electrode of a polyacrylamide slab gel (*see* Fig. 1). A linear temperature gradient is established perpendicular to the direction of the electric field, so that the molecules at the left side of the gel migrate at low temperatures, the molecules at the right side at high temperatures, and those at positions in between at corresponding intermediate temperatures. Each individual molecule, however, migrates during the whole electrophoretic run at constant temperature. In the example shown in Fig. 1 a double-stranded RNA is analyzed. The molecule undergoes two highly cooperative, temperature dependent conformational transitions: from the native, base-paired state to a partially

From: *Methods in Molecular Biology, Vol. 31: Protocols for Gene Analysis*
Edited by: A. J. Harwood

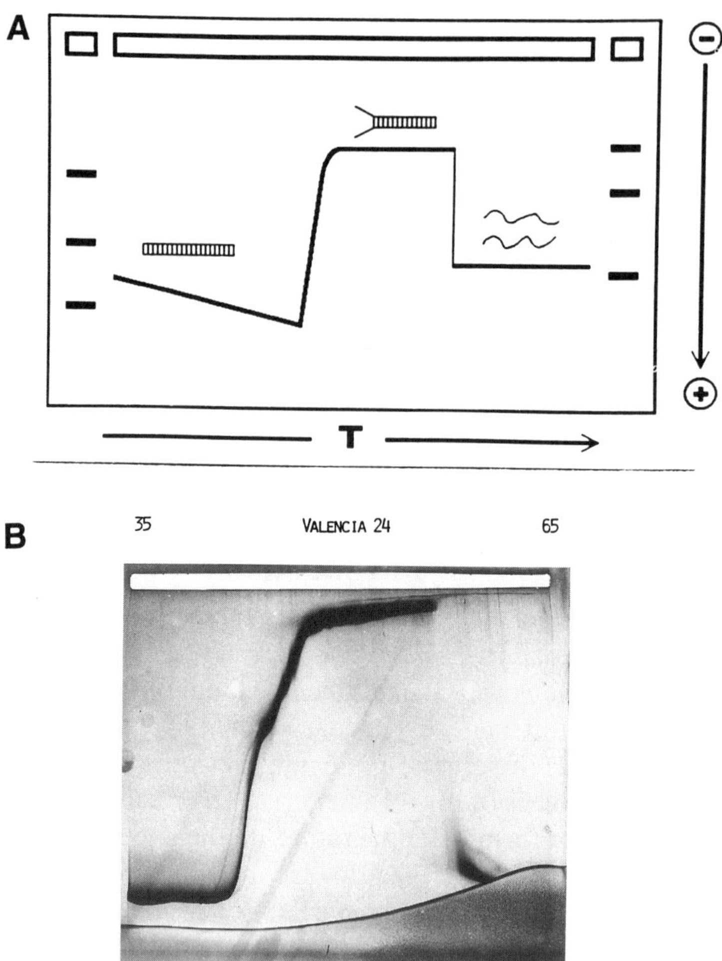

Fig. 1. Principle of TGGE. A linear temperature gradient is applied perpendicular to the electric field. The sample is applied to the long slot; small marker slots are on the left and right side of the gel. The mobility of a double-stranded nucleic acid is drastically decreased after a cooperative transition at T_m1 to a partially denatured molecule but is increased again after an irreversible transition of a complete strand separation at T_m2. A schematic representation is given in **(A),** and an experimental example in **(B).** In the experiment the double-stranded form of a satellite RNA of cucumber mosaic virus (strain Valencia 24) was analyzed. The lowest and highest temperatures are indicated. Two retardation transitions (40.5 and 44.2°C) and one dissociation transition (approx 58°C) are visible (from *7*).

denatured state, and then to the totally denatured state where the strands separate. The transitions occur at defined temperatures, T_m1 and T_m2. The molecules at temperatures above T_m1 migrate much slower than those at temperatures below T_m1. In the narrow temperature range of the transition the molecules switch between both states reversibly and assume a mobility averaged according to the degree of transition. After staining of the nucleic acid in the gel, the band exhibits a curve representative of the conformational transition. Commercial electrophoresis units are available (Fig. 2).

Similar considerations, as outlined above for TGGE, earlier led to the development of solvent-gradient gel electrophoresis. Mainly Lerman and coworkers *(1)* for nucleic acids and Creighton *(2)* for proteins developed gel electrophoresis techniques where a concentration gradient of a denaturing solvent was established in the gel perpendicular to the electric field. These techniques produce results similar to those of TGGE. Use of a denaturing solvent, rather than temperature, means, however, that the concentration gradient must be established during gel pouring. This adds to the complexity and reduces the versatility of the gel system. A temperature gradient can be established by use of a simple device of an otherwise normal horizontal gel and leads to better consistency of results; fewer restrictions in the choice of solution conditions and easier handling of the whole procedure *(3)*.

The perpendicular gradient system so far described is useful in establishing the transition temperature by providing complete denaturation curves. This approach is impractical if larger numbers of samples are to be investigated. Using the information from pilot perpendicular gradient gel experiments, electrophoresis can be carried out with a parallel temperature gradient (*see* Figs. 3 and 4B). In this case, samples are applied to small slots and run from low to high temperature. Their movement is retarded as they reach the first transition temperature. Provided that electrophoresis is not allowed to proceed to the second transition temperature, at which point the strands separate completely, the position of the bands is characteristic for the temperature of the transition.

As the T_m1 depends on the sequence of a given stretch of DNA, TGGE can be used to distinguish between mutant and wild-type DNAs

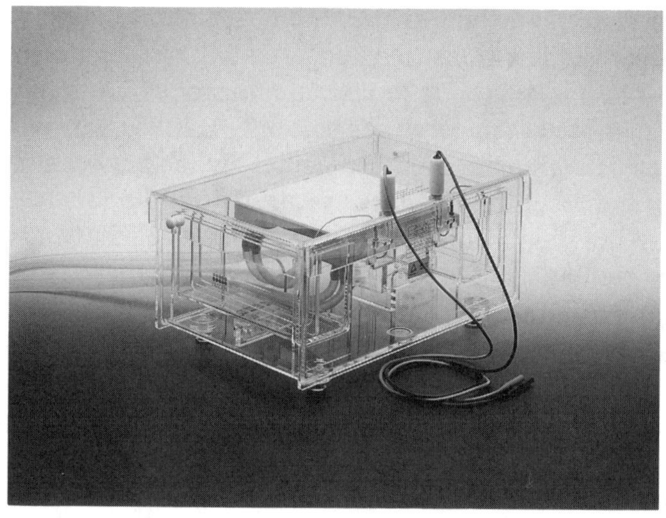

Fig. 2. Photograph of gel apparatus.

that differ by as little as a single nucleotide residue. These differences are best visualized by annealing wild-type and mutant DNAs to make a heteroduplex. The presence of the mismatch will tend to destabilize the helix, altering the T_m1 and resulting in a mobility difference during TGGE. In this chapter, in addition to the basic technique, we describe how the combination of PCR and TGGE can be used as a rapid and simple detection of point mutations in genomic DNA (*see* Fig. 3).

Although the above discussion relates to double-stranded nucleic acids, TGGE may also be used for the analysis of certain problems involving single-stranded RNA or DNA. All single-stranded nucleic acids will adopt a secondary structure at low temperature that melts as the temperature is increased. The partially melted form may have a reduced electrophoretic mobility. In some cases an increase of mobility has also been observed. Naturally occurring mutations or those introduced by, for example, site directed mutagenesis, may drastically influence the secondary structures of single-strands and hence their thermal stability and transition curves during TGGE. This approach has successfully been used to identify point mutations in transcripts *(4–6)*, and genomic sequence of various RNA viroids *(7,8)*.

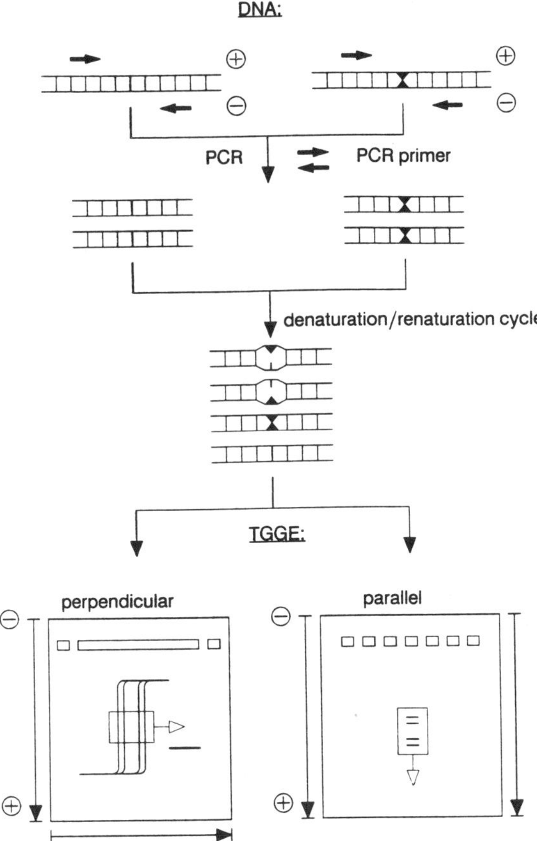

Fig. 3. Scheme for detecting mutations in genomic DNA. Samples are taken from a wild-type strain and from a mutant strain. The segments of the genomic DNA of interest are amplified by PCR, base pair exchanges are transferred into mismatches by mixing wild-type and mutant followed by the denaturation–renaturation cycle. The sample is analyzed in either perpendicular TGGE or in parallel TGGE.

In addition, TGGE increases the power of Single-Strand Conformation Polymorphism (SSCP; *9*). Whereas during conventional SSCP single-stranded conformers may only be analyzed at a single temperature, when SSCP is carried out on a temperature gradient mobility differences are visualized over a broad range of temperature, greatly improving the chance to detect a mutation (unpublished results).

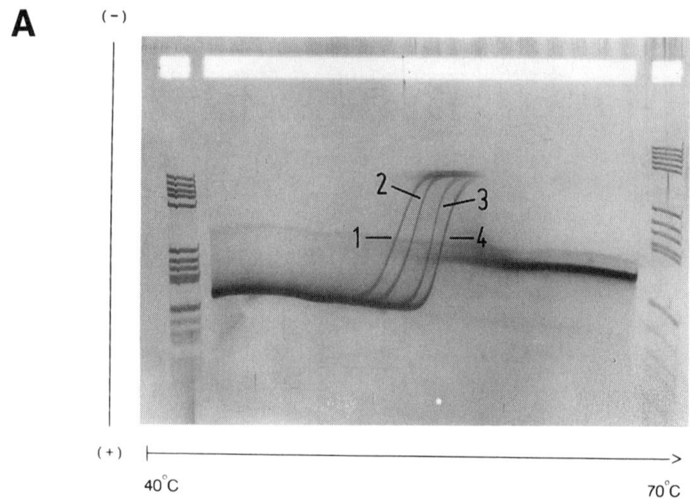

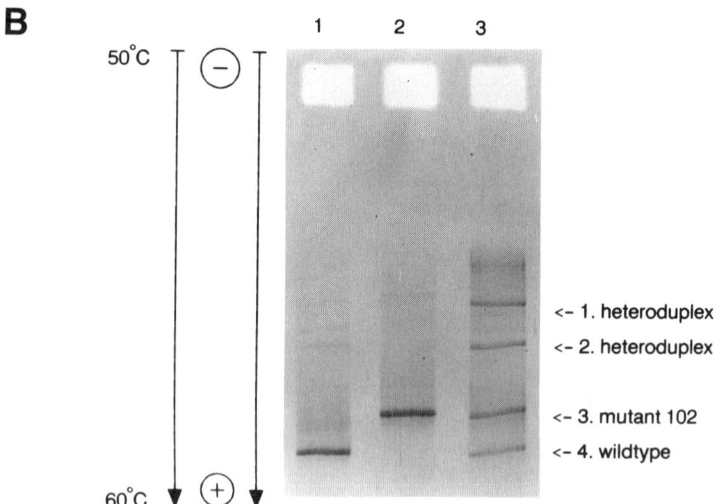

Fig. 4. Experimental analysis by TGGE of a C toT mutation in the human prion-protein in codon 102. The temperature gradient is indicated. (**A**) Perpendicular TGGE showing complete transition curves. Electrophoresis for 60 min at 300 V after establishing the temperature-gradient. (**B**) Parallel TGGE for a multisample analysis. Wild-type dsDNA *(4)*, lane 1; mutant dsDNA *(3)*, lane 2; mixture of wild-type and mutant dsDNA after a denaturation-renaturation cycle, lane 3; heteroduplex *(1)* contains the A:C mismatch, Heteroduplex *(2)* the G:T wobble-pair. Electrophoresis for 80 min at 300 V after establishing the temperature-gradient.

2. Materials

2.1. Basic Procedure

1. The TGGE gel apparatus (*see* Fig. 2): A conventional horizontal gel electrophoresis apparatus to which a linear temperature gradient is applied (*see* Note 1).

   The gradient is established by a metal plate connected to two thermostating baths with different temperatures on the opposite edges of the plate. An instrument consisting of a specially designed buffer tank and an aluminum plate electrically insulated to greater than 500 V by a sintered layer of epoxide is available commercially from Diagen (TGGE-System, Diagen GmbH, Hilden, Germany). The thermostating bath for the lower temperature should be a cryostate (e.g., Haake F3-C or Julabo F20-HC) with sufficient cooling power (at least 200 W at 20°C). For the higher temperature a small circulator, such as Haake F3-S or Julabo UC-5, is sufficient. To ensure ease of operation and control during electrophoresis, both thermostating baths should be equipped with digital temperature displays. The baths are connected to the edges of the gradient plate by large diameter water tubes shielded in neoprene tubes. To ensure accurate control of temperature the tubes should be as short as possible and the water pumped by strong pumps with a flow rate above 10 L/min. Air bubbles in tubes or channels of the gradient plate must be avoided (*see* Note 2).

   Good electrical contact between the gel and the electrodes is made through electrode contact cloths (well soaked rags) that are insulated in polyethylene foil in order to prevent drying out at a higher temperature, except where in direct contact with the gel. A number of gel compositions and buffers may be used in the TGGE system (*see* Note 3 for ones that differ from that described here).
2. Gel support films and gel casting molds (available from Diagen GmbH).
3. 10X Na-TAE buffer: 400 m*M* Tris (48.46 g/L), 1*M* NaOAc (82.03 g/L), and 10 m*M* EDTA (3.72 g/L), pH 8.4. Adjust the buffer with glacial acetic acid to pH 8.4 and autoclave. For electrophoresis dilute 50-fold.
4. Electrophoresis buffer: 0.2X Na-TAE.
5. 2X sample loading buffer: 400 µL of 10X Na-TAE, 120 µL of bromophenol blue stock solution (10 mg/mL), 120 µL of xylene cyanol FF stock solution (10 mg/mL), and 40 µL of Triton X-100; fill up to 10 mL wth autoclaved water.
6. PAA gel stock: A 30% acrylamide: 1% N'N'-methylene *bis*-acrylamide solution.

7. Gel solution for one TGGE (8% PAGE): Add 21.6 g of urea (8*M* in the gel) to 900 µL of 10X Na-TAE buffer and 12 mL of PAA gel stock (30:1). Fill up to 42 mL with distilled water. Dissolve the urea (sometimes briefly heating in a microwave oven may be necessary). Start polymerization with 75 µL of TEMED and 3 mL of 1% ammonium persulfate.
8. Buffers for silver staining: Buffer A: 100 mL of ethanol and 5 mL of acetic acid are filled up with distilled water to 1 L. Buffer B: Dissolve 1 g of $AgNO_3$ in 1 L of distilled water. One L Buffer B can be reused 5–10 times. Buffer C: Dissolve first 15 g of NaOH,and 0.1 g of $NaBH_4$ in 1 L of distilled water and then add 4 mL of formaldehyde (stock: 37% in water). This buffer must be freshly prepared immediately before use. Buffer D: Dissolve 7.5 g of Na_2CO_3 in 1 L water.

2.2. Detection of Mutations in Genomic DNA

9. Oligonucleotide primers: Appropriate primers for PCR (*see* Note 4) were synthesized on a DNA synthesizer (Applied Biosystems, Foster City, CA 380 A) and purified by PAGE. For PCR, 10 pmol/µL stock solutions were used.
10. Genomic DNA: This can be prepared using standard protocols. The concentration of DNA is determined spectrophotometrically. Store as 0.1–1 µg/µL stock at –20°C. Boil genomic DNA samples in water for 10 min and then quick-chill on ice before using in PCR.
11. Standard PCR 100 µL reaction mixture: This contains 200 µ*M* of each dNTP, 20 pmol of each primer, 20 m*M* Tris-HCl, pH 8.3, 1.5 m*M* $MgCl_2$, 50 m*M* KCl, 0.05% Tween 20, 100 µg/mL autoclaved gelatine and 1.8–2.5 U/100 µL reaction *Taq* DNA polymerase. *Taq* polymerases from different supplier shave been used successfully.
12. TNE-buffer: 100 m*M* Tris-HCl, 100 m*M* NaCl, 10 m*M* EDTA, pH 8.0. Make buffer with distilled water, adjust to pH 8.0, and autoclave.
13. Phenol: Use gloves and avoid any skin contact. Add 8-hydroxy quinoline to 0.1% (w/v) and saturate with 1*M* Tris-HCl, pH 8.0, overnight. Remove as much of the aqueous layer as possible and re-extract three times with 100 m*M* Tris-HCl, 10 m*M* EDTA, pH 8.0. Keep phenol in the dark and do not store for long periods *(10)*.
14. Phenol/chloroform: 24: 24:1 phenol:chloroform: isoamyl alcohol.
15. Chloroform: 24:1 chloroform: isoamyl alcohol.
16. 96% ethanol.
17. Hybridization-buffer: 10 m*M* Tris-HCl, 100 m*M* NaCl, 1 m*M* EDTA, pH 8.0. Make buffer with distilled water, adjust pH to 8.0, and autoclave.

3. Methods

3.1. Basic Procedure

This protocol describes a TGGE analysis with a perpendicular temperature gradient and should be used to optimize the parameters to obtain the maximum difference in mobility of different nucleic acids at the respective transition temperatures (*see* Notes 3–5).

3.1.1. Gel Casting

TGGE gels are polyacrylamide gels covalently bound to gel support films. Different slot forming plates are used for different TGGE applications: a single long slot for perpendicular TGGE and multiple slots for parallel TGGE (*see* Note 6).

1. Prior to gel casting clean the plates with distilled water, avoiding organic solvents. Put a gel support film with the hydrophobic side against the plain glass plate (*see* Note 7). Cover the hydrophilic side of the gel support film with the attached sheet of paper. Fix the gel support film by rubbing the film against the glass plate with the aid of a straight edge or another suitable device. Remove the protective paper.
2. Layer the U-shaped spacer and the slot forming plate onto the gel support film. The slot formers are oriented toward the bottom of the gel mold. Fix the gel mold with clamps.
3. Mix the gel solution and add the TEMED and ammonium persulfate to start polymerization. Transfer the gel solution into the gel mold. Avoid air bubbles.
4. Polymerize for 60 min and then remove the gel from the glass plates. To prevent leakage the gel mold may be placed into horizontal position.

3.1.2. Perpendicular TGGE Run

1. Fill each buffer tank with 1 L of electrophoresis buffer. Boil the synthetic electrode wicks in distilled water, wring, and put them into the buffer tanks. Soak the wicks with electrophoresis buffer. Spread 2 mL of 0.1% Triton X-100 dropwise onto the dry TGGE plate. Mount the gel with the gel support film facing the plate. Avoid trapping air bubbles by "rolling" the gel onto the plate. The long slot of the perpendicular TGGE gel is oriented toward the cathode (black). Establish the buffer bridges to the gel by a double layer of electrode wicks. Place one side of each wick on the gel while the other side is submerged in the buffer inside the tank. Ensure that the cathode wick does not approach the

slot closer than 5 mm, otherwise your sample might be absorbed by the wick.

2. Load 150–200 µL of sample into the long slot. Cover the gel with the glass plate. Close the safety lid and start electrophoresis at 300 V (constant voltage). Stop after 30 min.
3. Cover the gel with protective plastic wrap (SaranWrap™) to prevent evaporation (*see* Note 8). Establish the temperature gradient in the gel by switching on the thermostat and the cryostat. The standard temperature gradient suitable for most DNAs (GC content approx 50–70%) is 20°C at the left end of the TGGE plate and 60°C at the right edge (*see* Note 9). When the heating and cooling devices have reached the desired temperatures, wait a further 10 min for equilibration.
4. Once equilibrated, run the gel at 300 V for 1 h (*see* Note 10). Stop electrophoresis. Take the gel out of the TGGE-System and remove the protective foil. Carefully clean the Triton X-100 from the back of the gel before staining (*see* Note 11).
5. This protocol is designed for silver staining of DNA and RNA in polyacrylamide gels, especially TGGE gels. For all incubations shake the gel gently while submerged in approx 300 mL of the appropriate solution. Put the gel into a plastic tray. The gel support film is oriented toward the bottom of the tray. Cover the gel with buffer A and incubate for 3 min. Discard the buffer. Repeat the incubation with buffer A. Incubate in buffer B for 10 min. Restore this buffer because 1 L of buffer B may be reused for 5–10 gels. Wash twice with distilled water for 10 s. Incubate the gel in buffer C for 20 min. Discard the solution. Incubate in buffer D for 5–10 min. Discard the solution.

3.1.3. Parallel TGGE Run

Electrophoresis is carried out parallel to the temperature gradient. The standard temperature gradient is 20°C (cathode side) to 60°C (anode side). For special applications a different temperature gradient with a smaller temperature range may be favorable (*see* Note 12; for detailed handling instructions *see* Section 3.1.2.).

1. Pour a gel with multiple slots according to the protocol given under Section 3.1.1.
2. Orient the plate in the gel apparatus with the cold end toward the cathode and the heated end toward the anode side.
3. Fill each buffer tank with 1 L of electrophoresis buffer. Boil the synthetic electrode wicks in distilled water, wring, and put them into the buffer tanks. Soak the wicks with electrophoresis buffer. Spread 2 mL

of 0.1% Triton-X100 dropwise onto the dry TGGE plate. Mount the gel with the gel support film facing the plate. Avoid trapping air bubbles by "rolling" the gel onto the plate. The multiple slots of the parallel TGGE gel are oriented toward the cathode (black). Establish the buffer bridges to the gel by a double layer of electrode wicks. Place one side of each wick on the gel while the other side is submerged into the buffer inside the tank. Ensure that the cathode wick does not approach the slot closer than 5 mm. Otherwise your sample might be absorbed by the wick.

4. Cover the gel except the slots with the protective plastic foil to prevent evaporation and establish the temperature gradient in the gel by switching on the thermostat and the cryostat. When the heating and cooling devices have reached the desired temperatures, wait 10 min for equilibration.
5. Load the 6 μL samples into each slot. Cover the gel with the glass plate and loosen the safety lid. Run the gel at a constant 300 V for the desired time (*see* Note 12).
6. Stop electrophoresis and silver stain the gel as in Section 3.1.2.

3.2. Detection of Mutations in Genomic DNA

TGGE provides a powerful means with which to detect mutations in DNA from all sources. Mismatches in double-stranded DNA decrease the melting temperature, and hence lead to larger mobility shifts in the melting temperature, T_m1. By forming a hybrid between the wildtype and mutant sequences, the presence of mutations can readily be detected.

The detection of defined mutations in genomic DNA is of particular interest in genetics and in medicine. Since total genomic DNA usually cannot be analyzed directly, target sequences are amplified by polymerase chain reaction (PCR). If the genomic DNA is extracted from a cell heterozygous for a mutant allele, the mutation is detected by amplifying both alleles together, followed by a denaturation–renaturation cycle to form a hybrid and subsequent analysis by TGGE. If no wild-type copy of the target is present in the genome it can be added exogeneously. The principle of the analysis is shown in Fig. 3. The operation may first be defined by analysis of a constructed "optimal sample" (*see* Note 4).

1. Use 0.2–1 μg of boiled DNA for each PCR reaction. To prevent evaporation of liquid during thermal cycling, layer 100 μL of mineral oil on top of PCR solution. Perform 35–40 cycles of PCR using the following

temperature profile: an initial single cycle of 94°C for 180 s; followed by 33–38 cycles of 94°C (denaturation) for 60 s, 55°C (primer annealing) for 60 s, 72°C (primer extension) for 90 s, and finally a single cycle where the length of the primer extension is increased to 240 s.
2. Transfer the liquid solutions into a new Eppendorf tube. If no wild-type copy of the target is present in the genome, 0.5–1 µg of wild-type PCR product or a cloned fragment should be added to the PCR product. Add 100 µL TNE-buffer.
3. Mix the sample with an equal volume of phenol/chloroform and centrifuge for 3–5 min. Transfer the aqueous layer (top) to a new centrifuge tube and extract in a similar manner with 2 vol (300–400 µL) chloroform:isoamyl alcohol (24:1).
4. Transfer the supernatant into a screwed Eppendorf tube and precipitate the DNA with 2.5–3 vol of cold 96% ethanol. Place at –20°C for 30 min. Precipitate the nucleic acid by centrifugation at 12,000*g* for 20 min.
5. Dissolve the nucleic acid pellet in 15 µL of hybridization-buffer. It is important to redissolve the pellet in the hybridization-buffer (*see* Note 13).
6. Transfer the sample into a beaker containing boiling (>95°C) water and boil for 3 min. Remove the beaker from the heat and let it cool at room temperature for 4–5 h. This gives a slow renaturation of the DNA samples in a temperature range from 98–40°C, allowing a perfect reassociation of double strands even when the optimal temperature for renaturation is unknown (*see* Note 14).
7. Add 60 µL of autoclaved distilled water to the sample and mix with 75 µL (an equal volume) of sample loading buffer.
8. Use either all of the 150 µL sample-solution for one perpendicular TGGE or 6 µL for each slot for parallel TGGE. Proceed with the protocol as described in Section 3.1.

Figure 4 illustrates the use of TGGE to detect the C to T mutation in the human prion-protein gene at codon 102 that results in Gerstmann-Sträussler-syndrome (GSS; *11,12*).

4. Notes

1. A vertical TGGE apparatus using two gradient-forming plates, was described by Thatcher and Hodson *(13)* and later by Wartell et al. *(14)*. In our experience, the horizontal arrangement with only one metal plate has several advantages: The temperature gradient is highly reproducible; handling is easy; observation of the gel, i.e., the running of the dye-markers, is possible during electrophoresis, and the change from parallel to perpendicular gradients is easy to carry out.

2. The linearity of the temperature gradient may be confirmed experimentally by measuring the temperature over the whole plate with a calibrated thermistor.
3. The following gel and buffer systems have been applied successfully:
   a. The MOPS buffer system for analysis of dsDNA: Gel conditions: 20 m*M* MOPS, pH 8.0, 1 m*M* EDTA, 8*M* urea, 2% glycerol, 8% polyacrylamide (acrylamide:*bis* = 30:0.5), 0.17% TEMED (v/v), 0.03% (w/v) ammonium persulfate for starting the polymerization. The pH-value should be adjusted by adding NaOH to free MOPS acid. 10 m*M* MOPS instead of 20 m*M* lowers T_m by ~ 5°C, and allows higher voltage for faster migration of the nucleic acid but leads sometimes to irreversible conformational transitions of the nucleic acids. Using 4*M* urea instead of 8*M* urea (for low G:C and high number of mismatches) increases the T_m by 10–12°C. Each added mole of urea (for high G:C) lowers the T_m in general by 2.5–3°C for dsDNA and 3–4°C for dsRNA. Omission of glycerol increases the cooperativity of the transitions; higher concentrations of glycerol lead to less cooperative, i.e., broader transitions. For fragments with more than 500 base pairs use 5% polyacrylamide, and with more than 2000 base pairs use 3–4% polyacrylamide. In order to achieve a reproducible degree of polymerization, higher starter concentrations should be avoided. Electrophoresis buffer: 20 m*M* MOPS, pH 8.0, 1 m*M* EDTA.
   b. For analysis of viroids and secondary structure of RNAs: 5% polyacrylamide (acrylamide:Bis = 30:1), 17.8 m*M* Tris, 17.8 m*M* boric acid, and 0.4 m*M* EDTA (pH 8.3). Add 0.07% ammonium persulfate and 0.17% TEMED to start polymerization. To alter the transition temperatures the buffer concentration may be changed by a factor of 2 to higher or to lower concentrations. Five percent polyacrylamide with a ratio of polyacrylamide:*bis* of 30:1 is optimal for chain lengths of more than 400 nucleotides. Eight to twelve percent polyacrylamide with the same ratio of polyacrylamide:Bis should be used for chain lengths in the range of 150–300 nucleotides, up to 20% polyacrylamide for studies on oligonucleotides *(15,16)*.
   c. For analysis of secondary structures of mRNAs: 10 m*M* sodium phosphate, pH 6.0, with or without 1 m*M* $MgCl_2$. In addition amounts of 3–30 m*M* sodium acetate may be added to vary the ionic strength *(16)*.
4. The optimal sample: If the site of a mutation is known, an optimal sample may be constructed for routine diagnosis of this particular mutation. For the design several empirical rules are combined in order to obtain large shifts in TGGE:

a. The mutation should be located in that part of the amplified segment that denatures at the lowest temperature and in a continuous transition.
b. Different transitions of the amplified segment should be well resolved on the temperature scale.
c. If necessary, the highest temperature transition may be further stabilized by additional G:C base pairs (a GC clamp).

Calculation of the transition curves of DNA segments as analyzed by TGGE can be carried out with several computer programs as described in the literature (*7,17*; program for a PC available from Diagen, GmbH, Dusseldorf, Germany). Those programs are based on the algorithm of Poland *(18)*. Applying such a program the influence of every potential mutation on the transition curve may be predicted. These calculations are very helpful when determining the location of primers for PCR. Calculation of the transition curves and selection to the abovementioned rules is strongly recommended in order to achieve an up to 100% rate of detectability for unknown point mutations inside the amplified fragment. For further details of the theory see more detailed articles *(7)*.

5. The variables described below may be used to optimize the mobility differences between nucleic acids seen by TGGE:
   a. The pore size of the gel matrix is varied by the concentration of polyacrylamide. A compromise has to be found between a large change in electrophoretic mobility and an acceptable migration velocity.
   b. The electric field tends to stretch the molecules. These effects are very sensitive to the charge distribution, the conformation, and the flexibility of the molecule. The mobility changes of different conformational transitions in the same molecule exhibit different dependencies upon the electric field. High voltage (>500V), although inducing large changes, may be disadvantageous because of the large electric current.
   c. The ionic strength of the recommended buffers is related to T_m1 of the conformational transition, but cannot be varied independently from other features of TGGE. For nucleic acids one may keep in mind that as a general rule changes in electrophoretic mobility are greater in low ionic strength buffers. High ionic strength always improves the reversibility of a transition, and reversible transitions may be evaluated more easily and more accurately than irreversible or discontinuous transitions. The increase in the transition temperature owing to higher ionic strength may be compensated by the addition of urea. Each molar increase in urea concentration decreases the temperature in gereral by 2.5–3°C for double-stranded DNA.
6. The slot-forming plates have to be silanized by 7% dichlorodimethylsilane in chloroform in order to prevent sticking of the gel to the

plates. This procedure should be repeated after 10–20 runs. Keep in mind that dichlorodimethylsilane is very toxic. Therefore, all work during the silanization procedure has to be performed in a hood.

7. Gel support films normally consist of one hydrophobic (not sticking to the gel) and one hydrophilic side (sticking to the gel). Both sides can be easily distinguished by placing a drop of water on the film surface.
8. This procedure and the sealing of the electrode contact wicks in polyethylene foil provide good protection against drying out and therefore guarantee a uniform electric current. For extreme protection against evaporation at high temperatures the whole gel may be covered with an additional layer of SaranWrap™.
9. The range of the temperature gradient should be chosen carefully to ensure optimal results. The extreme temperatures of the gradient should be large enough to enclose the desired part of the transition curve but may be restricted by the limited power of the thermostating baths. For example, a temperature difference of 70°C may lead to an inconstant temperature at the low temperature side owing to insufficient heat removal. Temperatures above 80°C at the high temperature side can cause trouble because of drying out of the gel. Drying out can be avoided partially by sufficient buffer transport through the electrophoresis wicks and by shielding the gel against evaporation of water. The nearly closed Diagen TGGE-System suppresses evaporation well. If the transition temperatures of the biopolymer are too high under the conditions of the electrophoresis, a denaturing agent like urea may be added in uniform concentration throughout the gel to lower the respective transition temperatures (*see* Note 5).
10. Several experiments with TGGE with different values of the voltage between 300–500 V should be tested since different structural transitions may be detected by varying the voltage. In runs with higher ionic strength the voltage should be lowered to avoid uncontrolled heating.
11. After TGGE, the band or the whole transition curve of the biopolymer can be visualized either directly in the gel or by blotting. For blotting you should polymerize the gel to the hydrophobic side of the gel support film (not sticking to the gel). All procedures that are suitable for normal gel electrophoresis can be used after TGGE. These are staining, hybridization, and in specific cases, e.g., double-stranded RNA, antibody binding *(19)*.

    Silver staining of nucleic acids is carried out according to a method applied to conventional polyacrylamide gel electrophoresis (PAGE) *(20,21)*. Typically in perpendicular gradients 100–300 ng of a single nucleic acid species yields a well stained transition curve throughout the gel, whereas the limit of detection of a single band in parallel gradients is approx 100 pg. If a new band shows up in the analysis of a particular gene locus amplified by PCR it can be directly analyzed in sequencing by puncturing the gel even after silver staining followed by elution of the DNA and reamplification of the particular variant for direct sequencing *(22)*.

For the detection of a specific nucleic acid molecule within a mixture of other nucleic acids TGGE may be combined with hybridization. In order to transfer the nucleic acid onto the nylon-membrane, secondary and tertiary structures have to be dissociated. This denaturation has to be carried out after TGGE, of course, and may be achieved by chemical modification with glyoxal. The entire gel is incubated for 2 min in 10% ethanol, 0.5% acetic acid at room temperature for fixation of the nucleic acids and subsequently incubated in 6% glyoxal, 10% sodium phosphate, pH 6.5, at 50°C for 15 min. Transfer of the nucleic acid to the membrane and hybridization may be carried out afterward using standard protocols. A specific example for viroid RNA was described in the literature *(3)*.

12. Using the information from pilot perpendicular gradient gel experiments, electrophoresis can be carried out with a parallel temperature gradient. In this case, samples are applied to small slots and run from low to high temperature. Their movement is retarded as they reach the first transition temperature, T_m1. Provided that electrophoresis is not allowed to proceed to the second transition temperature, T_m2, at which point the strands separate, the position of the bands is characteristic for the temperature of the transition. The optimal running time for each sample has to be optimized for parallel TGGE.
13. The buffer system (200 m*M* MOPS, 4*M* urea, 5 m*M* EDTA, pH 8.0) may also be used.
14. Optimal renaturation conditions are known from hybridization experiments in solution:
    a. The ionic strength should be at least 100 m*M*. It can be achieved by 100 m*M* NaCl or 200 m*M* MOPS, pH 8.0; one sample volume of 200 m*M* NaCl or 400 m*M* MOPS should be added at the high temperature.
    b. The optimal temperature for renaturation to receive homo- and heteroduplexes is about 20°C below the average T_m value of the DNA, which is approximated according to the following formula:

$$T_m = 81.5 + 0.41 \times \%(G:C) + 16.6 \log [Na^+] - (300 + 2000 [Na^+])/L$$

where T_m is given in °C, % (G:C) is the percentage of G:C content, and L is the length of the DNA fragment in base pairs. Each increase in urea concentration lowers the T_m value by 2.5–3°C/1*M* urea for dsDNA.

c. The time $t_{0.5}$ for 50% of double-strand formation is approx

$$t_{0.5} = N \times \text{In } 2 / (3.5 \times 10^5 \times (L \times C_o)^{0.5})$$

where $t_{0.5}$ is given in seconds, N is the length of the nonrepetitive sequence (usually $L = N$), and C_o is the molar concentration of nucleotides. Renaturation should be allowed for at least $10 \times t_{0.5}$, leading to renaturation times between 15 min and several hours. A simple means of varying the hybridization time is to use different volumes of boiling water.

References

1. Lerman, L. S., Fischer, S. G., Hurley, I., Silverstein, K., and Lumelsky, N. (1984) Sequence-determined DNA separations. *Ann. Rev. Biophys. Bioeng.* **13,** 399–423.
2. Creighton, T. J. (1979) Electrophoretic analysis of the unfolding of proteins by urea. *Mol. Biol.* **179,** 235–264.
3. Rosenbaum, V. and Riesner, D. (1987) Temperature-gradient gel electrophoresis: thermodynamic analysis of nucleic acids and proteins in purified form and in cellular extracts. *Biophys. Chem.* **26,** 235–246.
4. Loss, P., Steger, G., Schmitz, M., and Riesner, D. (1991) Formation of a thermodynamically metastable structure containing hairpin II is critical for infectivity of potato spindle tuber viroid RNA. *EMBO J.* **10,** 719–727.
5. Steger, G., Baumstark, T., Mörchen, M., Tabler, M., Tsagris, M., Sänger, H. L., and Riesner, D., submitted. Structural requirements for viroid processing: correlation between processing data, thermodynamic studies, and model calculations.
6. Riesner, D. (1990) Structure of viroids and their replication intermediates. Are thermodynamic domains also functional domains? *Seminars in Virology* **1,** 83–99.
7. Riesner, D., Henco, K., and Steger, G. (1991) Temperature-Gradient Gel Electrophoresis: a method for the analysis of conformational transitions and mutations in nucleic acids and proteins, in *Advances in Electrophoresis,* Vol. 4 (Chrambach, A., Dunn, M. J., and Radola, B. J., eds.), VCH, Weinheim, pp. 169–250.
8. Zimmat, R., Gruner, R., Hecker, R., Steger, G., and Riesner, D. (1990) Analysis of mutations in viroid RNA by non-denaturing and temperature-gradient gel electrophoresis, in *Proceedings of the 6th Conversation in Biomolecular Stereodynamics,* (Sarma, R. and Sarma, M., eds.), Adenine, vol. 3, 339–357.
9. Orita, M., Iwahana, H., Kanazawa, H., Hayashi, K., and Sekiya, T. (1989) Detection of polymorphisms of human DNA by gel electrophoresis as single-strand conformation polymorphism. *Proc. Natl. Acad. Sci. USA* **86,** 2766–2770.
10. Sambrook, J., Fritsch, E. F., and Maniatis, T. (1989) *Molecular Cloning: A Laboratory Manual.* Cold Spring Harbor Laboratory, Cold Spring Harbor, NY.

11. Prusiner, S. B. (1991) Molecular biology of prion diseases. *Science* **252,** 1515–1522.
12. Hsiao, K., Baker, H. F., Crow, T. J., Poulter, M., Owen, F., Terwilliger, J. D. Westaway, D., Ott, J., and Prusiner, S. B. (1989) Linkage of prion protein missense variant to Gerstmann-Sträussler syndrome. *Nature* **338,** 342–345.
13. Thatcher, D. and Hodson, B. (1981) Denaturation of proteins and nucleic acids by thermal-gradient electrophoresis. *Biochem. J.* **197,** 105–109.
14. Wartell, R. M., Hosseini, S. H., and Moran, C. P., Jr. (1990) Detecting base pair substitutions in DNA fragments by temperature-gradient gel electrophoresis. *Nucleic Acids Res.* **18,** 2699–2705.
15. Hecker, R., Wang, Z., Steger, G., and Riesner, D. (1988) Analysis of RNA structure by temperature-gradient gel electrophoresis: viroid replication and processing. *Gene* **72,** 59–74.
16. Rosenbaum, V., Klahn, T., Lundberg, U., von Gabain, A., and Riesner, D., submitted. Co-existing structures of an mRNA stability determinant: the 5' region of the *Escherichia coli* and *Serratia marcescens* ompA mRNA.
17. Lerman, L. S. and Silverstein, K. (1987) Computational simulation of DNA melting and its application to denaturing gradient gel electrophoresis. *Meth. Enzymol.* **155,** 482–501.
18. Poland, D. (1978) *Cooperative Equilibria in Physical Biochemistry.* Clarendon, Oxford, England.
19. Schönborn, J., Oberstrass, J., Breyel, E., Tittgen, J., Schumacher, J., and Lukacs, N. (1991) Monoclonal antibodies to double-stranded RNA as probes of RNA structure in crude nuleic acid extracts. *Nucleic Acids Res.* **19,** 2993–3000.
20. Schumacher, J., Meyer, N., Riesner, D., and Weidemann, H. L. J. (1986) Diagnostic procedure for detection of viroids and viruses with circular RNAs by "return"-gel electrophoresis. *Phytopathology* **115,** 332–343.
21. Follett, E. A. C. and Desselberger, U., J. (1983) Cocirculation of different rotavirus strains in a local outbreak of infantile gastro-nenterilis: monitoring by rapid and sensitive nucleic acid analysis. *Med. Virol.* **11,** 39–52.
22. Meyer, C. G., Tannich, E., Harders, J., Henco, K., and Horstmann, R. D. (1991) Direct sequencing of variable HLA gene segments after in vitro amplification and allele separation by temperature-gradient gel electrophoresis. *J. Immunol. Meth.* **142,** 251–256.

CHAPTER 21

TGGE in Quantitative PCR of DNA and RNA

Jie Kang, Jutta Harders, Detlev Riesner, and Karsten Henco

1. Introduction

The copy numbers of DNA sequences (templates) can be determined quantitatively by a combination of PCR and temperature gradient gel electrophoresis (TGGE; *1,2;* Chapter 20) as illustrated in Fig. 1. Briefly, quantitation is performed by addition of a calibrated amount of DNA "standard" to the template that is subsequently amplified by PCR. The standard is identical to the template except for a single base substitution. Because of the identical priming sites and almost identical sequence composition, both template and standard are coamplified homogeneously; the relative amounts of each product type at any time point accurately reflect the original ratio prior to amplification. This holds true for the exponential amplification phase as well as for the stationary phase.

Subsequent to PCR, a small amount of the labeled standard is added to an aliquot of the PCR reaction. After denaturation and then renaturation, the labeled standard forms homoduplexes with amplified standard and heteroduplexes with amplified template. The ratio is identical with that of standard and template before amplification. A sample of this mixture is analyzed by TGGE with a parallel temperature gradient. Since a heteroduplex partially denatures at a lower temperature, its migration is drastically retarded in comparison to the homoduplex. After separation by TGGE, homo- and heteroduplex bands are quantified densitometrically.

From: *Methods in Molecular Biology, Vol. 31: Protocols for Gene Analysis*
Edited by: A. J. Harwood

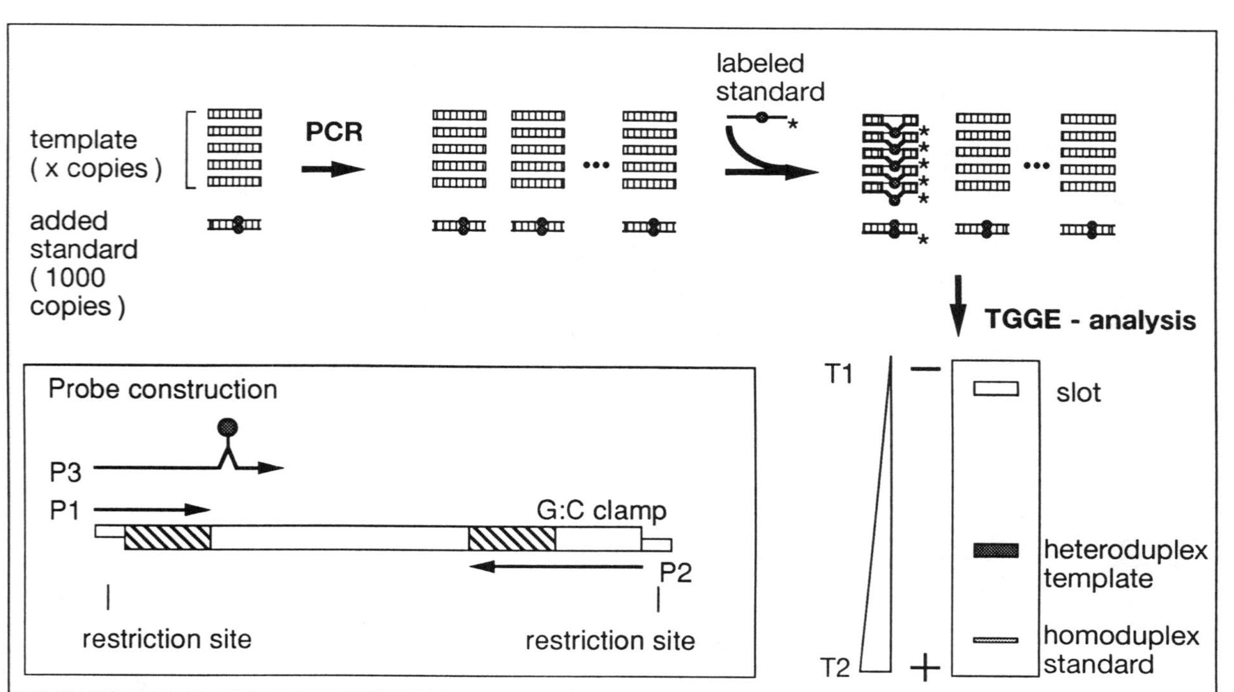

Fig. 1. Schematic protocol for PCR/TGGE quantitation. The insert shows the construction of the primers for template and standard amplification, respectively.

The protocol described in this chapter is also applicable to quantitation of RNA, such as virus RNA or mRNA, if the RNA is reverse transcribed prior to PCR. In this case a RNA runoff product is used as the standard. In addition, although a radioactive label and autoradiography is generally used, a fluorescent label may be used as an alternative.

2. Materials

2.1. Labeling of the DNA Standard

PCR primers and DNA standard: The standard DNA has to be constructed prior to the quantification experiment (*see* Fig. 2). The best way to do this is to PCR amplify a comparable DNA sequence to that under test and clone it into a suitable cloning vector. The PCR is carried out using primers that are almost identical to those for amplifying the template (P1 and P2; *see* Fig. 1 and Note 1) but substitute a third primer, P3, for P1. P3 should have an almost identical sequence to P1 except contain a single nucleotide substitution that will incorporate a point mutation 3–5 nucleotides before the 3'-end. The PCR primers should also be designed to use a different restriction enzyme site at each end, enabling the standard to be labeled. In the protocol described here we use *Eco*RI and *Sal*I sites. After preparation, dissolve in TE and store as a stock solution of 250 ng/µL at –20°C.

In addition to labeling, the DNA standard is used unlabeled in an appropriate concentration for quantitation. For this purpose, prepare a dilution series in TE-buffer. Store at –20°C.

1. 10X Buffer-E: 500 m*M* Tris-HCl, pH 8.0, 1*M* NaCl, 100 m*M* $MgCl_2$ (*see* Note 2).
2. 10X Buffer-S: 1*M* Tris-HCl, pH 7.6, 1.5*M* NaCl, 100 m*M* $MgCl_2$ (*see* Note 2).
3. Restriction enzymes: *Eco*RI, *Sal*I, 10 U/µL each (*see* Note 2). Store at –20°C.
4. dNTP Mix: dTTP, dGTP, and dCTP dissolved in TE buffer. Note the absence of the radioactive labeled dNTP, in this case dATP. Prepare as a 1 m*M* stock solution. Store at –20°C.
5. Radioactive label: In this protocol we use α-[^{32}P]-dATP (3000 Ci/mmol). Store at –20°C.
6. Klenow: The large fragment of *E. coli* DNA polymerase I. Store as a stock at 2 U/µL at –20°C.
7. Spin columns: These can either be made in the laboratory or are available commercially. We use QIAprep columns from Diagen GmbH, Hilden, Germany.
8. TE Buffer: 10 m*M* Tris-HCl, pH 8.0, 1 m*M* EDTA.

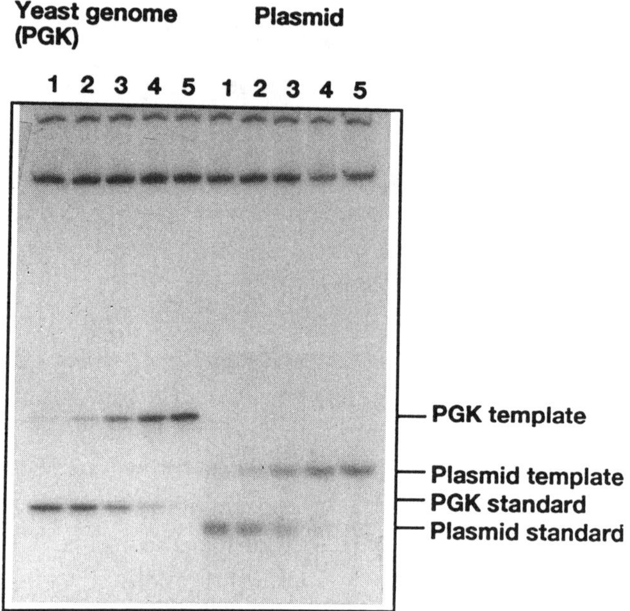

Fig. 2. Experimental data for the template quantification of plasmid copy numbers per yeast genome copy number during fermentation. The yeast genome is quantified by a PCR fragment from a single copy number gene (Phospho-glycerate-kinase, PGK); the plasmid copy number by a PCR fragment of the cloned insert of the plasmid. Gel conditions: *see* Methods. Temperature gradient: 35–60°C, 2.5 h running time. Copy numbers of PGK standard fragments: 1) 1.35×10^6; 2) 4.5×10^5; 3) 1.5×10^5; 4) 5×10^4; 5) 1.67×10^4. Copy numbers of plasmid standard fragments: 1) 1.35×10^8; 2) 4.5×10^7; 3) 1.5×10^7; 4) 5×10^6; 5) 1.67×10^6. Result: The sample contained 1.58×10^5 copies of yeast genome, 1.82×10^7 copies of plasmid. One genome copy corresponds to 115 plasmid copies.

2.2. PCR

9. PCR reaction mixture (100 µL) 10 pmol of each primer P1 and P2 (*see* Note 1); 200 µ*M* of each of the four dNTPs; 2.5 units of *Taq* polymerase; 10 m*M* Tris-HCl, pH 8.3 (at 72°C), 50 m*M* KCl, 0.1% gelatin, and 1.5 m*M* $MgCl_2$.
10. Denaturation/renaturation buffer: 8*M* urea, 400 m*M* MOPS, pH 8.0, 20 m*M* EDTA, bromophenol blue, xylene cyanol FF. Mix 4.8 g of urea, 4

mL of 50X ME buffer (*see* Section 2.2.12.), 30 µL of bromophenol blue stock solution (10 mg/mL), and 30 µL of xylene cyanol FF stock solution (10 mg/mL), fill up to 10 mL with distilled water.
11. TGGE apparatus: *see* Chapter 20.
12. 50X ME buffer: 1*M* MOPS, 50 m*M* EDTA, pH 8.0. Dissolve 293 g of MOPS (3-[N-morpholino]propanesulfonic acid), 18.6 g of EDTA·2 H_2O, 36 g of NaOH in 815 mL of distilled H_2O; this gives a total of 1 L. Filter sterilize.
13. Buffer A: 10% ethanol, 0.5% acetic acid in distilled water.
14. 2% glycerol.

3. Methods

3.1. Labeling of the Standard

1. Mix 2 µL of the DNA standard with 1.5 µL of 10X Buffer-E, 0.5 µL of *Eco*RI and 11 µL of distilled H_2O. Incubate at 37°C for 1 h and heat at 70°C for 10 min.
2. Directly add 2 µL of dNTP Mix, 1 µL α-[^{32}P]-dATP, 0.5 µL of Klenow, and 2 µL of distilled H_2O to the digested DNA. Incubate at room temperature for 1 h. Add 180 µL of TE.
3. Remove the unincorporated dNTPs from the labeled DNA by using standard methods or QIAprep columns according to the manufacturer's instruction (Diagen GmbH, Hilden, Germany).
4. Add 22 µL of 10X Buffer-S and 1 µL of *Sal*I (10 U). Incubate at 37°C for 1 h and store at –20°C.

3.2. PCR

1. Mix the template DNA (of unknown copy number) with a defined amount of unlabeled standard DNA (*see* Note 4) in a 100 µL PCR reaction. Denature for 5 min at 93°C and then PCR amplify for 35 cycles as follows: 92°C for 1 min; 55°C for 1 min; 72°C for 1 min.
2. Transfer 4 µL of the PCR mixture into a microfuge tube. Add 1 µL of labeled standard and 5 µL of denaturation/renaturation buffer.
3. Denature by heating at 98°C for 5 min and then renature by cooling at 50°C for 15 min.
4. Cast a parallel TGGE using a MOPS/EDTA buffer system, as described in Chapter 20. Set up a parallel temperature gradient in the TGGE system. Gradients routinely used for PCR/TGGE quantification are: 35°C (cathode) to 60°C (anode) and 30–60°C (*see* Note 5).
5. Load 6 µL of each sample and run a parallel TGGE gel for approx 2.5 h at 300 V.
6. Stop the electrophoresis and remove the gel. Incubate in buffer A for 15 min.

7. Incubate the gel with 2% glycerol for 10 min, cover with a sheet of cellophane, and dry at 50°C for 3 h (*see* Note 6).
8. Expose the gel to a Kodak X-ray film and evaluate the developed film densitometrically.

The initial template copy number is easily calculated according to the following formula:

template copy number = (intensity heteroduplex/ intensity homoduplex) × number of initial standard copies

4. Notes

1. Selection of the primer P1, P2: Preferably, a segment of 100–300 bp is chosen as target for the amplification using primers P1 and P2. Optimal results are obtained if P1 is located in a region of significant higher A:T-content compared to the region of P2 in order to generate a fragment with two independent melting domains. The artificial introduction of a 20–30 bp G:C-clamp by means of primer P2 is recommended if no naturally coded G:C-clamp can be used.
2. For radioactive or fluorescent labeling, we prefer to construct our standard DNA with a restriction site as a 5' overhang, such as *Eco*RI or *Sal*I that can easily be labeled by a filling-in reaction with α-labeled nucleotide triphosphates *(3)*. To use other restriction sites, appropriate changes must be made to 10X Buffer-E and 10X buffer-C. In the situation described in this protocol, 10X Buffer-E and 10X Buffer-S are the 10X concentrated enzyme buffers for *Eco*RI and *Sal*I, respectively.
3. Usually, a specific activity of 10^6 cpm ^{32}P/pmol end is achieved. The resulting volume of the probe (approx 220 µL) is sufficient for 220 TGGE samples. The DNA concentration of the probe is approx 1 ng/µL.
4. The best results are obtained if the ratio of copy number template: copy number standard is approx 1:1, thus for exact quantitation a dilution series of standard DNA is recommended. One particular standard concentration allows a quantitative evaluation with an error deviation of <20% if the evaluated signal intensities differ not more than a factor of 10. As a consequence one standard concentration allows quantification within a range of two orders of magnitude.
5. Since the DNA fragments selected for quantification of different templates vary in length and thermal stability, their mobility in the temperature gradient gel may be different. The optimal separation of the heteroduplex and the homoduplex should be determinated empirically. Ready to use protocols are available from Diagen, Hilden, Germany, for the quantitation of CMV-DNA, human β-globin DNA (single

copy gene marker), β actin mRNA, TNF alpha mRNA, IL2 mRNA, and others.

6. The cellophane layer should also be incubated in 2% glycerol for 1–2 min. Avoid air bubbles between cellophane and gel. For easier handling place the gel on top of a glass plate. Fix the cellophane layer with clamps at the glass plate.

References

1. Henco, K. and Heibey, M. (1990) Quantitative PCR: the determination of template copy numbers by temperature gradient gel electrophoresis (TGGE). *Nucleic Acids Res.* **19,** 6733–6734.
2. Kang, J., Immelmann, A., Welters, S., and Henco, K. (1991) Quality control in the fermentation of recombinant cells. *Biotech. Forum Europe* **10,** 590–593.
3. Sambrook, J., Fritsch, E. F., and Maniatis, T. (1989) *Molecular Cloning: A Laboratory Manual.* Cold Spring Harbor Laboratory, Cold Spring Harbor, NY.

CHAPTER 22

The PGK–PCR Clonality Assay (PPCA)

Lambert Busque and D. Gary Gilliland

1. Introduction

The clonal analysis of cell populations in neoplastic disorders and in solid tumors is an important tool in understanding the origin and progression of disease. The use of glucose-6-phosphate dehydrogenase (G6PD) isozymes in females *(1)* has proven to be extremely informative in these studies *(2)* but its use is limited by the low frequency of G6PD polymorphisms in females. X-linked restriction fragment length polymorphisms (RFLP) of both the phosphoglycerate kinase gene (PGK) and the hypoxanthine phosphoribosyl transferase gene (HPRT), used in conjunction with methylation sensitive enzymes that distinguish the active from inactive X chromosome, has dramatically increased the number of informative samples that can be analyzed for clonality *(3)*. However, RFLP analysis may be limited by the quantity of DNA available. The polymerase chain reaction (PCR) can obviate this problem. The PGK–PCR Clonality Assay (PPCA; *4,5*) can be utilized for clonality determination on as few as 100 cells. PPCA opens the way to the analysis of cells isolated by fluorescence activated cell sorting (FACS), as well as isolated hematopoietic colonies grown in semisolid media, such as methylcellulose. PPCA can also facilitate the study of clonality of blood elements in leukopenic states or any situation where the quantity of DNA is limited. Therefore, there is a broad scope of application for PPCA. The technique is particularly valuable in the study of acquired clonal hematopoietic disorders where the analysis of hematopoietic colonies or FACS sorted cells may give new insight in the pathogenesis of malignancy.

From: *Methods in Molecular Biology, Vol. 31: Protocols for Gene Analysis*
Edited by: A. J. Harwood Copyright ©1994 Humana Press Inc., Totowa, NJ

The principles used in the PPCA are similar to those used in RFLP clonality studies where:

1. In informative heterozygotes, the maternal X chromosome can be distinguished from the paternal X chromosome by RFLP;
2. During female embryogenesis, there is random inactivation of one X chromosome in each cell *(6)* and the pattern of X inactivation is faithfully maintained in all progeny of that cell *(7)*, even after malignant transformation;
3. Inactivation of the X chromosome is accompanied by changes in methylation patterns *(8)* that can be delineated with methylation-sensitive restriction endonucleases, such as *Hpa*II.

In a similar way, PPCA relies on a polymorphic *Bst*XI site at the PGK locus to differentiate maternal from paternal X chromosomes (Fig. 1). The polymorphic *Bst*XI site has been found to be present on 34% of the 450 females screened in our laboratory. To differentiate the active from the inactive X chromosome, PPCA takes advantage of two variably methylated cytosine residues close to the polymorphic *Bst*XI site. At the PGK locus, the two *Hpa*II sites shown in Fig. 1 are invariably methylated on the inactive X chromosome and unmethylated on the active one *(9)*.

PPCA is a rapid and easy technique to perform in a molecular biology laboratory. The method consists of digesting DNA with *Hpa*II prior to amplification with primers 1A and 1B that flank the differentially methylated *Hpa*II sites and the polymorphic *Bst*XI site. *Hpa*II cleavage precludes amplification from the active X chromosome since the uncut, methylated allele will be the only product amplified. Digestion of amplified DNA with *Bst*XI will then differentiate the maternal from the paternal X chromosome.

The possible results are shown schematically in Fig. 2. There are two types of results for a clonal population of cells. In type 1 clonally derived cells, the *Bst*XI site is absent on all inactive X chromosomes, and PPCA gives only one band of 530 bp after *Bst*XI digestion. In type 2 clonally derived cells, the *Bst*XI site is present on all inactive X chromosomes, and PPCA will give a smaller 433-bp band after *Bst*XI digestion. In a polyclonal population both the type 1 and type 2 cell populations coexist, and both the 530- and the 433-bp alleles are present in equal proportions reflecting random Lyonization in the population. In practice, however, because of the formation of hetero-

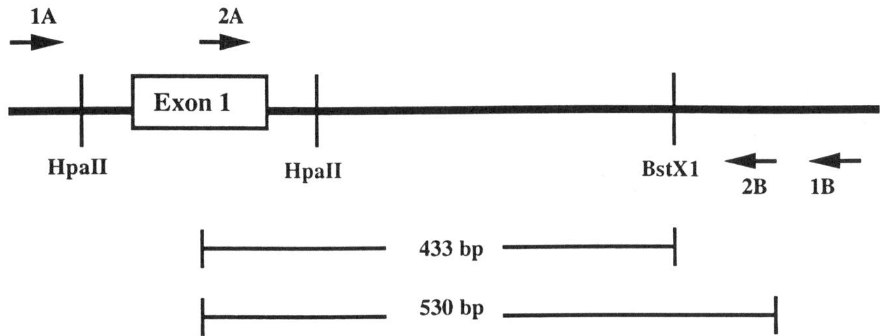

Fig. 1. PGK locus. *Bst*XI site is present on 34% of tested females. The *Hpa*II sites are unmethylated on active X chromosome and methylated on the inactive one. Amplification is carried with primer 1A and 1B and nested with primers 2A and 2B. Postcutting with *Bst*XI will give a band of 433 bp for the allele with the *Bst*XI site and 530-bp band for the allele without the *Bst*XI site.

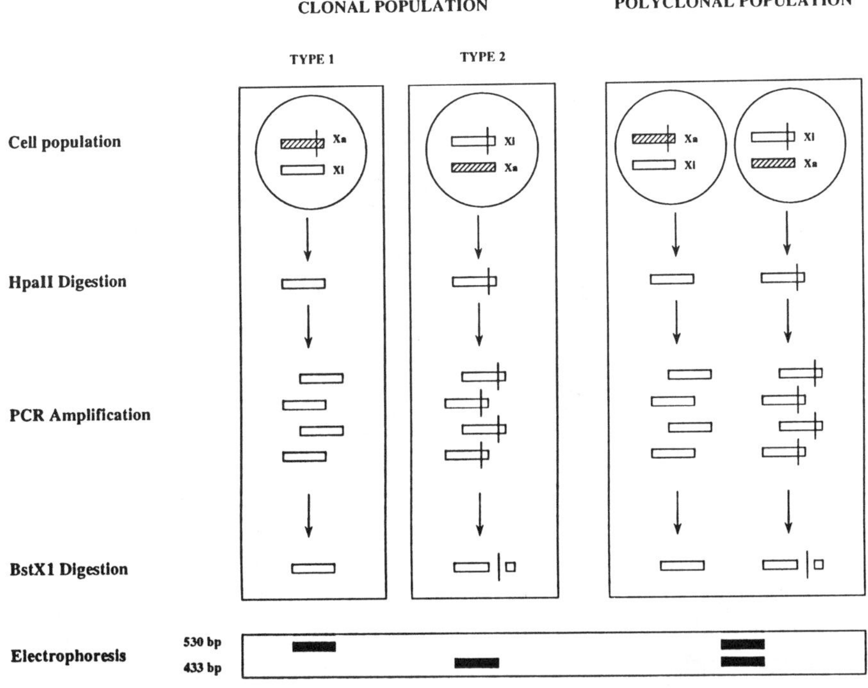

Fig. 2. Principle of PPCA and possible results for clonal and polyclonal cell populations (*see* text for comments). Xa: active X chromosome. Xi: inactive X chromosome.

duplexes between the two alleles during the PCR, ratios are shifted 3:1 in favor of the 530-bp band.

The interpretation of PPCA results for both clonal and polyclonal populations of cells is straightforward. In those circumstances, it is not necessary to analyze somatic tissue to determine the extent of Lyonization. However, the study of oligoclonal populations merit special considerations. A deviation from the 3:1 band ratio should not automatically be interpreted as evidence for partial clonal involvement since the Lyonization ratio can vary significantly from the theoretical equal distribution of the two alleles *(1,10,11)*. A PPCA reference of somatic tissue ratio must be established for an individual female that then can be compared to the specimen of interest. The PPCA protocol may easily be adapted to give more precise evaluations of the band ratios by either using densitometry or addition of trace labeled α-[^{32}P]dCTP to the PCR mix and quantification of each band by scintillation counting.

The power to study small populations of cells and the rapid and easily interpretable results makes PPCA one of the most useful techniques to study clonality of human tissue.

2. Materials

2.1. DNA Isolation

1. PBS: Phosphate buffered saline. 140 m*M* NaCl, 2.6 m*M* KCl, 10 m*M* Na_2HPO_4, 1.8 m*M* KH_2PO_4, pH 7.4. Store at 4°C.
2. IMDM: Iscove's modified Dubecco's medium. Store at 4°C. PBS, RPMI, MEM, or other isotonic media may be substituted.
3. Ficoll-Paque: Pharmacia (Piscataway, NJ), store at 4°C.
4. IMDM–20% Fetal calf serum (FCS): Store at 4°C.
5. Lux plates: Tissue culture dishes (100 × 15 mm).
6. Triton X-100 lysis buffer: 0.32 mol/L sucrose, 10 m*M* Tris-HCl, pH 7.5, 5 m*M* $MgCl_2$, 1% by vol Triton X-100. Store at 4°C.
7. Proteinase K buffer: 75 m*M* NaCl, 24 m*M* EDTA. Store at room temperature.
8. Proteinase K/SDS: 2 mg/mL proteinase K, 5% SDS. Two-milliliter aliquots stored at –20°C.
9. Ammonium salts (NaOAc): 3*M* NaOAc; 3 mol/L. Store at room temperature.
10. Tris-EDTA (TE) buffer: 10 m*M* Tris-HCl, 1 m*M* EDTA, pH 7.4. Store at room temperature.

2.2. BstXI Heterozygosity Screening

All reagents should be made in dH_2O and free of contaminating DNA.

11. 10X PCR buffer: 500 m*M* KCl, 100 m*M* Tris-HCl pH 8.3, 15 m*M* $MgCl_2$, 0.01% gelatin. One-milliliter aliquots are stored at –20°C.
12. dNTPs: 12.5X stock solution: 2.5 m*M* of each in dH_2O. One-milliliter aliquots stored at –20°C. May be prepared from 0.1*M* stocks.
13. Primers: Synthetic primers are made using 1.0-μ*M* columns. Primers are purified over NAP-10 columns in dH_2O according to manufacturer's protocol and diluted to 5 pmol/μL. Aliquots of 0.5 mL are stored at –20°C. The following primers are used for amplification:
    1A: 5' -CTGTTCCTGCCCGCGCGGTGTTCCGCATTC- 3'
    1B: 5'-ACGCCTGTTACGTAAGCTCTGCAGGCCTCC- 3'
    2A: 5'-AGCTGGACGTTAAAGGGAAGCGGGTCGTTA- 3'
    2B: 5'-TACTCCTGAAGTTAAATCAACATCCTCTTG- 3'
14. *Taq* polymerase: Amplitac (250 U/50 μL) is purchased from Perkin-Elmer/Cetus (Norwalk, CT) and stored at –20°C.
15. *Bst*XI change buffer: 2.5*M* NaCl; 0.125*M* $MgCl_2$.
16. *Bst*XI: *Bst*XI at 10 U/μL. Store at –20°C.
17. *Hpa*II: *Hpa*II at 40 U/μL. Store at –20°C.
18. α-[^{32}P] dCTP: Labeled dCTP with a specific activity of 3000 Ci/mmol.

3. Methods

3.1. DNA Processing

A number of different methods of DNA extraction are employed depending on the source.

1. Single hematopoietic colony: Pluck a single colony with a drawn glass microcapillary. Suspend in 100 μL of H_2O and boil for 3 min. Centrifuge at 14,000*g* in a microcentrifuge for 5 s to bring down lysed cells. Use the supernatant as the DNA source.
2. FACS sorted cells: Transfer the cells to a 1.5-mL plastic microfuge tube. Centrifuge cells for 5 s at 1000*g* in a microcentrifuge. Wash two times with PBS. Resuspend in 100 μL of dH_2O. Boil 3 min and centrifuge for 5 s at 14,000*g*.
3. Buccal epithelial cells (*see* Note 1): Buccal epithelial cells can be recovered easily with a cotton swab stick. Cells are rinsed from the stick to a 1.5-mL microfuge tube with PBS. Proceed as described in Section 3.1.2.
4. Peripheral blood:
    a. Dilute peripheral blood samples with an equal amount of IMDM. Layer over Ficoll-Paque; centrifuge at 250*g* for 40 min. Recover the

monolayer; wash the cells three times with IMDM. Resuspend in 5 mL of IMDM–20% FCS.

b. Plate the cells on a Lux dish; incubate at 37°C in a CO_2 incubator overnight. Recover the nonadherent cells (lymphocyte rich) in a 15-mL conical centrifuge tube. Centrifuge cells at 250*g* for 10 min, discard supernatant (*see* Note 2).

c. Add 15 mL of the Triton X-100 lysis buffer. Centrifuge at 600*g* for 10 min, discard supernatant, and add 2.25 mL of proteinase K digestion buffer and 250 µL of proteinase K/SDS. Incubate 24–48 h at 37°C in a water bath.

d. Extract with an equal volume of phenol, followed by 2 extractions with chloroform:isoamylalcohol 24:1. Precipitate with 1/10 vol of ammonium salts and 2 vol of absolute ethanol. Resuspend in TE buffer.

3.2. BstXI Heterozygosity Screening

Since not all females are heterozygous for the *Bst*XI polymorphism in this section we describe a rapid screen for finding the informative females.

1. In a 0.5-mL microfuge tube add: 100 ng–1 µg of genomic DNA (2 µL of extracted DNA or 10 µL of cell lysate supernatant), 10 µL of 10X PCR buffer, 8 µL of dNTP, 4 µL of primer 1A, 4 µL of primer 1B, 0.2 µL of *Taq* polymerase, and H_2O to 100 µL total volume (*see* Note 3). Overlay with 50 µL (2 drops) of mineral oil.
2. Denature at 94°C for 3 min, then perform 50 cycles of PCR with the following parameters: 1 min at 94°C; 1 min 45 s at 58°C; 2 min 45 s at 72°C.
3. To carry out a nested amplification (*see* Note 4), add to a fresh microfuge tube the following: 5 µL of the first PCR reaction; 9.5 µL of 10X PCR buffer; 4 µL of primer 2A; 4 µL of primer 2B; 8 µL of dNTP; 0.2 µL of *Taq* polymerase, and 69.5 µL of H_2O. Amplify as in step 2.
4. To verifiy the amplification, run 15 µL of each PCR reaction on 2% agarose gel containing 50 µg/100 mL of ethidium bromide at 12 V/cm^2 (*see* Note 5).
5. Then digest positives samples with *Bst*XI restriction enzyme by adding 1 µL of *Bst*XI change buffer and 1 µL of *Bst*XI to 23 µL of the PCR reaction. Digest at 45°C for 4–12 h.
6. Run the 25 µL digestion on a 2% agarose gel as in step 4. Informative heterozygotes will have both the 530 and 433 bp bands (*see* Fig. 3, lane 2).

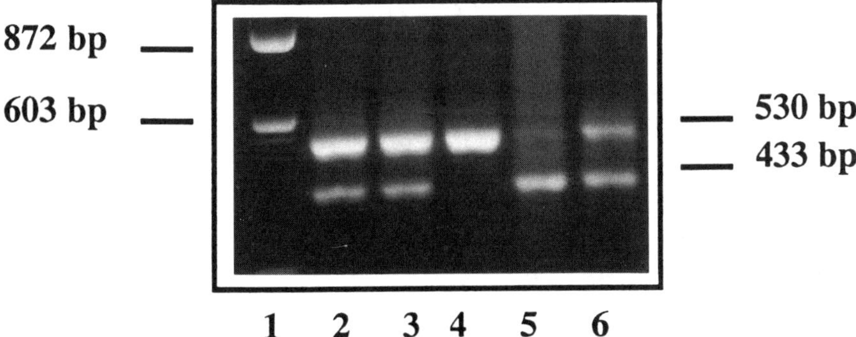

Fig. 3. Two percent agarose gel of different PPCA assays. Lane 1: molecular markers: ϕ X174/*Hea*III, λ/*Hind*III. Lane 2: *Bst*XI screening (DNA is not precut with *Hpa*II) of an heterozygote female. Lane 3: PPCA (DNA precut with *Hpa*II) of a normal heterozygote female. Lane 4: PPCA of a type 1 clonal population of cells. Lane 5: PPCA of a type 2 clonal population of cells. Lane 6: PPCA of type 2 oligoclonal population of cells, ratio between the 530 bp and 433 bp is 1:2 instead of 3:1.

3.3. Clonality Assay

The clonality assay involves precutting the DNA of known *Bst*XI heterozygotes with *Hpa*II, followed by the same amplification as for the *Bst*XI screening. In addition to the samples under test, both positive and negative controls should be included (*see* Notes 6 and 7).

1. Prepare two tubes for each informative female; one will serve as an auto-control. To each microfuge tube add: genomic DNA (2 µL of extracted DNA or 10 µL of cell lysate supernatant); 2 µL of PCR 10X buffer and H_2O to 19.5 µL total volume. Add 0.5 µL of *Hpa*II to the experimental sample but omit from the control tube. Digest at 37°C for 4–12 h (*see* Note 8).
2. Centrifuge the tubes for 1 s at 14,000 rpm in a microfuge. Heat microfuge tubes at 100°C for 4 min to inactivate the *Hpa*II (*see* Note 9). Repeat the centrifugation process.
3. Prepare a PCR master mix that contains: 8 µL of 10X PCR buffer, 8 µL of dNTP, 4 µL of primer 1A, 4 µL of primer 1B, 0.2 µL of AmpliTaq, and 56 µL of dH_2O per reaction. Add 80 µL to each microfuge tube from step 3.3.2.
4. Amplify as described for the *Bst*XI screening (steps 3.2.2.–3.2.6.).

3.4. Band Quantification

To quantify the ratio of the 530- and 433-bp bands, proceed with the amplification as describe in Section 3.3. but add 5 µL of α-[^{32}P] dCTP to the PCR mix at the nested amplification step (3.2.2.). After migration in agarose, cut the two bands individually, transfer to scintillation vials, and count. Alternatively, the agarose gel may be dried, autoradiographed, and the band intensities measured with a densitometer (*see* Note 10).

4. Notes

1. Epithelial cells recovered from the buccal mucosa provide enough DNA to permit PPCA. This is a less invasive way to study Lyonization status than skin biopsy. Hair follicles may be used as well, however, multiple follicles from different sites on the scalp are necessary. Any single follicle may be embryologically derived from a single cell, and will not reflect true Lyonization of somatic tissue.
2. For monocytes, scrape the plate to recover adherent cells and proceed as for lymphocytes. For polymorphonuclear cells, lyse the erythrocyte phase using Triton X-100 buffer in a 20:1 ratio with the erythrocyte fraction. Repetitive phenol extraction may be necessary to eliminate hemoglobin degradation products that inhibit *Taq* polymerase.
3. For multiple samples we prepare a master mix to minimize contamination.
4. The use of nested primers is essential for small amounts of DNA (e.g., colonies or FACS sorted cells). When using larger quantity of highly purified DNA, only one amplification, with primers 2A and 2B, is necessary.
5. Even in the best conditions working with sufficient quantities of highly purified DNA, the amplification of the PGK locus may be difficult. We have tested more than 20 primer pairs in this region before choosing the optimal primers that we described in this chapter. Nonetheless, we have found amplification of PGK from genomic DNA to be more difficult than for other genes. In general, only 50–70% of DNA samples amplify. If a sample does not amplify, reextract and precipitate the DNA. After this further treatment our overal is success rate increases to between 80–90%. When working with FACS sorted cells or impure DNA, the amplification is more difficult; inclusion of dimethylsulfoxide to 5% in the PCR reaction has overcome these problems.
6. For positive controls, we use DNA from a type 2 clonal patient and from a male hemizygote for the *Bst*XI site. Clonal DNA controls for the efficiency of the *Hpa*II precutting, whereas the male DNA is a good control for the *Bst*XI postcutting.

7. To monitor contamination, we concurrently run several water controls. For the clonality assay we use two water controls, one containing *Hpa*II and one without *Hpa*II. Water controls are nested like the other samples. We also start another water control with the second amplification. The three water controls will help to pinpoint the source of contamination:

*Hpa*II	*Hpa*II -	Second amplification	Possible contamination
+	+	+	Water, dNTP, buffer, Primer 2A or 2B
+	–	–	*Hpa*II
+	+	–	Primer 1A or 1B

8. The *Hpa*II digestion must be carried out prior to amplification since PCR removes cytosine methylation.
9. Heat inactivation of *Hpa*II is important. If *Hpa*II activity is present during the first ramp to denaturing temperature, then inactive alleles can be digested from the lack of fidelity of methylation during the amplification process.
10. We have shown using mixing experiments with clonally derived DNAs that the PPCA can detect a clonal population of cells that comprise 20% of the total *(4)*.

Acknowledgments

L. Busque is supported by the Samuel Mclaughlin Foundation. Our research is supported in part by a grant from the American Cancer Society.

References

1. Fialkow, P. (1973) Primordial cell pool size and lineage relationships of five human cell types. *Ann. Hum. Genet.* **37,** 39–48.
2. Raskind, W. and Fialkow, P. J. (1987) The use of cell markers in the study of human neoplasia. *Adv. Cancer Res.* **49,** 127–167.
3. Vogelstein, B., Fearon, E. R., Hamilton, S. R., and Feinberg, A. P. (1985) Use of restriction fragment length polymorphisms to determine the clonal origin of human tumors. *Science* **227,** 642–645.
4. Gilliland, D. G., Levy, J., Perrin, S., Blanchard, K., and Bunn, H. F. (1989) Determination of clonality in myeloproliferative diseases using polymerase chain reaction. *Clin. Res.* **37,** 601A.
5. Gilliland, D. G., Blanchard, K., Levy, J., Perrin, S., and Bunn, H. F. (1991) Clonality in myeloproliferative disorders: analysis by means of the polymerase chain reaction. *Proc. Natl. Acad. Sci. USA* **88,** 6848–6852.
6. Lyon, M. F. (1988) The William Allan Memorial Award address: X-chromosome inactivation and the location and expression of X-linked gene. *Am. J. Hum. Gen.* **42,** 8–16.

7. Riggs, A. D. and Pfeifer, G. (1992) X-chromosome inactivation and cell memory trends. *Genetics* **8,** 169–174.
8. Grant, S. G. and Chapman, V. M. (1988) Mechanisms of X-chromosome regulation. *Annu. Rev. Genet.* **22,** 199–233.
9. Pfeifer, G. P., Tanguay, R. L., Steigerwald, S. D., and Riggs, A. D. (1990) In vivo footprint and methylation analysis by PCR-aided genomic sequencing: comparison of active and inactive X-chromosome DNA at CpG island and promoter of human PGK-1. *Gene Dev.* **4,** 1277–1287.
10. Vogelstein, B., Fearon, E. R., Hamilton, S. R., Preisinger, A. C., Huntington, F. W., Michelson, A. M., Riggs, A. D., and Orkin, S. H. (1987) Clonal analysis using recombinant DNA probes from the X-chromosome. *Cancer Res.* **47,** 4806–4813.
11. Gale, R. H., Wheadon, H., and Linch, D. C. (1991) X-chromosome inactivation patterns using HPRT and the PGK polymorphisms in haematologically normal and post-chemotherapy females. *Br. J. Haem.* **79,** 193–197.

CHAPTER 23

Direct Sequencing of PCR Products

Geraldine A. Phear and Janet Harwood

1. Introduction

The development of the polymerase chain reaction (PCR) has allowed the rapid isolation of DNA sequences utilizing the hybridization of two oligonucleotide primers and subsequent amplification of the intervening sequences by *Taq* polymerase. There are many applications of this technique. One of the most useful is the screening of large numbers of samples in the search for mutations at a defined locus, for example in clinical studies or in the analysis of cultured cell lines *(1–3)*. In the absence of PCR this can only be achieved by isolating DNA from each individual and making and screening a library. The use of PCR means that whereas previously it may have taken a month to examine each individual sample it is now possible to examine many samples in a few days. In addition, by using redundant primers it is even possible to isolate related novel genes.

To make full use of this technique it is necessary to be able to sequence the amplified fragments rapidly. There is, however, a major technical problem encountered when sequencing PCR products. The oligonucleotide primers used for the PCR reaction also act as primers in the sequencing reaction. This makes the sequencing gels unreadable because of the overlaying of multiple sequences. A number of methods have been used to address this problem. Initially, PCR products were cloned into a vector suitable for sequencing *(4)*. This, however, has two disadvantages: It is time consuming and each clone represents only a single molecule from the population of PCR prod-

From: *Methods in Molecular Biology, Vol. 31: Protocols for Gene Analysis*
Edited by: A. J. Harwood

ucts, therefore several clones must be sequenced to avoid artifacts owing to the infidelity of *Taq* polymerase. By sequencing the PCR product directly these errors are minimized since the template comprises a sample of the whole population of molecules generated by the PCR reaction.

Two general approaches to direct sequencing have been taken. First, the double-stranded template may be purified from the oligonucleotide primers. A number of methods have been devised to do this. They include gel purification and isolation by the use of size exclusion of DNA molecules using materials such as GeneClean™ or Centricon™ columns. Second, the double-stranded PCR product can be used as a template to make single-stranded DNA. This can be achieved by asymmetric PCR *(5,6)*. Alternatively, the PCR reaction may be modified so that one of the primers is biotinylated and then the single-stranded template can be purified from it by binding the biotinylated strand to streptavidin coated magnetic beads *(7)*. Although many of these methods have been used successfully, they all involve additional manipulations to the PCR template prior to sequencing.

The protocol presented here is a simple and rapid way of sequencing PCR products directly. It is based on two previously described methods *(4,8)*. It is ideal for sequencing PCR products that are less than 2 kb in length, where an internal sequencing primer is available. The double-stranded template is denatured in the presence of an end-labeled primer by boiling, then snap-chilled in ice/water. This prevents reassociation of the template, favoring primer-template annealing. Subsequent sequencing is a modification of the dideoxy-nucleotide sequencing technique *(9)*.

The end-labeled oligonucleotide used as a sequencing primer is situated internally to the PCR primers, i.e., is a "nested" primer. This has several advantages.

1. There is no need to remove the PCR primers from the template. The products of the sequencing reaction primed by the excess of PCR primers will not be detected since the PCR primers are not radioactively labeled. This minimal purification of the template material reduces loss of material and is sufficient for sequencing purposes.
2. The sequencing primer is a third oligonucleotide which is "nested" to the PCR primer and specific to the desired PCR product. It will not

anneal to nonspecific PCR products, and therefore sequence specificity is maximized.

3. The sensitivity of the technique is such that only 150 ng of template is required per sequencing reaction.

The one drawback of this protocol is the requirement of a nested sequencing primer. The method, however, can be adapted to sequence PCR products using one of the PCR primers, provided the oligonucleotides used in the PCR reaction are removed from the template first.

This protocol is optimized to sequence a PCR product of approx 250 bp using a 20-mer as a primer. It is applicable for different lengths of PCR products and primers, however, the quantity of material used must be adjusted to maintain the molar ratio of 1 pmol of template to 3 pmol of primer (*see* Note 1).

2. Materials

All solutions should be made to the standard required for molecular biology. Use molecular biology grade reagents and sterile distilled water. The reagents for sequencing are available commercially as kits.

2.1. End-Labeling Oligonucleotide Primers

1. γ-[^{32}P] ATP: 5000 Ci/mmol, Amersham (Arlington Heights, IL). This is a high energy β emitter, therefore adequate safety precautions must be taken (*see* Note 2).
2. 10X polynucleotide kinase buffer: 0.5*M* Tris-HCl, pH 7.6, 0.1*M* $MgCl_2$, 50 m*M* DTT, 1 m*M* Spermidine,1 m*M* EDTA, pH 8.0. Store in aliquots at –20°C.
3. Nested oligonucleotide primer specific to PCR sequence: 0.1 µg/µL in water (*see* Note 3).
4. Polynucleotide kinase (10,000 U/mL, New England Biolabs, Beverly, MA).
5. TE: 10 m*M* Tris-HCl, 1 m*M* EDTA, pH 8.0.
6. Sephadex G50 (Pharmacia, Piscataway, NJ) column in TE (*see* Note 4).

2.2. Purification and Sequencing of the PCR Product

7. 3*M* Sodium acetate, pH 5.2.
8. 100% Ethanol.
9. 5X Sequenase buffer: 200 m*M* Tris-HCl, pH 7.5, 100 m*M* $MgCl_2$, 250 m*M* NaCl.
10. 0.1*M* DTT. Store at –20°C.

11. Labeling mix: A 7.5 µ*M* solution of each deoxynucleotide triphosphate (dNTP). To make up the stock labeling mix:

1 m*M* dATP	7.5 µL
1 m*M* dCTP	7.5 µL
1 m*M* dGTP	7.5 µL
1 m*M* dTTP	7.5 µL
water	970 µL

Store at –20°C. Dilute eightfold in water for the working labeling mix.
12. Sequenase (Version 2.0 USB, 14 U/µL). Dilute eightfold to 1.75 U/µL in cold TE just before use.
13. Termination mixes: A termination mix is required for each nucleotide (i.e., A, C, G, and T). These comprise solutions of 80 µ*M* of each dNTP, 8 µ*M* of the respective dideoxynucleotide (ddNTP) in 50 m*M* NaCl.
    The termination mixes can be made as follows:

	A mix	C mix	G mix	T mix
10 m*M* dATP	8 µL	8 µL	8 µL	8 µL
10 m*M* dCTP	8 µL	8 µL	8 µL	8 µL
10 m*M* dGTP	8 µL	8 µL	8 µL	8 µL
10 m*M* dTTP	8 µL	8 µL	8 µL	8 µL
1 m*M* ddATP	8 µL	—	—	—
1 m*M* ddCTP	—	8 µL	—	—
1 m*M* ddGTP	—	—	8 µL	—
1 m*M* ddTTP	—	—	—	8 µL
1*M* NaCl	50 µL	50 µL	50 µL	50 µL
water	910 µL	910 µL	910 µL	910 µL

14. Formamide dye mix: 95% Formamide (v/v), 20 m*M* EDTA, 0.05% bromophenol blue (w/v), 0.05% xylene cyanol (w/v).
15. Sequencing gel: A 0.4-mm thick, 6% denaturing polyacrylamide gel.
16. Fix solution: 10% acetic acid/10% methanol (v/v). The fix can be stored at room temperature and reused 4–5 times.
17. Autoradiographic film: e.g., Kodak XAR5.

3. Methods

3.1. End-Labeling the Sequencing Primer

We describe labeling of 200 ng of oligonucleotide. This is sufficient for ten sequencing reactions. The labeling reaction can be adapted for 100 ng (*see* Note 5).

1. Dry down 100 µCi γ-[^{32}P] ATP in a 1.5-mL Eppendorf tube.
2. Place on ice and add sequentially 5 µL of water, 1 µL of 10X polynucleotide kinase buffer, 200 ng (2 µL) of oligonucleotide, and 2 µL of polynucleotide kinase (20 U).

3. Incubate at 37°C for 1 h. Add 90 µL of TE to the reaction. Spin the labeled oligonucleotide through a Sephadex G50 spun column, and collect it in a 1.5-mL Eppendorf tube.
4. Dry the sample down and resuspend in water to 10 ng/µL. Freeze the sample at –20°C until required.

The labeled oligonucleotide is stable for several weeks at –20°C but its use is limited by its radioactive half-life.

3.2. Preparation and Sequencing of the PCR Product

The yield of the PCR product should be estimated on an agarose gel before sequencing the template. At least 150 ng of PCR product is required for each sequencing reaction.

1. Spin the whole PCR reaction through a Sephadex G50 spun column in TE. Take care not to load any paraffin oil. If necessary this can be removed by chloroform extraction. Collect the eluate in a 1.5-mL Eppendorf tube. This procedure removes the excess of dNTPs from the PCR reaction.
2. Measure the volume of the reaction and add 0.1 vol of 3*M* sodium acetate, pH 5.2, and 2.5 vol of ethanol to precipitate the PCR product. Cool on dry ice for 15 min or leave at –20°C overnight.
3. Spin in a microfuge at 12,000*g* for 15 min at 4°C. Remove the supernatant and briefly dry the pellet. Resuspend the pellet in sterile distilled water to give a final concentration of 150 ng/µL.
4. Mix 1 µL (150 ng) of prepared PCR product, 2 µL (20 ng) of labeled primer, 2 µL of 5X sequenase buffer and 5 µL of water in a 0.5 mL Eppendorf tube.
5. Anneal the template to the oligonucleotide by boiling the tube for 5 min and then immediately plunging it into ice/water (*see* Notes 6 and 7). Leave in ice/water for 5 min.
6. Spin for a few seconds in a microfuge to collect any liquid on the tube walls. Add 1 µL of 0.1*M* DTT, 2 µL of working labeling mix, and 2 µL of diluted sequenase. Incubate at room temperature for 5 min (*see* Note 8).
7. While the labeling reaction is proceeding, label four 1.5-mL Eppendorf tubes, one for each termination mix, i.e., A, C, G, and T. Add 2.5 µL of the relevant termination mix to each tube.
8. At the end of the first 5-min incubation add 3.5 µL of the labeling reaction to each termination mix. Incubate at 37°C for 5 min.
9. Add 4 µL of formamide dye mix. Store at –20°C until required. These reactions are fairly stable for a week at –20°C.

10. Heat the sequencing reactions to 95°C for 3 min before loading onto a sequencing gel. Run the gel until the bromophenol blue has just run off the bottom; this will allow the excess of labeled primer to run off the gel but it will be possible to read the sequence within 20 bp of the primer.
11. Transfer the gel to a piece of 3MM paper (Whatman, Clifton, NJ); alternatively it may be fixed on the glass sequencing gel plate. Place the gel in fix solution for at least 10 min. Remove the gel from the fix solution. Cover the surface of the gel in cling film and dry.
12. Autoradiograph at –70°C overnight. As a rough guide, if the reaction has worked well it should be possible to obtain a reading of 50 cps on a dry gel using a series-900 mini-monitor (Mini Instruments, Ltd.) at the position where the xylene cyanol runs.

A typical result is shown in Fig. 1A. A number of common problems that may arise are discussed in Notes 9–14.

4. Notes

1. The limit in template size for sequencing using this method is about 2 kb. It is important to alter the amount of template when using larger templates to keep the molar ratio of template-to-primer constant. For example, 1 pmol of a 250-bp template corresponds to 150 ng but for a 1.6-kb template this would be 1 µg.
2. It is possible to end-label the sequencing primer using $[^{35}S]$-γ-ATP. This gives improved resolution toward the top of the gel and reduces the radiation exposure to the experimenter. The end-labeling reaction with $[^{35}S]$-γ-ATP takes 4 h and the reduced signal intensity means that autoradiography may take several days.
3. It is important that the sequencing primer does not contain sequences that are likely to form any secondary structure, that it does not contain a long run of G-residues, and that there are no mismatches at the 3' end.
4. Sephadex G50 spun columns can be made quickly and simply. Add Sephadex G50 to TE, pH 8.0, and allow it to stand at room temperature overnight. Plug a 1-mL disposable syringe with polymer wool, fill it with the Sephadex G50, and allow the TE to drain out (the syringe should be full to the top with Sephadex G50 before it is spun). Put the syringe into a centrifuge tube (a 15-mL Falcon tube 2095 will do). Spin for 3 min at 200*g* (1000 rpm in a bench centrifuge). The column should pack to 1 mL volume. Load the sample on to the top of the column and spin it again (at the same speed for 3 min) collecting the sample in a decapped Eppendorf tube.

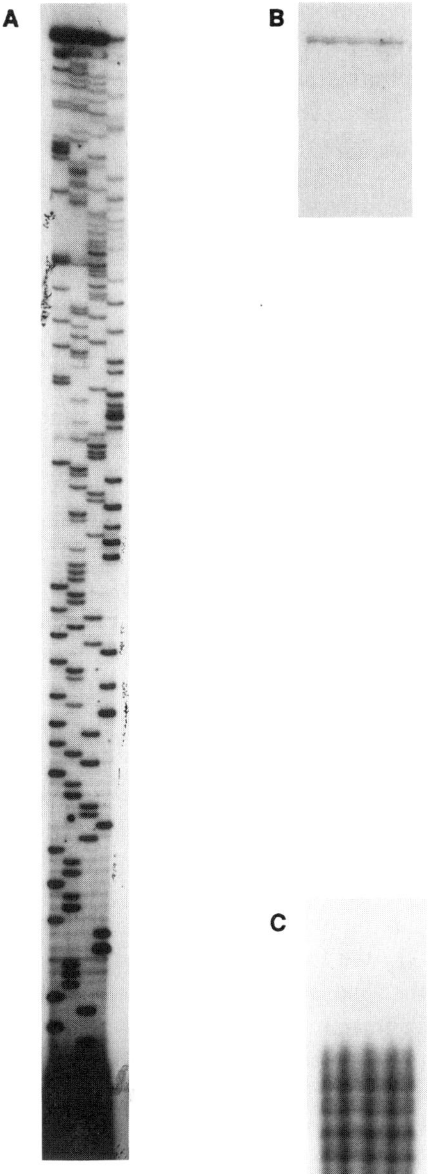

Fig. 1. (**A**) Autoradiograph of a PCR product sequenced using a ^{32}P-labeled nested primer. (**B**) The upper region of the gel showing the fully extended product produced by primer annealing and extension but not termination. (**C**) The bottom of the sequencing gel showing the free end-labeled primer that has not extended.

5. To label 100 ng (15 pmol) of oligonucleotide primer, use 50 µCi of γATP. It is not necessary to dry the label down in this case. The final reaction volume remains at 10 µL but the amount of polynucleotide kinase should be reduced to 1 µL (10 U).
6. For most PCR products 5 min boiling is sufficient, but in some cases, for example human DNA and DNA rich in guanine and cytosine residues, better denaturation may be achieved by boiling for as long as 10 min.
7. It is essential to plunge the annealing reaction into ice water without delay since slow cooling of the primer–template mix will allow the template strands to reanneal.
8. An eightfold dilution of the labeling mix will allow the sequence to be read as close as 20 bp to the primer. Addition of manganese buffer (0.15*M* sodium isocitrate, 0.1*M* $MnCl_2$) to the annealing reaction allows the sequence to be read very close to the primer. One microliter of this buffer is added to the annealed primer–template mix at step 3.2.4. To read further from the primer, a more concentrated labeling mix may be used.
9. Loss of sequence but a strong band at the top of the gel results from extension but no termination of the template. This is likely to be owing to degradation of old termination mixes or their improper dilution, therefore fresh solutions should be made (Fig. 1B).
10. Loss of sequence but a strong band at the bottom of the gel is caused by failure of the primer to anneal to the template (Fig. 1C). The strong band is the free end-labeled primer. This may result from mutations in the template where the primer anneals. Try another primer. An alternative explanation is that the template is not fully denatured; *see* Notes 6 and 7.
11. Faint sequences may be caused by insufficient template, or poor labeling of the primer. In the latter case the γ-[^{32}P] ATP may be too old or the efficiency of the kinase reaction may not be high enough to achieve the required incorporation (10^8 cpm/µg).
12. Secondary structure may cause premature termination of the sequence in all lanes (Fig. 2A). This may be solved by incorporating nucleotide analogs such as dITP, or 7-deaza-2'dGTP *(10)* into the sequencing mixes, or by sequencing using *Taq* polymerase *(11,12)*. More recently a method has been reported that couples PCR and sequencing using end-labeled primers *(13)* that may prove to be a valuable advance in this respect.
13. The sequence may become too faint to read after a short distance. This is caused by either the labeling mix being too dilute or the concentration of dideoxynucleotide in the termination mixes being too high.

A

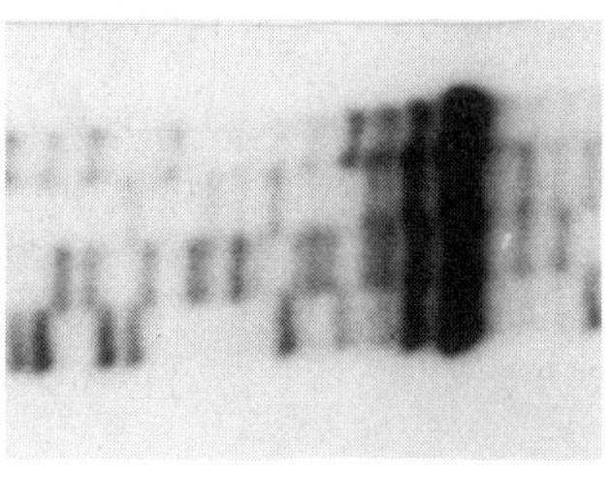

B

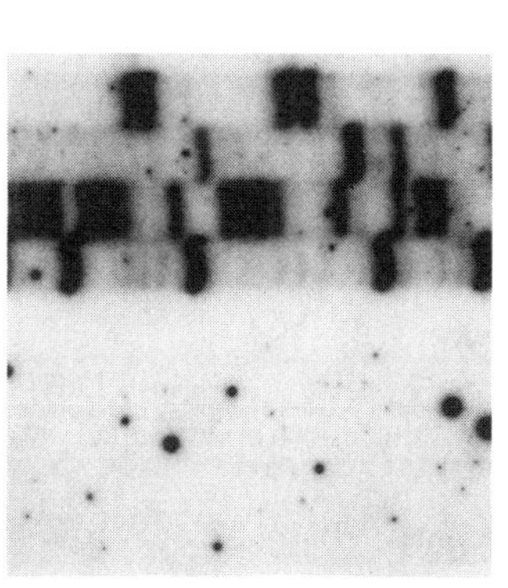

C

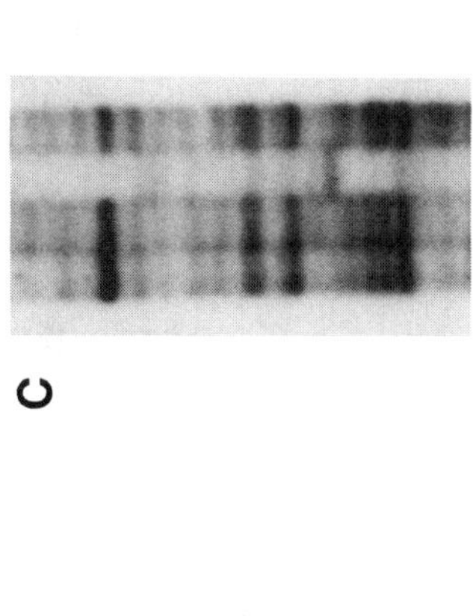

Fig. 2. **(A)** Bands appear in all four lanes. This is probably because the polymerase halts because of a region of secondary structure in the template. **(B)** Dots on the sequencing gel may be caused by degradation of the end-labeled primer. The left panel shows the extreme case where the primer is degraded to such an extent that it does not prime the sequencing reaction. The right panel shows the case where it is still possible to read the sequence. **(C)** The result of using an end-labeled PCR primer for sequencing rather than a nested primer. The picture shows an unreadable sequence since multiple sequences are superimposed on each other. This is generated by the end-labeled primer annealing to nonspecific PCR-products.

14. Black spots on the autoradiograph (Fig. 2B) are usually caused by degradation of the primer, probably occurring during the labeling reaction. This is not a serious problem unless the primer is degraded to such an extent that it is very short and does not prime the sequencing reaction.

References

1. Harwood, J., Tachibana, A., and Meuth, M. (1991) Multiple dispersed spontaneous mutations: a novel pathway of mutation in a malignant human cell line. *Mol. Cell Biol.* **11,** 3163–3170.
2. Phear, G. A., Armstrong, W., and Meuth, M. (1989) Molecular basis of spontaneous mutation at the *aprt* locus of hamster cells. *J. Mol. Biol.* **209,** 577–582.
3. Gibbs, R. A., Nguyen, P.-N., McBride, L. J., Koepf, S. M., and Caskey, C. T. (1989) Identification of mutations leading to the Lesch-Nyhan syndrome by automated DNA sequencing of *in vitro* amplified cDNA. *Proc. Natl. Acad. Sci. USA* **86,** 1919–1923.
4. Scharf, S. J., Horn, G. T., and Ehrlich, H. A. (1986) Direct cloning and sequence analysis of enzymatically amplified genomic sequences. *Science* **233,** 1076–1078.
5. Ward, M. A., Skandalis, A., Glickman, B. W., and Grosovsky, A. J. (1989) Rapid generation of a specific single-stranded template DNA from PCR-amplified material. *Nucleic Acids Res.* **20,** 8394.
6. Loh, E. Y., Elliott, J. F., Cwirla, S., Lanier, L. L., and Davis, M. M. (1989) Polymerase chain reaction with single-sided specificity: analysis of T cell receptor δ chain. *Science* **243,** 217.
7. Hultman, T., Stahl, S., Hornes, E., and Uhlen, M. (1989) Direct solid phase sequencing of genomic and plasmid DNA using magnetic beads as solid support. *Nucleic Acids Res.* **17 (13)** 4936–4937.
8. Wong, C., Dowling, C. E., Saiki, R. K., Higuchi, R. G., Ehrlich, H. A., and Kazazian, H. H. (1987) Characterisation of β-thalassemia mutations using direct genomic sequencing of amplified genomic DNA. *Nature* **330,** 384.
9. Sanger, F., Nicklen, S., and Coulson, A. R. (1977) DNA sequencing with chain-terminating inhibitors. *Proc. Natl. Acad. Sci. USA* **74,** 5463–5467.
10. Barnes, W. M., Bevan, M., and Son, P. H. (1983) Kilo sequencing: creation of an ordered nest of asymmetric deletions across a large target sequence carried on phage M13. *Meth. Enzymol.* **101,** 98–122.
11. Innis, M. A., Myambo, K. B., Gelfand, D. H., and Brow, M. A. D. (1988) DNA sequencing with *Thermus aquaticus* polymerase and direct sequencing of polymerase chain reaction-amplified DNA. *Proc. Natl. Acad. Sci. USA* **85,** 9436–9440.
12. Peterson, G. (1988) DNA sequencing using *Taq* polymerase. *Nucleic Acids Res.* **22,** 10915.
13. Ruano, G. and Kidd, K. K. (1991) Coupled amplification and sequencing of genomic DNA. *Proc. Natl. Acad. Sci. USA* **88,** 2815–2819.

PART V

Gene Expression

CHAPTER 24

The Use of Riboprobes for the Analysis of Gene Expression

Dominique Belin

1. Introduction

The isolation and characterization of RNA polymerases from the *Salmonella* phage SP6 and the *E. coli* phages T7 and T3 has revolutionized all aspects of the study of RNA metabolism *(1–6)*. Indeed, it is now possible to generate unlimited quantities of virtually any RNA molecule in a chemically pure form. This technology is based on a number of properties of the viral transcription units. First, and in contrast to their cellular counterparts, the enzymes are single-chain proteins that are easily purified from phage-infected cells and are now produced by recombinant DNA technology. Second, they very specifically recognize their own promoters, which are contiguous 17–20 bp long sequences rarely encountered in bacterial, plasmid, or eukaryotic sequences. Third, the enzymes are highly processive, allowing the efficient synthesis of very long transcripts from DNA templates. In this chapter, I will discuss the preparation of the DNA templates, the transcription from the template of ^{32}P-labeled synthetic RNA molecules, commonly called riboprobes, and their use in Northern and RNase protection assays.

2. Materials

These protocols require the use of standard molecular biological materials and methods for carrying out subcloning, polymerase chain reaction (PCR), and gel electrophoresis, in addition to those listed

From: *Methods in Molecular Biology, Vol. 31: Protocols for Gene Analysis*
Edited by: A. J. Harwood

below. More rigorous precautions are required for working with RNA than commonly used for most DNA studies, because it is important to eliminate RNase contamination. The two major sources of unwanted RNases are the skin of investigators and microbial contamination of solutions. Most RNases do not require divalent cations and are not irreversibly denatured by autoclaving. Gloves should be worn and frequently changed. Sterile plasticware should be used, although some mechanically manufactured tubes and pipet tips have been used successfully without sterilization. Glassware should be incubated at 180°C in a dried baking oven for several hours.

2.1. Preparation of Riboprobes

1. Water: Although a number of protocols recommend treatment of the water used for all the solutions with diethylpyrocarbonate, I find this to be unnecessary. Double-distilled water is used for the preparation of stock solutions that can be autoclaved (121°C, 15–30 min), and sterilized water is used otherwise.
2. TE: 10 m*M* Tris-HCl, pH 8.1, and 1 m*M* EDTA.
3. 10X TB: 0.4*M* Tris-HCl, pH 7.4, 0.2*M* NaCl, 60 m*M* $MgCl_2$, and 20 m*M* spermidine. If ribonucleotides are used at concentrations >0.5 m*M* each, the $MgCl_2$ concentration should be increased to provide a free magnesium concentration of 4 m*M*.
4. 0.2*M* DTT: The solution is stored in small aliquots at –20°C. Aliquots are used only once. EDTA can be included at 0.5 m*M* to stabilize DTT solutions.
5. Ribonucleotides: Neutralized solutions of ribonucleotides are commercially available, or can be made up from dry powder (*see* Note 1). Ribonucleotide solutions can be stored at –20°C for a few months.
6. RNA polymerase stocks: the three RNA polymerases, T3, T7, and SP6, are available commercially. Store at –20°C.
7. RNA polymerase dilution buffer: 50 m*M* Tris-HCl, pH 8.1, 1 m*M* DTT, 0.1 m*M* EDTA, 500 µg/mL BSA, and 5% glycerol. Diluted enzyme is unstable and should be stored on ice for no more than a few hours.
8. Stop-mix: 1% SDS, 10 m*M* EDTA, and 1 mg/mL tRNA. The tRNA may be omitted.
9. TEN: 10 m*M* Tris-HCl, pH 8.1, 1 m*M* EDTA, and 100 m*M* NaCl.
10. Sample buffer (polyacrylamide/urea gels): 80% deionized formamide, 2*M* urea, 0.1X TBE (8.9 m*M* Tris-HCl, 8.9 m*M* boric acid, 0.2 m*M* EDTA) and 0.01% each of xylene cyanol and bromophenol blue. Use 1–2 µL/µL of RNA in aqueous solution.

2.2. Northern Blot Hybridization

1. Formamide: Pure formamide is slowly hydrolyzed by water vapor to ammonium formate and therefore must be dieonized (*see* Note 2). Store at –20°C.
2. Glyoxal: A 30% glyoxal solution (6*M*) is deionized by several incubations at room temperature with a mixed bed resin (AG501-X8, Bio-Rad, Richmond, CA), until the pH is 6–7 and the conductivity of 3% glyoxal in water is below 30 μSiemens. Store at –20°C in small aliquots.
3. Northern sample buffer: 1*M* glyoxal in 50% DMSO and 10 m*M* Na_2 HPO_4, pH 6.8.
4. 5X RSB: 50% glycerol, 10 m*M* Na_2HPO_4, pH 6.8, and 0.4% bromophenol blue. Autoclave and store at –20°C in small aliquots.
5. Transfer membrane: Nitrocellulose (Schleicher and Schuell, Keene, NH) and nylon membranes (Hybond N, Amersham, Arlington Heights, IL or Biodyne A, Pall, Glen Cove, NY) have been used successfully.
6. 50X Denhardt's solution: Dissolve 2 g of bovine serum albumin (fraction V) in 80 mL of sterile water. Bring the pH to 3.0 with 2*N* HCl, boil for 15 min, and cool on ice for 10 min. Bring the pH to 7 with 2*N* NaOH and add sterile water to 100 mL. Autoclave a 100-mL solution of 2% polyvinylpyrrolidone (K90, Fluka, Buchs, Switzerland) and 2% Ficoll 400 (Pharmacia, Piscataway, NJ), and add to the 2% albumin solution.
7. Hybridization solution: 50% deionized formamide, 0.8*M* NaCl, 50 m*M* Na-PIPES, pH 6.8, 2 m*M* EDTA, 0.1% SDS, 2.5X Denhardt's solution, and 0.1 mg/mL sonicated and heat-denatured salmon sperm DNA. The hybridization solution is heated, filtered over 0.45 μm Nalgene sterilization units or Millipore nitrocellulose filters, and kept at the hybridization temperature during prehybridization. The filtration step decreases nonspecific attachment of the probe to the membrane and degasses the solution.
8. 20X SSC: 3*M* NaCl, and 0.3*M* Na/citrate, pH 7.

2.3. RNase Protection

1. Hybridization mixture for RNase protection: 80% deionized formamide, 0.4*M* NaCl, 40 m*M* Na-PIPES, pH 6.8, and 1 mM EDTA.
2. RNase digestion buffer: 300 mM NaCl, 10 mM Tris-HCl, pH 7.4, and 4 m*M* EDTA.
3. Pancreatic RNase: Make a 10 mg/mL solution in TE containing 10 m*M* NaCl. Vials containing lyophilized RNases should be carefully open in a ventilated hood to avoid contamination. Boil for 15 min and slowly cool to room temperature. Store in aliquots at –20°C.

4. T1 RNase: Make a 1 mg/mL solution in TE. Adjust the pH to 7. Store in aliquots at –20°C.
5. Proteinase K: Dissolve the enzyme at 20 mg/mL in water and store in aliquots at –20°C.

3. Methods

3.1. Preparation of Riboprobes

3.1.1. Linearized Plasmid Templates for Runoff Transcription

1. Subclone the desired gene fragment into an appropriate transcription vector (*see* Note 3).
2. Isolate plasmid DNA by alkaline lysis from a 30–100-mL saturated culture. Plasmid DNAs are purified by CsCl/ethidium bromide centrifugation or by precipitation with polyethyleneglycol (*see* Note 4).
3. Linearize 2–20 µg of plasmid DNA with an appropriate restriction enzyme (*see* Note 5). Verify the extent of digestion by electrophoresis of an aliquot (0.2–0.5 µg of DNA) on an agarose minigel in the presence of ethidium bromide (0.5 µg/mL) (*see* Note 6).
4. Purify the restricted DNA by 2–3 extractions with phenol/chloroform, and remove residual phenol by one extraction with chloroform or 3–4 extractions with ether. Precipitate the DNA with ethanol. If little DNA is present (below 5 µg), add 10 µg of glycogen as a carrier; this carrier has no adverse effect in the transcription reactions. After washing the ethanol pellet, the DNA is dried in air and resuspended in TE at 1 µg/µL.

3.1.2. Synthetic and PCR-Derived Templates

The major limitation in using restriction enzymes to clone inserts and to linearize plasmid templates is that appropriate sites are not always available. Furthermore, the transcripts will almost always contain 5' and 3' portions that differ from those of the endogenous RNA. One possibility to circumvent these difficulties is based on the transcription of small DNA fragments obtained by annealing of synthetic oligodeoxynucleotides *(6,7)*. I present an alternative and more general approach by generating the transcription templates via PCR amplification of plasmid DNA. In theory, such templates could direct the synthesis of virtually any RNA sequence.

1. Design the 5' primer, which has a composite sequence: Its 5' portion is constituted by a minimal T7 promoter, and its 3' portion corresponds to the beginning of the transcript (*see* Note 7). Six to ten nucleotides are usually sufficient to prime DNA synthesis on the plasmid template.

2. Design the 3' primer, which is usually 17–20 nucleotides long and defines the 3' end of the transcript (*see* Note 8).
3. PCR amplify 2–50 ng of plasmid DNA in a total volume of 100 µL. Verify that a DNA fragment of the expected size has been amplified, and estimate the amount of DNA by comparison with known standards.
4. Purify the DNA as described in Section 3.1.1. and resuspend in TE at an appropriate concentration. The PCR-derived templates are transcribed at lower DNA concentrations than plasmids to maintain the molar ratio of enzyme to promoter; I use 3 µg/mL for a 100-bp fragment.

3.1.3. Basic Transcription Protocol for Radioactive Probes

1. Assemble the transcription mixture to a total volume of 20 µL by adding in the following order: water (as required), 2 µL of 10X TB, 1 µL of bovine serum albumin (2 mg/mL), 1 µL of 0.2*M* DTT, 0.25 µL of placental RNase inhibitor (40 U/µL), 2 µL of a 5 m*M* solution of each ribonucleotide (i.e., ATP,GTP,CTP), 5 µL of α-[^{32}P]-UTP (400 Ci/mmol, 10 mCi/mL), 1 µL of restricted plasmid DNA template and 0.25–0.75 µL of RNA polymerase (*see* Notes 9 and 10).
2. Incubate 40 min at 37°C for T7 and T3 RNA polymerases or 40°C for SP6 RNA polymerase. After addition of the same number of units of enzyme (*see* Note 10), incubate for a further 40 min.
3. Degrade the template with RNase-free DNase (1 U/µg of DNA) for 20 min at 37°C. Stop the reaction by adding 40 µL of stop-mix.
4. After 2 extractions with phenol/chloroform, in which the organic phases are back-extracted with 50 µL of TEN, the combined aqueous phases are purified from unincorporated ribonucleotides by spun-column centrifugation. The spun-column is prepared by filling disposable columns (QS-Q, Biolab, Hörth, Germany) with a sterile 50% slurry of G50 Sephadex (Pharmacia) in TEN and followed by centrifugation for 5 min at $200g_{av}$. The samples are carefully deposited on top of the dried resin, and the column is placed in a sterile conical centrifuge tube. After centrifugation for a further 5 min at 200*g*, 200 µL of TEN is deposited on top of the resin and the column recentrifuged. The eluted RNA, approx 200–400 µL, is ethanol precipitated and resuspended in water (*see* Note 11).
5. Measure the incorporation efficiency by counting an aliquot (*see* Note 12). The size of the transcript may be verified by electrophoresis on polyacrylamide/urea gels (*see* Note 13).

3.2. Northern Blot Hybridization

The use of riboprobes in RNA blot hybridizations follows the same general principles as that for DNA probes. The major disadvantage in using double-stranded DNA probes results from self-annealing, which decreases the availability of DNA probes to bind to the immobilized target; this is particularly critical with heterologous probes, where the reannealed probe may displace incompletely matched hybrids. Self-annealing, of course, does not occur with single-stranded RNA probes. Most difficulties encountered with riboprobes stem from the increased thermal stability of RNA:RNA hybrids. Thus, crosshybridization of GC-rich probes to rRNAs can generate unacceptable backgrounds. This problem is often solved by increasing the stringency of hybridization, as illustrated in Fig. 1B. Alternatively, the template may have to be shortened to remove GC-rich regions from the probe.

1. Denature the sample RNA in 8 µL of Northern sample buffer for 15–30 min at 50°C. Add 2 µL of 5X RSB and electrophorese in 0.7–2% agarose gels in 10 m*M* Na_2HPO_4, pH 6.8. The buffer should be circularized with a peristaltic pump, so that the pH near the electrodes remains neutral.
2. Transfer the RNAs by capillarity onto a membrane (*see* Note 14). Fix the RNA by incubating the blots at 80°C under vacuum. This step is essential to remove glyoxal covalently fixed to guanine residues in the RNA. UV-crosslinking is often used to improve RNA retention, although irradiation may increase background hybridization (*see* Fig. 1A and Note 15).
3. Prehybridize the blots for 4–12 h in the hybridization solution. Use 200 µL of hybridization solution per cm^2 of membrane.
4. Dilute the probe in hybridization solution (25–50 µL/cm^2 of membrane), then add to the membrane. I frequently hybridize 2–3 filters per bag. Hybridize at the appropriate temperature (58–68°C, *see* Fig. 1B and Note 16), usually for 12–18 h (*see* Note 17).
5. Wash the membranes twice for 10–20 min at the hybridization temperature with 100 µL/cm^2 of 3X SSC and 2X Denhardt's solution, and then 3 times with 0.2X SSC, 0.1% SDS, and 0.1% Na pyrophosphate at the appropriate temperature (*see* Fig. 1B and Note 17).
6. Expose to autoradiographic film. As long as the membranes are not allowed to dry, they can be further washed at increased stringency to reduce background.

3.3. RNase Protection

This assay is based on solution hybridization and on the resistance of RNA:RNA hybrids to single-strand specific RNases. A ^{32}P-labeled probe is synthesised that is complementary to the target RNA sequence. It is hybridized in excess to the target so that all comple-

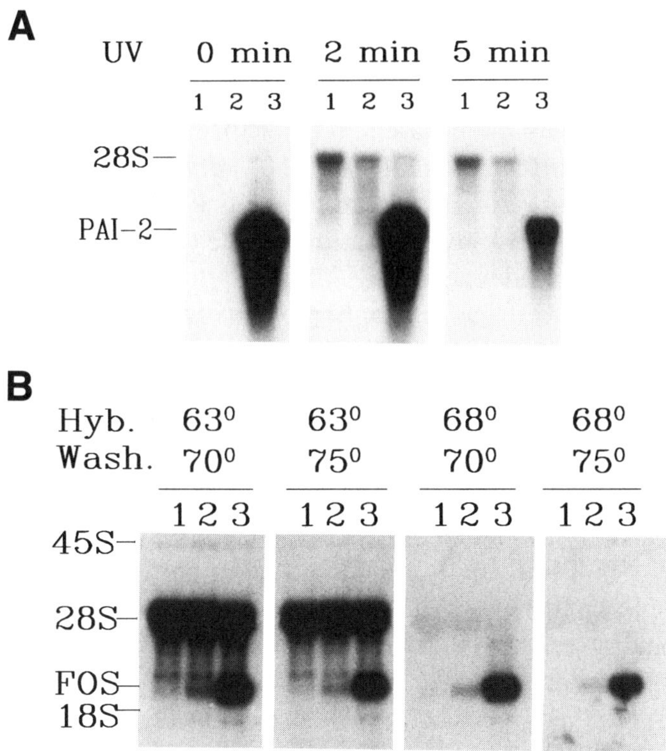

Fig. 1. Northern blot hybridization with riboprobes. (**A**) Effect of UV crosslinking. Northern blot hybridization of PAI-2 mRNA in murine total cellular RNA with a homologous cRNA probe. Lanes 1 and 2: 5 µg of placental RNA (15.5 and 18.5 d of gestation) that do not contain detectable levels of PAI-2 mRNA. Lane 3: 1 µg of LPS-induced macrophage RNA, an abundant source of PAI-2 mRNA *(16)*. All samples were electrophoresed and transferred together. After cutting the membrane, each filter was UV-treated as described. The filters were hybridized together at 58°C, washed at 70°C, and exposed together. Crosshybridization of the probe to 28S rRNA is more pronounced after UV irradiation, and specific hybridization is decreased after 5 min of UV exposure. (**B**) Effect of hybridization temperature. Northern blot hybridization of *c-fos* mRNA in rat total cellular RNA with a murine *v-fos* cRNA probe. Lane 1: uninduced cells. Lane 2: partially induced cells. Lane 3: fully induced cells (M. Prentki and D. Belin, unpublished). All samples were electrophoresed and transferred together. The filters, which were not UV crosslinked, were hybridized and washed in parallel at the indicated temperatures. The four filters were exposed together. Crosshybridization of the probe to 28S rRNA was essentially abolished by hybridizing at 68°C. Some specific signal was lost with the 75°C stringency wash.

mentary sequences are driven into the labeled RNA:RNA hybrid. Unhybridized probe, or any single-stranded regions of the hybridizied probe are then removed by RNase digestion. The "protected" probe is then detected and quantitated on a denaturing polyacrylamide gel. It can be used to map the ends of RNA molecules or exon–intron boundaries. It also provides an attractive and highly sensitive alternative to Northern blot hybridization for the quantitative determination of mRNA abundance.

RNase protection has a number of advantages. First, solution hybridization tolerates high RNA input (up to 60 µg of total RNA), and is not affected by the efficiency of transfer on membranes or by the availability of membrane-bound RNAs. Second, the signal-to-noise ratio is much more favorable, since crosshybridizing RNAs yield only short protected fragments. Third, a significant fraction of mRNAs is often partially degraded during RNA isolation; in Northern blots, this generates a trail of shorter hybridizing species, which reduces the sensitivity of detection. Finally, the detection of hybridized probes on sequencing gels is much more sensitive because the width of the bands can be reduced to a tenth of those used in agarose gels. Only two features of Northern blots are lost in RNase protection assays: complete size determination of target RNAs and multiple use of each sample.

1. Linearize the plasmid DNA template as described in Section 3.1.1. (*see* Note 18). Transcribe the template as described in Section 3.1.3.; although the amount of labeled/ribonucleotide may be varied (*see* Note 19).
2. An optional step is to purify the full length transcripts by electrophoresis (*see* Note 20). Separate the transcript on a preparative 5–6% polyacrylamide/urea gel (gel thickness: 0.4–1.5 mm). Cover the wet gel with SaranWrap and expose for 30 s to 5 min at room temperature to localize the full length transcript. Cut the exposed band on the film with a razor blade. After aligning the cut film on the gel, excise the gel band with a sterile blade. The cut gel should be re-exposed to verify that the correct band has been excised. Elute the RNA from the gel either by electroelution or soaking.
   a. Electroelute for 1–2 h at 30 V/cm in 0.1X TBE in a sterile dialysis bag, after which invert the polarity for 30 s, to detach the eluted RNA from the membrane. Purify the eluate by two phenol/chloroform extractions and ethanol precipitation with a known amount of tRNA carrier (*see* Note 21). This procedure is very sensitive to RNase degradation.

b. Incubate the gel fragment in an Eppendorf tube in 500 µL of 0.5*M* ammonium acetate, 1% SDS, and 20 µg/mL tRNA for 1–3 h at 37°C. The eluate and residual gel can be counted to ensure that more than 60% of the RNA is eluted. After two extractions with phenol/chloroform, recover the eluted RNA by ethanol precipitation.

3. Resuspend the probe in water at 1–2 ng/µL. Add 1 µL of probe to 29 µL of hybridization mixture for each assay. The probe is resuspended in water and constitutes 3% of the final volume; the exact amount of probe is not critical, since it is in excess of its specific target (*see* Note 22).
4. Lyophilize or ethanol precipitate 10 µg of the sample RNA. Resuspend in 30 µL of complete hybridization mixture (including probe). Heat for 2 min at 90°C, and incubate overnight, usually at 45°C (*see* Notes 19 and 23).
5. Cool the samples on ice and add 300 µL of RNase digestion buffer. Digest for 1 h at 25°C with pancreatic RNase, which cleaves after uracil and cytosine residues, with T1 RNase, which cleaves after guanine residues, or with both RNases (*see* Notes 24 and 25).
6. Add 20 µL of 10% SDS, and degrade the enzyme(s) with 0.5 µL (10 µg) of proteinase K for 10–20 min at 37°C. Extract twice with phenol/chloroform, and precipitate the RNAs with ethanol with 10 µg of carrier tRNA.
7. Resuspend the RNAs in sample buffer, denature the hybrids for 2 min at 90°C, and electrophorese in polyacrylamide/urea sequencing gels. Fix the gels with 20% ethanol and 10% acetic acid to remove the urea, dry, and autoradiograph (*see* Note 26).

4. Notes

1. Powdered ribonucleotides should be resuspended in water, neutralized to pH 6–8 with 1*M* NaOH or HCl, and adjusted to the desired concentration by measuring the UV absorbance of appropriate dilutions:

   100 m*M* ATP: 1540 absorbance units at 259 nm.
   100 m*M* GTP: 1370 absorbance units at 253 nm.
   100 m*M* CTP: 910 absorbance units at 271 nm.
   100 m*M* UTP: 1000 absorbance units at 262 nm.

   The integrity of ribonucleotide solutions can be verified by thin layer chromatography on PEI-cellulose (PEI–CEL300). The resin is first washed with water by ascending chromatography to remove residual UV-absorbing material and dried. 10–30 nmol of ribonucleotides are deposited on the resin, which is then resolved by ascending chromatography with 0.5*M* KH_2PO_4, pH 3.5 with H_3PO_4. After drying, the ribonucleotides are detected by UV-shadowing at 254 nm.

2. To deionize formamide, incubate at –80°C until 75–90% of the solution has crystallized. Discard the liquid phase, thaw, and incubate at 4°C for several hours with a mixed bed resin (AG501-X8, Bio-Rad). Use a Teflon-covered magnet that has been freed of RNase by treatment with 0.1*M* NaOH for 10 min, and rinsed with water and with crude formamide. Check the conductivity, which should be below 20 µSiemens. Filter the solution onto sterile paper (Whatman, LS-14) over a sterile funnel. The absorbance at 270 nm should be below 0.2.
3. All the vectors that are commercially available consist of high copy number *E. coli* plasmids derived from ColE1. The original plasmids (pSP64 and pSP65, Promega, Madison, WI) contained only one SP6 promoter located upstream of multiple cloning sites (MCS; *2*). In the second generation of plasmids, two promoters in opposite orientation flank the MCS allowing transcription of both strands of inserted DNA fragments. The pGEM series (Promega) contain SP6 and T7 promoters (*5*), while the pBluescript™ series (Stratagene, La Jolla, CA) contain T7 and T3 promoters.

   The choice of plasmid is mostly a matter of personal preference, although it can be influenced by the properties of individual sequences. For instance, we have frequently observed premature termination with SP6 transcripts. The nature of the termination signals is not completely understood *(4,8)*, and their efficiency can be more pronounced when one ribonucleotide is present at suboptimal concentration. The problem is often solved by recloning the inserts in front of a T7 or T3 promoter. The partial recognition of T7 promoters by T3 polymerases may result, however, in the transcription of both strands when the ratio of enzyme to promoter is not carefully controlled. This can be a source of artifacts, particularly when the templates are linearized within the cloned inserts.
4. It is possible to use plasmid DNA from "minipreps," although transcription efficiency can be reduced, particularly with SP6 polymerase. The RNA present in the "minipreps" is digested with pancreatic RNase (20 µg/mL), which is then removed during purification of the linearized templates. Spun-column centrifugation of the digested "minipreps" can improve transcription efficiency.
5. Since RNA polymerases can initiate transcription unspecifically from 3' protruding ends, restriction enzymes that generate 5' protruding or blunt ends are usually preferred *(4,9)*. If the only available site generates 3' protruding ends, the DNA can be blunt-ended by exonucleolytic digestion using T4 DNA polymerase or the Klenow fragment of DNA polymerase I. Restriction with enzymes that cut the plasmids more than

once may also be used, provided that the promoter is not separated from the insert.

6. The plasmids must be linearized as extensively as possible. Since circular plasmids are efficient templates, their transcription may yield RNA molecules that can be up to 20 kb long and thus incorporate a significant portion of the available ribonucleotide.
7. There are additional constraints on the 5' sequence of transcripts since it has been shown that the sequence immediately downstream of the start site may affect the efficiency of elongation mode. The first 6 nt have a strong influence on promoter efficiency. In particular, the presence of uracil residues is usually detrimental *(6,7)*. It may be necessary, therefore, to include in the 5' end of the transcripts 5–6 bases that differ from those present in natural RNA. A similar case has been reported for SP6 polymerase *(10,11)*. I have used a number of composite T7 promoters, whose efficiency is summarized below. In addition to the promoter sequence (5'-TAATACGACTCACTATA) at the 5' end, the first 6 nt of the templates are:

Efficient promoters	Inefficient promoters
GGGAGA (T7 consensus)	GTTGGG (5% efficiency)
GGGCGA (pBS plasmids)	GCTTTG (1% efficiency)
GCCGAA	

8. The 3' end of the transcripts should be exactly defined by the 5' end of the downstream primer. However, during transcription template-independent addition of 1–2 nt usually generates populations of RNA with differing 3' ends. The proportion of each residue at the 3' end may depend on context and is influenced by the relative concentration of each ribonucleotide, a limiting ribonucleotide being less frequently incorporated *(2,6,7,12)*.
9. The order of addition of components may be changed. Remember that dilutions of RNasin are very unstable in the absence of DTT, and that DNA should not be added to undiluted 10X TB. The temperature of all components must be at least at 25°C to avoid precipitation of the DNA:spermidine complex. Since the radioactive nucleotides, which constitute half of the reactions, are provided in well-insulated vials, they can take up to 10–15 min after thawing to reach an acceptable temperature. I routinely incubate all components, except the RNasin and the polymerase, at 30°C for 15–20 min. When more than one probe is to be synthesized, a reaction mixture with all components is added to the DNA template. It has also been possible to reduce the total volume to as small as 8 µL in order to conserve materials; however, since the en-

zymes are sensitive to surface denaturation, these incubations are done in 400-μL vials.

10. For SP6 polymerase, use 4–6 U/μg of plasmid DNA (size: 3–4 kbp). For T7 and T3 polymerases, use 10–20 U/μg of plasmid DNA.
11. The purification step (Section 3.1.3., step 4) may not be required, and some investigators use the transcription mixtures directly in hybridization assays. However, low backgrounds, consistency, and quantitation of the newly synthesized RNAs may justify the additional time and effort.
12. More than 50% of the labeled ribonucleotide is routinely incorporated and often greater than 80%. An input of 50 μCi may therefore yield up to 4×10^7 Cerenkov-cpm of RNA. This represents 100 pmol of UMP and, assuming no sequence bias (i.e., 25% of residues are uracil residues), 130 ng of RNA (specific activity: 3×10^8 Cerenkov-cpm/μg).

    Similar results are obtained with labeled CTP and GTP, although GTP rapidly loses its incorporation efficiency on storage. ATP is not routinely used because of its higher K_m *(2)*. UTP is particularly stable and can be used even after several weeks, once radioactive decay is taken into account. With each labeled/ribonucleotide, the initial concentration must be >12.5 μ*M* to ensure efficient incorporation and to prevent polymerase pausing. ^{3}H-labeled probes may also be made for use in *in situ* hybridization (*see* ref. *13* for a detailed protocol).

    The T4 gene *32* protein can increase transcription efficiency when added at 10 μg/μg of template DNA (D. Caput, personal communication). It has been recently reported that a reduction of premature termination and an increase synthesis of large full length transcripts (size >1 kb) can be obtained by performing the transcription at 30°C or at room temperature *(4,14)*.
13. Large transcripts that do not enter the polyacrylamide gel are diagnostic of incompletely linearized templates. Small transcripts are indicative of extensive pausing or premature termination.
14. Hybridization of dot-blots is often used to quantitate mRNA levels with large numbers of samples. To accurately quantitate specific hybrids, it is necessary to include as negative control total RNA from cells that do not express the transcript under examination.
15. The intensity provided by transilluminators, a common source of UV light, varies with time, and excessive crosslinking can severely reduce hybridization efficiency (Fig. 1A). Furthermore, I have often observed that UV-crosslinking can increase crosshybridization to rRNAs, possibly by inducing covalent crosslinking with the probe. UV irradiation should thus be limited to membranes subjected to multiple rounds of hybridization.

16. The major variables in optimizing signal-to-noise ratio are the temperature of hybridization and the temperature of the stringency washes (Fig. 1B). Most probes can be hybridized at 58°C, although the temperature must be increased to 68°C for certain pairs of probe and target RNAs. Stringency washes are usually performed at 70 or 75°C. It is difficult to increase stringency much further, since most waterbaths cannot maintain accurate temperatures at or above 80°C. For the detection of DNA targets, hybridization is usually performed at 42°C and the stringency washes are at 65°C.
17. I use RNA probes that are 200–800 nt long. With 600 nt long RNA probes, a hybridization plateau is achieved in 20 h at 58°C with 2.5 ng/mL of probe. This represents an input of 7.5×10^5 Cerenkov/mL. The probe concentration can be increased to 10 ng/mL without increasing background.
18. The probes should be 100–400 nt long, and should include at least 10 nt that are not complementary to the target RNA. Residual template DNA generally produces a trace of full length protected probe that must be distinguishable from the fragment protected by the target RNA (Fig. 2).
19. It is often useful to decrease the specific activity of the probe; since more RNA is synthesized at the resulting higher ribonucleotide concentration, the probes are less susceptible to radiolysis, and less radioactivity is used. The following guideline can be used to alter the specific activity of the probes according to the sensitivity required:

Target RNA abundance	Unlabeled UTP	[^{32}P]-UTP	Probe/sample*
High	100 μ*M*	2.5 μ*M*, 10 μCi	6–12 Kcpm
Moderate	10 μ*M*	2.5 μ*M*, 10 μCi	60–120 Kcpm
Low	—	12.5 μ*M*, 50 μCi	300–600 Kcpm

20. Purification is often necessary for maximal sensitivity or for mapping purposes. The transcription reaction can be directly loaded on the preparative gel after DNase digestion of the template, provided that enough EDTA is present in the sample buffer to chelate all the magnesium. Omitting the DNase digestion of the template results in higher amounts of fully protected probe in the assay.
21. To facilitate the RNase digestion step, each sample should contain the same amount of total RNA. Inequalities may be eliminated by addition of tRNA. A negative control sample, containing only tRNA, is always included (Fig. 2A, lane 4, Fig. 2B, lane 3).

*The amount of probe required per sample for the detection of the target (*see* Note 22).

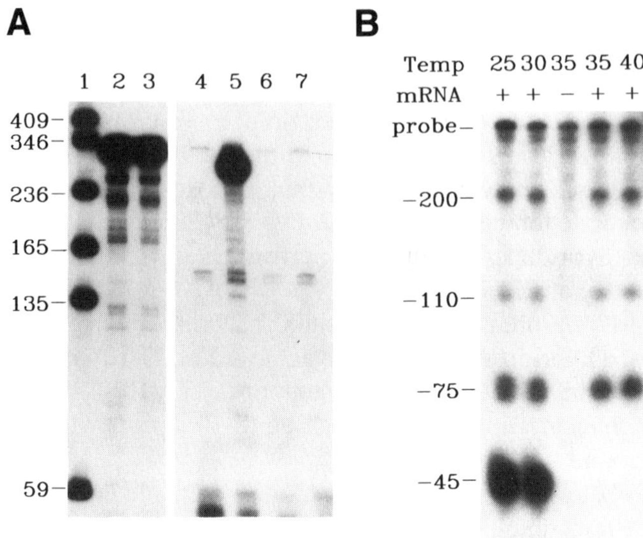

Fig. 2. RNase protection assay. **(A)** Discrimination between target-specific signal and complete probe protection by residual DNA. Detection of PN-I mRNA in total RNA from murine tissues *(17)*. The probe (310 nt long) was gel purified. Lane 1: Size markers. Lane 2: purified probe. Lane 3: probe hybridized and processed without RNase digestion. Lane 4: control hybridization with 10 μg of tRNA; traces of fully protected probe are visible. Lane 5: 10 μg of RNA from seminal vesicles, an abundant source of PN-I mRNA; the specific protected fragment is 260 nt long. Lane 6: 10 μg of liver RNA, which does not contain detectable levels of PN-I mRNA. Lane 7: 10 μg of testis RNA, which contains trace levels of PN-I mRNA. **(B)** Effect of hybridization temperature on the detection of short complementary RNAs. The 5' ends of phage T4 gene 32 transcripts in total RNA from bacteria carrying a gene 32 expression cassette were mapped by hybridization to a cRNA probe containing 400 nt of gene 32 upstream sequences *(15)*. The probe was not gel purified, and hybridization was performed at the indicated temperatures. Traces of fully protected probe result from incomplete DNase digestion of the template; they are also visible in the control hybridization without target RNA. The 44-nt protected fragment is no longer detected above 30°C.

22. The probe must be in excess over the target RNA. Since initially it is not possible to know the abundance of the target, this may have to be calculated empirically. The figures presented in Note 19 may be used as a guide. In addition, for very low abundance target RNAs, the amount of sample RNA may be increased up to 60 μg.

    An input of 1–2 ng of a 300-nt probe in a total volume of 30 μL will drive the hybridization of target RNA to completion (4–8 $T_{1/2}$) in approx

16 h. Shorter hybridizations can be performed but require higher probe input (R_0) to achieve the same extent of saturation, i.e., to maintain the $R_0 \times T_{1/2}$ value.

23. The temperature of hybridization must be reduced to detect small or very AU-rich protected fragments. For instance, a 44-nt fragment of a phage T4 gene 32 transcript (containing 35 A/U and 9 C/G) was only protected by performing the hybridization at 25–30°C *(15)*(Fig. 2B).
24. The amount of RNase is determined by the total amount of RNA present in the samples, including that contributed by the probe. I usually add 0.5 µg of pancreatic RNase and/or 0.25 µg of T1 RNase/µg of RNA. In most cases, digestion with pancreatic RNase alone is sufficient. When the probe and the target RNAs are from different species, the extent of homology can be sufficient to generate discrete protected fragments, particularly if digestion is performed with RNase T1 only. The temperature of digestion can be increased to 30–37°C, although this often leads to partial cleavages within the RNA:RNA hybrids.
25. To ensure that the probe remains intact during the hybridization, it may be useful to include a parallel control that is hybridized and processed without RNase treatment (Fig. 2A, lanes 2 and 3).
26. Minor shorter protected fragments are often detected, and they may complicate the interpretation of mapping assays. To distinguish between digestion artifacts and rare target RNAs that are only partially complementary to the probe, it may be useful to use as a control target RNA a synthetic sense transcript fully complementary to the probe *(16)*.

Acknowledgments

I thank P. Vassalli for his early encouragement to use riboprobes for detecting rare mRNAs. Over the last few years, many collegues, students and technicians have contributed to the methods outlined in this chapter, including M. Collart, N. Busso, J.-D. Vassalli, H. Krisch, S. Clarkson, J. Huarte, S. Strickland, P. Sappino, M. Pepper, A. Stutz, G. Moreau, D. Caput, M. Prentki, P. Gubler, F. Silva, V. Monney, D. Gay-Ducrest, and N. Sappino. Research is supported by grants from the Swiss National Science Foundation. The author is affiliated with the University of Geneva Medical School, Department of Pathology, Geneva, Switzerland.

References

1. Butler, E. T. and Chamberlin, M. J. (1984) Bacteriophage SP6-specific RNA polymerase. *J. Biol. Chem.* **257,** 5772–5788.

2. Melton, D. A., Krieg, P. A., Rebagliati, M. R., Maniatis, T., Zinn, K., and Green, M. R. (1984) Efficient in vitro synthesis of biologically active RNA and RNA hybridization probes from plasmids containing a bacteriophage SP6 promoter. *Nucleic Acids Res.* **12,** 7035–7056.
3. Davanloo, P., Rosenberg, A. H., Dunn, J. J., and Studier, F. W. (1984) Cloning and expression of the gene for bacteriophage T7 RNA polymerase. *Proc. Natl. Acad. Sci. USA* **81,** 2035–2039.
4. Krieg, P. A. and Melton, D. A. (1987) In vitro RNA synthesis with SP6 RNA polymerase. *Meth. Enzymol.* **155,** 397–415.
5. Yisraeli, J. K. and Melton, D. A. (1989) Synthesis of long, capped transcripts in vitro by SP6 and T7 RNA polymerases. *Meth. Enzymol.* **180,** 42–50.
6. Milligan, J. F. and Uhlenbeck, O. C. (1989) Synthesis of small RNAs using T7 RNA polymerase. *Meth. Enzymol.* **180,** 51–62.
7. Milligan, J. F., Groebe, D. R., Witherell, G. W., and Uhlenbeck, O. C. (1987) Oligoribonucleotide synthesis using T7 RNA polymerase and synthetic DNA templates. *Nucleic Acids Res.* **15,** 8783–8798.
8. Roitsch, T. and Lehle, L. (1989) Requirements for efficient in vitro transcription and translation: a study using yeast invertase as a probe. *Biochem. Biophys. Acta* **1009,** 19–26.
9. Schenbon, E. T. and Mierendorf, R. C. (1985) A novel transcription property of SP6 and T7 RNA polymerases: dependence on template structure. *Nucleic Acids Res.* **13,** 6223–6234.
10. Nam, S. C. and Kang, C. (1988) Transcription initiation site selection and abortive initiation cycling of phage SP6 RNA polymerase. *J. Biol. Chem.* **263,** 18123–18127.
11. Solazzo, M., Spinelli, L., and Cesareni, G. (1987) SP6 RNA polymerase: sequence requirements downstream from the transcription start site. *Focus* **10,** 11–12.
12. Moreau, G. (1991) RNA binding properties of the Xenopus LA proteins. Ph.D. thesis, University of Geneva.
13. Sappino, A.-P., Huarte, J., Belin, D., and Vassalli, J.-D. (1989) Plasminogen activators in tissue remodeling and invasion: mRNA localization in mouse ovaries and implanting embryos. *J. Cell Biol.* **109,** 2471–2479.
14. Krieg, P. A. (1991) Improved synthesis of full length RNA probe at reduced incubation temperatures. *Nucleic Acids Res.* **18,** 6463.
15. Belin, D., Mudd, E. A., Prentki, P., Yi-Yi, Y., and Krisch, H. M. (1987) Sense and antisense transcription of bacteriophage T4 gene 32. *J. Mol. Biol.* **194,** 231–243.
16. Belin, D., Wohlwend, A., Schleuning, W.-D., Kruithof, E. K. O., and Vassalli, J.-D. (1989) Facultative polypeptide translocation allows a single mRNA to encode the secreted and cytosolic forms of plasminogen activators inhibitor 2. *EMBO J.* **8,** 3287–3294.
17. Vassalli, J.-D., Huarte, J., Bosco, D., Sappino, A.-P., Sappino, N., Velardi, A., Wohlwend, A., Erno, H., Monard, D., and Belin, D. (1993) Protease-nexin I as an androgen-dependent secretory product of the murine seminal vesicle. *EMBO J.* **12,** 1871–1878.

CHAPTER 25

Quantification of Absolute Amounts of Cellular Messenger RNA by RNA-Excess Solution Hybridization

Rai Ajit K. Srivastava and Gustav Schonfeld

1. Introduction

The quantification of mRNA in mammalian tissues or cell culture is an important part of analysis of gene expression and is required to fully understand many biological processes, such as cell replication, growth, and differentiation or the effects of hormonal, dietary, and genetical factors. The measurement of specific mRNAs relies on the hybridization of nucleic acids with labeled probes that have complementary sequences. The most common analysis utilizes Northern or slot blotting. For Northern blot analysis, RNA is resolved by agarose gel electrophoresis, transferred to a nitrocellulose or nylon membrane and probed either with cDNA *(1)* or riboprobe *(2,3)*. Slot blotting is performed without prior electrophoresis. Both methods are semiquantitative, unreliable for detecting small changes in the message levels, and provide only relative values of mRNA. By contrast, DNA-excess solution hybridization *(4)* can be used to quantify absolute levels of mRNA. In this method a cDNA fragment of a specific message is first subcloned into a M13 vector and single-stranded DNA prepared. The single-stranded DNA is used to synthesize a complementary cDNA probe, which is then excised from the vector with appropriate restriction enzymes and isolated by denaturing polyacrylamide gel electrophoresis *(4)*. A standard curve is generated by performing hybridization of the cDNA probe with the single-stranded M13 template. The RNA to be analyzed for the specific message is also hybridized and

From: *Methods in Molecular Biology, Vol. 31: Protocols for Gene Analysis*
Edited by: A. J. Harwood

the absolute levels of RNA calculated by comparison of the counts obtained to the standard curve. Although DNA-excess hybridization is a great improvement over blotting methods, the disadvantage is that it requires lengthy preparation times and the quantification values are based on a standard curve derived from the DNA:DNA-hybridization of the cDNA probe to the single-stranded DNA template. Furthermore, we have found that single-stranded templates made from phagemids, such as pGem3Zf(+), are always contaminated with other DNAs, which affect the accuracy of quantification of the single-stranded DNA standard.

RNA-excess solution hybridization represents an advance over DNA excess hybridization because it is simpler, more sensitive, and reproducible. Furthermore, this method relies solely on RNA:RNA hybridization kinetics. The desired cDNA is subcloned into the polylinker of a ribroprobe vector, such as pGem3Zf(+), which is flanked by T7 and SP6 RNA polymerase promoters. This enables complementary RNA to be synthesized from either of the cDNA strands. The recombinant vector is then used to prepare a ^{32}P-labeled RNA probe (riboprobe) from one strand and an RNA standard from the other. Total cellular RNA is hybridized with excess of a riboprobe complementary to a specific mRNA and the remaining unhybridized probe digested with RNase A and T1. The hybridized probe is collected on a glass fiber filter by TCA precipitation and counted in a liquid scintillation counter. This method has several advantages:

1. The hybridization is carried out in solution so that probe and mRNA are available for hybridization.
2. It relies on only one kinetic component, RNA:RNA hybridization.
3. It is highly sensitive and provides absolute levels of mRNA.
4. It is much simpler and quicker than DNA-excess solution hybridization and blotting techniques.

We have routinely used RNA-excess solution hybridization for the quantification of mRNA for mouse apolipoproteins and their receptors.

2. Materials

In addition to the solutions specifically mentioned in this section and Section 3., the following are required: appropriate restriction enzymes and buffers, dithiothreitol (DTT), ribonucloetides (NTPs), ribonuclease inhibitor (RNasin or human placental ribonuclease inhibitor), T7 and SP6 RNA polymerases, α-[^{32}P]UTP (600 mCi/m*M*),

RNase-free DNase I, DEPC (diethyl pyrocarbonate), phenol/chloroform/isoamylalcohol (50:49:1), 3*M* sodium acetate (pH 4.8), ethanol, TE (20 m*M* Tris-HCl, pH 7.5, 1 m*M* EDTA), deionized formamide, paraffin oil, yeast tRNA, GF/C glass fiber filters (Schleicher and Schuell, Keene, NH) and a Sephadex G-50 spin column.

All glassware must be treated overnight with 0.05% DEPC-water and autoclaved for 1 h. All reagents, except those containing Tris, are treated with 0.05% DEPC and autoclaved. Alternatively, as in the case of those containing Tris, they may be sterilized by filtration. Preparation of riboprobes should be avoided at working benches where ribonucleases are used.

2.1. Preparation of Riboprobe and RNA Standard

1. Template plasmid: The cDNA should be subcloned into a riboprobe vector (*see* Note 1). The DNA should be of high quality and free of contaminants (*see* Note 2). Before runoff transcription the vector should be linearized separately at two restriction enzyme sites, one for the RNA polymerase T7 directed transcript and the other for SP6 RNA polymerase directed transcript.
2. 5X transcription buffer: 200 m*M* Tris-HCl (pH 7.5), 30 m*M* $MgCl_2$, 50 m*M* NaCl, and 10 m*M* spermidine. Store at –20°C.
3. Nucleotide solutions: 10 m*M* solutions of each NTP. Store at –20°C.
4. Nucleotide mixes: Prepare nucleotide mixes as follows:
   Nucleotide mix A: A 1:1:1 ratio mix of ATP, CTP, and GTP solutions (if labeled UTP is used). Also dilute an aliquot of UTP mix to 100 µ*M*.
   Nucleotide mix B: A 1:1:1:1 mix of all 4 nucleotide solutions. Store in aliquots at –20°C.
5. TCA solution: 10% trichloroacetic acid, 1.5% sodium pyrophosphate. Store at 4°C.

2.2. Hybridization Assay

6. Hybridization buffer: 40% formamide (deionized with resins and stored at –20°C), 400 m*M* NaCl, 40 m*M* PIPES (piperazine-*N,N'*-*bis*[2-ethanesulfonic acid, pH 6.7), and 1 m*M* EDTA. Prepare fresh.
7. RNase buffer: 30 m*M* NaCl, 10 m*M* Tris-HCl, pH 7.5, 5 m*M* EDTA, 40 µg/mL RNase A (pancreatic), and 2 µg/mL RNase T1.

3. Methods

3.1. Preparation of Riboprobe and RNA Standard

1a. To prepare the riboprobe, mix the following at room temperature and in the order shown: 4 µL of 5X transcription buffer, 2 µL of DTT, 1 µL (40 U) of ribonuclease inhibitor, 3 µL of nucleotide mix A, 2 µL of 100 µ*M*

UTP, 1 μg of linearized template, 5 μL of α-[^{32}P]UTP, and 1 μL of T7 or SP6 polymerase (15–20 U). Make the final volume up to 20 μL with DEPC-treated water. Mix the contents by brief vortexing and centrifugation. Incubate for 1 h at 37°C.

1b. Prepare the RNA standard in a similar manner by mixing the following at room temperature and in the order shown: 20 μL of 5X transcription buffer, 10 μL of DTT, 1 μL (40 U) of ribonuclease inhibitor, 20 μL of nucleotide mix B, 2–3 μg of linearized template, and 2 μL of T7 of SP6 polymerase (30–40 U). Make the final volume up to 100 μL with DEPC-treated water. Mix the contents by brief vortexing and centrifugation. Incubate for 90 min at 37°C.

2. Stop the synthesis of both riboprobe and RNA standard by addition of RNase-free DNase I (5 U/mg of template DNA) and incubate for an additional 10 min at 37°C.
3. Add 5 μL of DEPC, vortex, and make the volume of the riboprobe up to 100 μL with water. Extract both with 100 μL of phenol/chloroform/isoamylalcohol (50:49:1). Spin and remove the aqueous layer to another tube. Precipitate by addition of 10 μL of 3*M* sodium acetate and 2.5 vol of absolute ethanol. Wash with 70% ethanol, dry briefly, and dissolve in 30 μL of TE.
5. Separate the riboprobe and RNA standard from free nucleotides using a spin column spun at 200*g* for 2 min.
6. Place 2 μL of the riboprobe and 2 μg of yeast tRNA in an Eppendorf tube. Add 1 mL of cold TCA and place on ice for 15 min. Collect the precipitate on a GF/C glass fiber filter assembled on a vacuum manifold. Wash the filter thoroughly with 10–20 mL of cold TCA solution and dry in an oven at 80°C for 1 h. Place the filter in a scintillation vial, add 4 mL of scintillation cocktail, and count in a liquid scintillation counter (*see* Note 3). Adjust the riboprobe volume to 50,000 cpm/μL.
7. The concentration of the RNA standard should be determined photospectromically at 260 nm in a 1-cm pathlength cuvet. Under these conditions 1.0 A_{260} is equivalent to a concentration of 40 μg/mL.

Both riboprobe and RNA standard should be stored at –70°C as 5-μL aliquots. Riboprobes should be used within a week and never older than 2 wk. The RNA standard can be stored up to 4 mo.

3.2. Hybridization Assay

For the hybridization two duplicate controls should be made in addition to the calibration curve. The first set contains the riboprobe and 5 μg of tRNA instead of RNA sample. This provides the background counts. The second contains only riboprobe and is not treated with RNase buffer. This gives the actual counts of the riboprobe available for hybridization.

1. Lyophilize a range of amounts (1–10 μg) of RNA samples (or appropriate amounts of the RNA standard mixed with 5 μg of yeast tRNA) in a 0.75-mL tube (*see* Notes 4 and 5). Redissolve in 28 μL of hybridization buffer plus 2 μL of riboprobe. Cover the tube contents with 15 μL of paraffin oil to prevent loss in volume. Perform the hybridization for 10 h at 63°C (*see* Note 6).
2. Add 300 μL of RNase buffer to each tube and incubate at 30°C for 1 h.
3. Transfer the tubes to an ice bath and add 1 mL of ice-cold TCA solution. Sit the tubes on ice for 15 min. A precipitate should form.
4. Collect the precipitate on a GF/C glass fiber filter placed on a manifold connected to a vacuum line. Wash each filter thoroughly with 10–20 mL of cold TCA solution and dry in an oven at 80°C for 1 h.
5. Place each filter in a scintillation vial and add 4 mL of scintillation cocktail to each count in a liquid scintillation counter.

The background counts should be subtracted and the results of both sample and RNA standard plotted. From the plot it is straightforward to calculate the absolute amount of specific RNA (*see* Notes 7–9). A typical result is shown in Fig. 1.

4. Notes

1. A cDNA fragment (between 250 and 500 bp) is subcloned into the polylinker region of the vector using general methods *(5)* and DNA is prepared *(6)*. Generally, it is necessary to sequence *(7)* the template to determine the orientation of the cDNA insert; this then determines which restriction enzymes are required to make either the riboprobe or the RNA standard transcripts.
2. Sufficiently pure templates are obtained after banding the plasmid in the presence of ethidium bromide by $CsCl_2$ ultracentrifugation *(8)*.
3. The extent of incorporation and quality of the synthesized RNA should be checked before the hybridization assay is set up. The extent of incorporation can be checked by comparing the counts obtained from the TCA precipitated riboprobe (*see* Section 3.1.; step 6) to the counts from 2 μL of a 1/10 dilution (in DEPC-treated water) of a sample of the probe removed before the spin column. The percentage incorporation can be calculated from these two figures. An estimate of the probe integrity can be obtained by comparing the counts of the TCA precipitated probe to a dilution of the riboprobe after the spin column step.

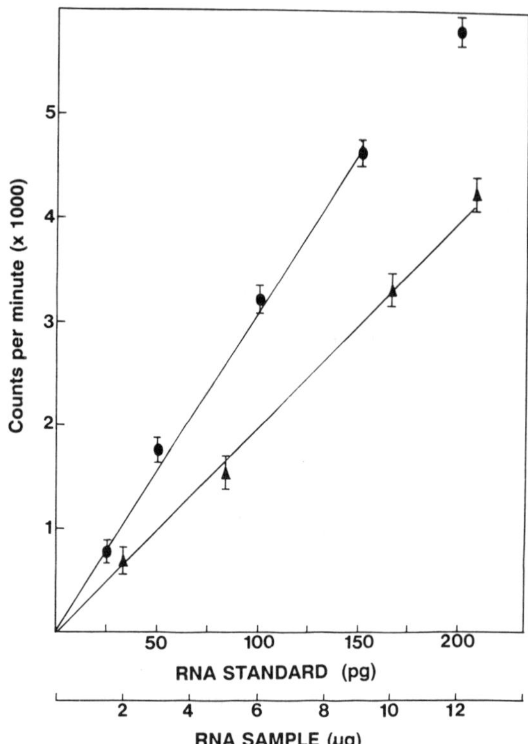

Fig. 1. A typical standard curve and result for a solution hybridization assay. Increasing concentrations of RNA standard synthesized in vitro were hybridized with a mouse apoAII riboprobe. Each assay was performed in triplicate. Increasing concentrations of mouse liver RNA were also hybridized in a similar way in order to assess the linearity of the response curve. (○), standard curve; (△), hybridization of riboprobe with RNA samples.

The integrity and size should also be determined by denaturing polyacrylamide gel electrophoresis.

4. We have used cellular RNA prepared either by commercially available kits (e.g., RNAZol^{TM}B *(9)*) or by Chirgwin's guanidinium isothiocyanate/$CsCl_2$ centrifugation method *(10)*. RNA samples are normally stored in aliquots in DEPC-treated water or TE at –70°C.
5. The amount of RNA to be used for the assay depends on the specific mRNA abundance. Therefore, a linear dose-response range must be determined. At least three different concentrations of cellular RNA in duplicate are hybridized to obtain the dose range of linear response.

6. The optimum hybridization temperature is determined for each riboprobe by performing hybridization over a range temperatures from 40–75°C.
7. The solution hybridization assay provides absolute mRNA values per microgram of total RNA to normalize these data to a per cell basis tissue DNA and RNA are determined colorimetrically *(4)*. The diploid DNA content per cell is required to convert RNA concentration from picogram to molecules per cell.
8. If the RNA standard contains extra nucleotides transcribed from the vector, then a correction needs to be made for the extra counts. A second correction should be made for the length of the riboprobe and the size of the mRNA to be quantified. For example, if the probe length is 500 nucleotides but the mRNA 1000 nucleotides in length, the final value is multiplied by 2.
9. To confirm the specificity of the riboprobe, Northern blotting and RNase protection analysis should be undertaken.

References

1. Twoney, T. A. and Krawetz, S. A. (1990) Parameters affecting hybridization of nucleic acids blotted onto nitrocellulose membrane. *Biotechniques* **8(5),** 478–481.
2. Srivastava, R. A. K., Pfleger, B., and Schonfeld, G. (1991) Expression of low density lipoprotein receptor, apolipoprotein B, apolipoprotein AI and apolipoprotein AIV mRNA in various mouse organs as determined by a novel RNA-excess solution hybridization assay. *Biochim. Biophys. Acta.* **1090,** 95–101.
3. Srivastava, R. A. K. and Schonfeld, G. (1991) Using riboprobes for northern blotting analysis. *Biotechniques* **11,** 584–587.
4. Williams, D. L., Newman, T. C., Shelnes, G. S., and Gordon, D. A. (1986) Measurements of apolipoprotein mRNA by DNA-excess solution hybridization with single stranded probes. *Meth. Enzymol.* **128,** 671–689.
5. Srivastava, R. A. K. and Schonfeld, G. (1990) A rapid method for screening large numbers of recombinant clones. *Biotechniques* **9(6),** 689–691.
6. Tiesman, J. and Rizzino, A. (1991) A rapid and reliable method for the purification of high quality plasmid DNA for double stranded sequencing. *Biotechniques* **10(3),** 326–328.
7. Mierendorf, R. C. and Pfeffer, D. C. (1987) Direct sequencing of denatured plasmid DNA. *Meth. Enzymol.* **152,** 556–562.
8. Davis, L. G., Dibner, M. D., and Battey, J. F. in *Basic Methods in Molecular Biology,* Elsevier, New York, pp. 99–101.
9. Chomczynski, P. (1989) Isolation of RNA by The RNAzol™B method. *Cinna Biotex Bulletin No. 3.*
10. Chirgwin, J. M., Przbyla, A. E., MacDonald, R. J., and Rutter, W. J. (1979) Isolation of biologically active ribonucleic acid from sources enriched in ribonuclease. *Biochemistry* **18,** 5294–5299.

CHAPTER 26

Measurements of Rate of Transcription in Isolated Nuclei by Nuclear "Run-Off" Assay

Rai Ajit K. Srivastava and Gustav Schonfeld

1. Introduction

Unlike gene expression in prokaryotic cells, which is primarily under transcriptional control, gene expression in eukaryotic cells is subject to both transcriptional and posttranscriptional controls. Since transcription and translation in eukaryotic cells are separated topographically, the regulation of mRNA metabolism can occur at multiple sites, within nuclei and in the cytoplasm. Nevertheless, transcription remains a critical locus of control of eukaryotic gene expression. Transcriptional regulation affects cellular mRNA abundance by affecting rates of transcription. Another important control mechanism that can affect mRNA abundance is the rate of mRNA decay. Thus, half-life of mRNA represents a balance between rates of transcription and intracellular degradation, e.g., an increase in the abundance of mRNA could result from decreased mRNA degradation, increased mRNA synthesis, or both.

Tissue specific expression and the level of expression of certain genes are routinely determined by measuring the abundance of the corresponding mRNA *(1–3)* and, frequently, mRNA levels are interpreted as reflecting rates of transcription. Although a safe interpretation in some cases *(4–8)*, this is not always true *(9–15)*. It is, therefore, important to measure the rate of transcription directly, in order to understand the contribution of transcription rates to "setting" the cellular level of any specific mRNA.

From: *Methods in Molecular Biology, Vol. 31: Protocols for Gene Analysis*
Edited by: A. J. Harwood

Relative rates of transcription are measured by nuclear "run-off" assays using isolated cell nuclei. The assay quantifies the elongations in vitro of nascent mRNA chains already initiated in vivo. Nuclei are isolated from homogenized tissues or cultured cells, and incubated in the presence of four ribonucleotides, one of them radiolabeled. The reaction is stopped by the addition of RNase-free DNase I, and total nuclear RNA that contains newly synthesized labeled RNA is extracted. The extracted RNA is hybridized to a denatured, immobilized cDNA corresponding to the mRNA being measured. After hybridization, the membrane is treated with RNase A and washed to remove the nonspecifically bound RNA. Bound mRNA are hydrolyzed from the filter and counted in a liquid scintillation counter. Nuclear transcription rates of given mRNAs are frequently compared to transcription rates of an "internal standard," such as, β-actin mRNA.

2. Materials

All glassware used for a nuclear run-off assay is treated with 0.05% DEPC (diethyl pyrocarbonate) overnight and autoclaved for 1 h. The reagents used are also treated in a similar way and either autoclaved or filter sterilized.

2.1. Isolation of Nuclei

2.1.1. Preparation of Nuclei from Tissues

1. Buffer A: 60 m*M* KCl, 15 m*M* NaCl, 0.15 m*M* spermine, 0.5 m*M* Spermidine, 14 m*M* β-mercaptoethanol, 0.5 m*M* EGTA, 2 m*M* EDTA, and 15 m*M* HEPES (pH 7.5). Store at 4°C.
2. Buffer B: Same as buffer A but with 0.1 m*M* of EGTA and EDTA. (Made in DEPC-treated water and sterilized by filtration through a sterile millipore filter.)
3. Sucrose solutions: Solution A: 0.3*M* sucrose in buffer A, solution B: 1*M* sucrose in buffer B, and solution C: 1.5*M* sucrose solution in buffer B. Store at 4°C.
4. Nuclei storage buffer: 20 m*M* Tris-HCl, pH 7.9, 75 m*M* NaCl, 0.5 m*M* EDTA 0.85 m*M* DTT, 0.125 m*M* PMSF 50% glycerol. Store at 4°C.

2.1.2. Preparation of Nuclei from Cultured Cells

5. Wash buffer: 20 m*M* Tris-HCl, pH 7.5, 15 m*M* NaCl, and 1.1 m*M* sucrose. Store at 4°C.

6. Hypotonic buffer: 20 m*M* Tris-HCl, pH 8.0, 4 m*M* $MgCl_2$, 6 m*M* $CaCl_2$, and 0.5 m*M* DTT. Store at 4°C.
7. Lysis buffer: 0.6*M* sucrose; 0.2% Nonidet P-40; and 0.5 m*M* DTT. Prepare fresh from stock solutions.

2.2. Elongation of Nascent Chains

8. Elongation buffer (2X): 200 m*M* Tris-HCl, pH 7.9, 100 m*M* NaCl, 0.8 m*M* EDTA, 0.2 m*M* phenylmethyl sulfonyl chloride, 2.4 m*M* DTT, 2 mg/mL heparin sulfate, 4 m*M* $MnCl_2$, 8 m*M* $MgCl_2$, and 20 m*M* creatine phosphate. Make fresh each time from the stock solutions.
9. Nucleotide mix: Prepare 100 m*M* solution of GTP, ATP, and CTP, and mix in 1:1:1 ratio. Store frozen in aliquots of 500 µL at –20°C.
10. Ribonuclease inhibitor: RNasin (Promega, Inc.) or human placental ribonuclease inhibitor at 40 U/µL.
11. α-[^{32}P] UTP: Labeled UTP with specific activity of 600 µCi/mmol.
12. DNase 1: RNase-free.
13. α-Omanitin: 2 pg/mL.
14. SET buffer: 5% sodium dodecyl sulfate, 50 m*M* EDTA, 100 m*M* Tris-HCl, pH 7.4. Prepare fresh.
15. Proteinase K: 10 mg/mL, store at 20°C.
16. Extraction buffer: 4*M* guanidinium thiocyanate, 25 m*M* sodium citrate, pH 7.0, 0.5% Sarkosyl, and 0.1*M* β-mercaptoethanol. Store at room temperature.
17. Phenol and chloroform: salt saturated phenol and a 49:1 chloroform:isoamylalcohol mix.
18. Yeast tRNA: 20 mg/mL solution.

2.3. Hybridization

19. 1*M* HEPES: N-[2-hydroxyethyl] piperazine-N'-[2-ethanesulfonic acid], pH 6.5.
20. Hybridization buffer: 20 m*M* PIPES (piperazine-*N,N'-bis*[2-ethanesulfonic acid]), pH 6.7, 50% formamide (deionized with mixed bed resins), 2 m*M* EDTA, 0.8*M* NaCl, 0.2% SDS, 0.02% Ficoll, 0.02% polyvinylpyrrolidone, 0.02% bovine serum albumin, and 500 µg/mL denatured salmon sperm DNA. Make fresh before use.
21. 2X SSC: A 1/10 dilution of 20X SCC. 20X SCC contains 3*M* NaCl and 0.3*M* *tri*-sodium citrate.
22. RNase A solution: A 10 pg/mL solution in 2X SCC.
23. Scintillation cocktail.

3. Methods

3.1. Isolation of Nuclei

3.1.1. Preparation of Nuclei from Tissues

1. Homogenize preweighed tissue in 10 vol of ice-cold sucrose solution A (10 mL solution A/1 g tissue). Perform all the steps of nuclei preparation at 0–4°C.
2. Filter the homogenate through either a 40-μ*M* nylon cloth or through 4 layers of cheesecloth, and layer over a 10-mL cushion of solution B. Spin for 20 min at 2500*g* and 4°C in a swinging bucket rotor of Beckman (Fullerton, CA) Centrifuge.
3. Resuspend pelleted crude nuclei in sucrose solution B, using 2–3 mL for up to 1 g of tissue sample, layer over 5 mL of solution C, and centrifuge at 45,000*g* in Beckman SW50 rotor at 4°C for 60 min.
4. Resuspend the pellet, containing clean nuclei, in nuclear storage buffer. This gives nuclei at a concentration of $\sim 10^6/\mu L$. Isolated clean nuclei are either used immediately after preparation or may be stored frozen in aliquots of about $1–5 \times 10^7$ at –70°C without loss in activity for up to 4 wk.

3.1.2. Preparation of Nuclei from Cultured Cells

1. Wash a confluent culture containing $1–2 \times 10^8$ cells with ice-cold wash solution and collect by centrifugation at 300*g*.
2. Resuspend the cells in 2.5 mL of ice-cold hypotonic buffer and allow to sit on ice for 5 min. Add 2.5 mL of lysis buffer.
3. Break the cells further with the tight-fitting pestle of a homogenizer. Usually 6–10 strokes are enough to break the cells.
4. Pellet nuclei by centrifugation at 1500*g* for 10 min and resuspend in 2 mL of sucrose solution A.
5. Layer the crude nuclear suspension over 2.5 mL of sucrose solution C and centrifuge for 1 h at 45,000*g*. Resuspend the nuclear pellet in a nuclear storage buffer and either use immediately for in vitro nuclear run-off assay or store at –70°C for up to 4 wk.

3.2. Elongation of Nascent mRNA Chains

1. To a total volume of 200 μL add 100 μL of 2X elongation buffer, 6 μL of nucleotide mix, 40 U (1 μL) RNase inhibitor, 10^7 nuclei (suspended in nuclei storage buffer), and 100 μCi ^{32}P[UTP].
2. Allow the transcription reaction to proceed at 26°C for 20 min and then stop by the addition of 100 U DNase I (RNase free). Incubate for an additional 5 min. To determine exclusively the RNA polymerase II

dependent transcription, transcription is also performed in presence of α-amanitin (2 μg/mL) and the counts obtained are subtracted from the total counts obtained.

3. Treat the samples with 2 μL of proteinase K (10 mg/ mL) and 20 μL of SET buffer for 30 min at 37°C. Add 400 μL of extraction buffer and 80 μL of sodium acetate (2.0*M*, pH 4.0) and vortex the contents for 10 s. Add 700 μL of salt saturated phenol and 150 μL of chloroform:isoamylalcohol, vortex for 10 s, and allow to sit on ice for 15 min.
4. Centrifuge at 4°C and 12,000*g* for 15 min. Transfer the aqueous (top) phase to a clean tube and add 20 μg of yeast tRNA and an equal volume of cold isopropanol. Mix tube contents and incubate at –20°C for 20 min. Centrifuge at 12,000*g* and 4°C for 15 min and wash the pellet with 70% ethanol. Lyophilize and redissolve in 100 μL of hybridization buffer (*see* Note 1).

3.3. Hybridization

The amount of specific mRNA synthesized in the nuclear run-off assay is determined by hybridizing the total synthesized RNA with the specific cDNA probe (*see* Notes 2 and 3). This is then compared to transcription of an "internal control," e.g., of β-actin mRNA.

1. The double-stranded recombinant plasmid containing the appropriate cDNA fragment (or β-actin cDNA) is denatured with 0.2*M* NaOH/2 m*M* EDTA by incubation for 15 min at 37°C. It is then neutralized by 1*M* HEPES.
2. Five micrograms of denatured plasmid DNA is applied to nitrocellulose paper using dot blot apparatus and baked for 2 h at 80°C in vacuum. The portions of the filter where plasmid DNA are spotted are cut out with the help of a sterile cork borer or sharp blade.
3. Prehybridize the membrane for 2 h in 400 μL of hybridization buffer. Remove buffer from the tube and replace with total RNA (10^5–10^6 cpm) dissolved in 250 μL of hybridization buffer. Cover the hybridization mixture plus membrane with 50–100 μL of mineral oil and hybridize for 50 h at 42°C.
4. After the hybridization wash the membranes twice with 500 μL of 2X SSC/0.1% SDS for 30 min at room temperature. In order to remove nonspecific binding, treat the membrane with 400 μL of RNase A solution (10 μg/mL in 2X SSC) at 37°C for 30 min. This will remove unhybridized RNA. Wash with 1 mL of 2X SSC twice at room temperature and expose to X-ray film (*see* Figs. 1 and 2).
5. To elute the hybridized RNA, incubate the membranes with 200 μL of 0.3*M* NaOH for 15 min at 65°C followed by the addition of 50 μL

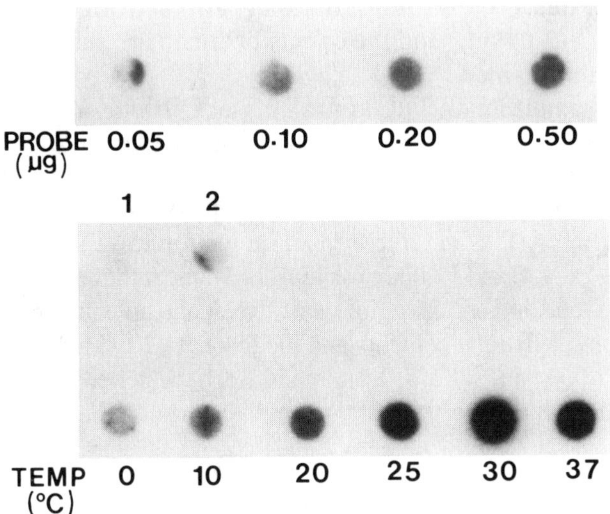

Fig. 1. Nuclear run-off assay, using 10^7 mouse liver nuclei per reaction, was performed as described in the text and probed for the rate of low density lipoprotein receptor mRNA transcription using 1.3 kb rat liver LDL-receptor cDNA. The top panel shows an increasing hybridization signal with increasing amounts of probe immobilized onto nitrocellulose membrane. Lower panel shows the effect of temperature on the LDL-receptor mRNA transcription. Lane 1, 5 µg plasmid DNA that does not contain LDL-receptor cDNA; Lane 2, transcription performed in the presence of 2 µg/mL α-amanitin.

glacial acetic acid and 4 mL of scintillation cocktail. Count ^{32}P radioactivity in a liquid scintillation counter. This value provides the relative rates of transcription of a specific mRNA (*see* Note 4).

4. Notes

1. The integrity of newly synthesized RNA and the extent of incorporation of label in the nascent RNA chains may be checked prior to proceeding for hybridization:
   a. To determine the extent of label incorporation, an aliquot of synthesized total nuclear RNA is diluted 10–20-fold depending on the label incorporation as judged by Geiger counter. An aliquot (usually 2 concentrations; 5 and 10 µL) of the diluted sample is counted in a liquid scintillation counter.
   b. To determine the quality of the synthesized RNA, an aliquot of newly synthesized RNA (50,000–100,000 cpm) are resolved in 6% denaturing polyacrylamide gel containing 7*M* urea. After the electrophore-

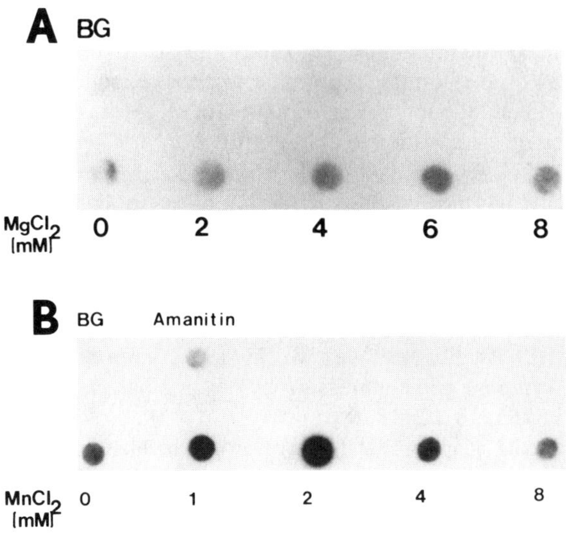

Fig. 2. LDL-receptor mRNA transcription on isolated mouse liver nuclei in the presence of different concentrations of $MgCl_2$ or $MnCl_2$. Each transcription assay was performed with 10^7 nuclei at 28°C for 20 min. The concentration of divalent cations are indicated in the figure.

sis, the gel is dried and exposed to X-ray film. If the transcribed RNAs are intact, one can see several distinct RNA bands all along the autoradiogram. If the RNAs are degraded, smaller fragments of RNA appear and there is "smudging" of the bands.

2. To ensure specificity of the probe, linear concentrations (0.1–1 μg) of recombinant plasmid are immobilized on one set of filters, and identical amounts of a plasmid that does not contain cDNA insert are bound to another set of nitrocellulose membranes. Hybridization is performed with the same amounts of ^{32}P[RNA] using both sets of filters. If the cDNA probe is specific, a linear increase in the hybridization signal is obtained with the recombinant plasmid but not with the nonrecombinant plasmid.
3. In another experiment, increasing amounts of ^{32}P[RNA] (10^5–10^6 cpm) are hybridized with constant amounts (5 μg) of the recombinant and the nonrecombinant plasmids. Here again one gets increasing hybridization signals with the recombinant plasmid but not with the nonrecombinant plasmid.
4. For background counts hybridization is carried with the nonrecombinant plasmid and the counts obtained are subtracted from the counts obtained by hybridization with the recombinant plasmid.

References

1. Sorci-Thomas, M., Wilson, M. D., Johnson, F. L, Williams, D. L., and Rudel, L. L. (1989) Studies on the expression of genes encoding apolipoproteins B-100 and B-48 and the low densiy lipoprotein receptor in non-human primates. Comparison of dietary fat and cholesterol. *J. Biol. Chem.* **264,** 9039–9045.
2. Elshourbagy, N. A., Boguski, M. S., Liao, W. S. L., Jefferson, L. S., Gordon, J. I., and Taylor, J. M. (1985) Expression of rat apolipoprotein A-IV and A-I genes: mRNA induction during development and in response to glucocorticoids and insulin. *Proc. Natl. Acad. Sci. USA* **82,** 8242–8246.
3. Karathanasis, S. K. and Yunis, S. (1986) Structure, evolution, and tissue-specific synthesis of human apolipoprotein AIV. *Biochemistry* **25,** 3962–3970.
4. Feng, B., Hilt, D. C., and Max, S. R. (1990) Transcriptional regulation of glutamine synthase gene expression by dexamethasone in L6 muscle cells. *J. Biol. Chem.* **265,** 18702–18706.
5. Brock, M. L. and Shapiro, D. J. (1983) Estrogen regulates the absolute rate of transcription of the *Xenopus laevis* vitellogenin genes. *J. Biol. Chem.* **258,** 5449–5455.
6. McKnight, G. S. and Palmiter, R. D. (1979) Transcriptional regulation of the ovalbumin and conalbumin genes by steroid hormones in chick oviduct. *J. Biol. Chem.* **254,** 9050–9058.
7. Chazenbalk, G. D., Wadsworth, H. L., and Rapoport, B. (1990) Transcriptional regulation of ferritin H messenger RNA levels in FRTL5 rat thyroid cells by thyrotropin. *J. Biol. Chem.* **265,** 666–670.
8. Chinsky, J. M., Maa, M. C., Ramamurthy, V., and Kellems, R. E. (1989) Adenosine deaminase gene expression. Tissue-dependent regulation of transcriptional elongation. *J. Biol. Chem.* **264,** 14,561–14,565.
9. Saini, K., Thomas, P., and Bhandari, B. (1990) Hormonal regulation of stability of glutamine synthetase mRNA in cultured 3T3-L1 adipocytes. *Biochem. J.* **267,** 241–244.
10. Brock, M. L. and Shapiro, D. J. (1983) Estrogen stabilizes vitellogenin mRNA against cytoplasmic degradation. *Cell* **34,** 207–214.
11. Antrast, J., Lasnier, F., and Pairault, J. (1991) Adipsin gene expression in 3T3-F442A adipocytes is post-transcriptionally down-regulated by retinoic acid. *J. Biol. Chem.* **266,** 1157–1161.
12. Jefferson, D. M., Clayton, D. F., Darnell, J. E., Jr., and Reid, L. M. (1986) Posttranscriptional modulation of gene expression in cultured rat hepatocytes. *Mol. Cell Biol.* **4,** 1929–1934.
13. Hod, Y. and Hanson, R. W. (1988) Cyclic AMP stabilizes the mRNA for phosphoenol pyruvate carboxykinase (GTP) against degradation. *J. Biol. Chem.* **263,** 7747–7752.
14. Gordon, D. A., Shelness, G. J., Nicosia, M., and Williams, D. L (1988) Estrogen-induced destabilization of yolk precursor protein mRNA in avian liver. *J. Biol. Chem.* **263,** 2625–2631.
15. Srivastava, R. A. K., Jiao, S., Tang, J., Pfleger, B., Kitchens, R. T., and Schonfeld, G. (1991) In vivo regulation of low density lipoprotein receptor and apolipoprotein B gene expresion by dietary fatty acids and dietary cholesterol in inbred strains of mice. *Biochim. Biophys. Acta* **1086,** 29–43.

CHAPTER 27

An RNA Polymerase II In Vitro Transcription System

Roberto Mantovani

1. Introduction

A great deal of attention in recent years has been focused on the mechanisms governing transcription. Repression and/or activation of specific genes, or sets of genes, represents a key regulatory step in such diverse processes as cell growth, differentiation, and response to external stimuli.

It is largely accepted that the genomic DNA sequences immediately preceding and surrounding the transcriptional initiation start site—the promoter—are responsible for the correct spatial and temporal activation of genes transcribed by RNA polymerase II. Sometimes this may be in conjunction with remote regulatory elements—the enhancer—often located several kb away *(1,2)*. Promoter and enhancer elements are known to bind proteins responsible for the activation and the correct positioning of transcription. Thanks to the recent spectacular advances of methods to study DNA-binding proteins and to clone the corresponding genes, it has been possible to gain a considerable amount of information concerning these important biological phenomena.

Among these techniques, the development of soluble transcription systems derived from isolated nuclei *(3–5)* or from whole cells *(6)* has been and will continue to be extremely useful. In vitro transcription has been used to:

1. Analyze rapidly DNA elements of importance in a given promoter *(5)*;
2. Correlate these elements to specific DNA-binding proteins *(7)*;

From: *Methods in Molecular Biology, Vol. 31: Protocols for Gene Analysis*
Edited by: A. J. Harwood

3. Dissect the interplay of different activator proteins *(8)*;
4. Purify proteins involved in transcription that do not bind to DNA *(9)*;
5. Perform structure–function studies on the role of a given transcription factor whose gene is cloned *(10)*; and
6. Explore the role of chromatin structure in such processes *(11)*.

I will describe here a method that has proven valuable in obtaining transcriptional extracts from many cell lines. I will also describe polyacrylamide gel electrophoresis and preparation for S1 analysis—two commonly used means of detecting the in vitro transcripts.

2. Materials

2.1. Preparation of Transcription Extracts

1. PBS: Phosphate buffered saline. Dissolve 8 g NaCl, 0.2 g KCl, 0.2 g KH_2PO_4, and 1.15 g Na_2HPO_4 in 1 L of H_2O.
2. PMSF: Phenylmethylsulfonylfluoride (Sigma, St. Louis, MO). Dissolve in isopropanol; prepare a 100 m*M* stock solution. Store at –20°C.
3. Protease inhibitors solution: Pepstatin and chymostatin (Sigma) are dissolved in ethanol; antipain, leupeptin, and aprotinin (Sigma) in water. Prepare a 2.5 mg/mL stock solution. Store at –20°C.
4. Hypotonic buffer (HB). 10 m*M* HEPES, pH 7.9, 0.75 m*M* spermidine, 0.15 m*M* spermine, 0.1 m*M* EDTA, 1 m*M* DTT, 10 m*M* KCl, 0.5 m*M* PMSF, and 2.5 µg/mL proteases inhibitors.
5. Sucrose restore buffer (SRB). Sucrose 67.5%, 5 m*M* HEPES, pH 7.9, 0.75 m*M* spermidine, 0.15 m*M* spermine, 10 m*M* KCl, 0.2 m*M* EDTA, 1 m*M* DTT, 0.5 m*M* PMSF, and 2.5 µg/mL protease inhibitors.
6. Nuclear resuspension buffer (NRB). 18 m*M* HEPES, pH 7.9, 0.675 m*M* spermidine, 0.135 m*M* spermine, 0.18 m*M* EDTA, 2 m*M* DTT, 22.5% glycerol, 0.42*M* NaCl, 0.5 m*M* PMSF, and 2.5 µg/mL proteases inhibitors.
7. Nuclear dialysis buffer (HDB). 20 m*M* HEPES, pH 7.9, 100 m*M* KCl, 0.2 m*M* EDTA, 2 m*M* DTT, and 20% glycerol. Store at 4°C. Prepare solutions 4–6 fresh; PMSF and DTT should be added to the solutions just before use.

2.2. In Vitro Transcription: Nuclear Runoff

8. Template plasmid DNA *(12)*: A plasmid template that contains a suitable promoter element and a means of detecting the generated transcript. Two common forms of detection are the G-less Cassette or the use of S1 analysis (*see* Notes 1 and 2 and Fig. 1).
9. 10X Transcription mix: 100 m*M* HEPES, pH 7.9, 300 m*M* KCl, 60 m*M* $MgCl_2$, 20 m*M* creatine phosphate, 6 m*M* ATP, 6 m*M* UTP, 200 µ*M* CTP, 2.5 m*M* 3'-O-Methyl GTP (*see* Note 3). Store at –20°C.

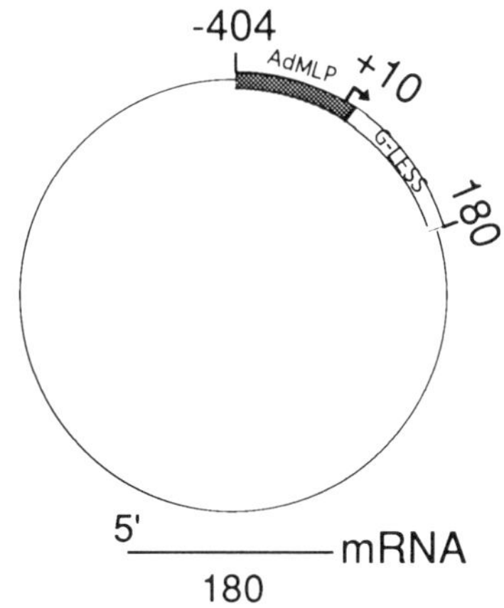

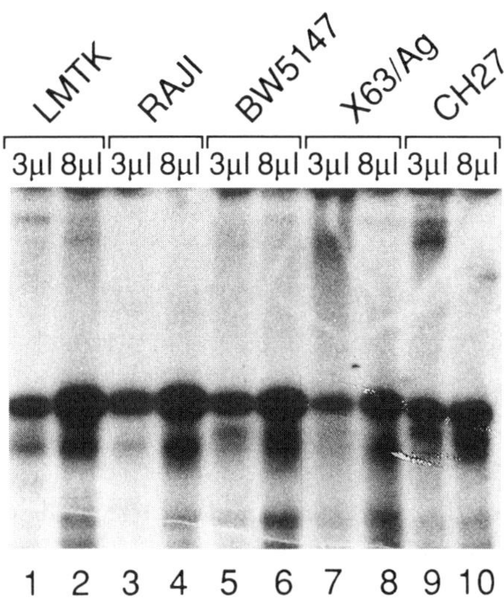

Fig. 1. Detection of in vitro transcription using extracts from different cell lines by the runoff method. The G-less Cassette vector contains a synthetic gene of 170 nucleotides with no G, driven by a full Adenovirus Major Late promoter from pos. –404 to pos. +10 *(17)*, generating a 180-nucleotides long transcript. Amounts of 3 and 8 µL of each extract were tested for activity on 400 ng of plasmid DNA. All extracts are equally active both at low and high concentrations.

10. Transcription stop mix: 0.25*M* NaCl, 20 m*M* HEPES, pH 7.9, 5 m*M* EDTA, 1% SDS.
11. α-[^{32}P] CTP: specific activity > 400 Ci/mmol.

12. RNasin: 25 U/µL supplied from Promega (Madison, WI). Store at –20°C.
13. Proteinase K: 10 mg/mL stock solution. Store at –20°C.
14. tRNA: 10 mg/mL stock solution. Store at –20°C.
15. Phenol buffered with Tris-HCl, pH 7.5. Store at –20°C.
16. Ether.
17. Ethanol: absolute.
18. 25X TBE: 270 g; Tris-base, 137 g; Boric acid, 232.5 g; EDTA, H_2O to 1 L.
19. Formamide dye: 96 µL of deionized formamide, 1 µL of a 10% bromophenol blue solution, 1 µL of a 10% Xylencyanol solution, 2 µL of 25X TBE. Store at –20°C.
20. 6% polyacrylamide/7*M* urea solution in 0.5X TBE. Store at 4°C for up to 3 mo.

2.3. *In Vitro Transcription: S1 Analysis*

21. 10X S1 transcription mix: 300 m*M* KCl, 100 m*M* HEPES, pH 7.9, 80 m*M* $MgCl_2$, 20 m*M* creatine phosphate. Store at –20°C.
22. 10X S1 NTP mix: 5 m*M* ATP, 5 m*M* CTP, 5 m*M* UTP, 5 m*M* GTP. Store at –20°C.
23. 10X S1 DNase I mix: 300 m*M* Tris-HCl, pH 7.9, 50 m*M* $MgCl_2$.
24. VRC: Vanadyl ribonucleoside complex (BRL). Store at –20°C.
25. RNase-free DNase I: *See* Note 4.

3. Methods

3.1. *Preparation of Transcription Extracts*

The whole procedure should be performed at 4°C and steps 3–5 should be completed as fast as possible (do not exceed 5 min) to minimize leakage of transcription factors from nuclei. This method is derived from that of Shapiro et al. *(4)*. Cells are grown in suspension whenever possible and harvested in the mid-logarithmic growth phase, between 6–8.1 × 10^5/mL depending on the cell lines (*see* Notes 5 and 6).

1. Pellet cells at 200*g* for 5 min at 4°C and wash once with 100 mL of cold PBS for every liter of cells used. Pellet cells and resuspend in 5 vol of HB. Incubate on ice for 10 min.
2. Pellet cells by centrifugation at 300*g* for 10 min. The supernatant gently discarded, resuspend the pelleted cells in 2 vol of HB. Homogenize with eight strokes of a Dounce homogenizer (type B pestle).
3. Immediately add 0.1 vol of SRB and homogenize the suspension with type A pestle (two strokes). Rapidly transfer the homogenate to a plas-

tic tube and spin down in a Sorvall HB4 rotor at 8000 rpm (10,000*g*) for 30 s (*see* Note 7).

4. Quickly resuspend the pelleted nuclei in NRB and rock for 30 min at 4°C. Centrifuge the resulting viscous material at 150,000*g* for 45 min (e.g., 38,000 rpm in a SW41 rotor).
5. Transfer the supernatant to a new tube. Add 0.3 g of solid $(NH_4)_2SO_4$ for every mL and gently shake the tube until the salt is dissolved completely. Place on ice for 20 min. Recover the precipitate by centrifugation at 100,000*g* for 20 min (e.g., 28,000 rpm in a SW41 rotor).
6. Resuspend the protein pellet in 0.6 mL of NDB for every 1×10^9 cells. Dialyze against >250 vol of the same buffer (the proteases inhibitors may be omitted) for 4 h with one buffer change.
7. During the dialysis a cloudy precipitate will form. Remove by centrifugating for 10 min in an Eppendorf centrifuge. The supernatant is batched as 100-µL aliquots, quick frozen, and stored in liquid nitrogen (*see* Note 8). The final protein concentration should be 8–12 µg/µL.

3.2. *In Vitro* Transcription: Nuclear Runoff

1. Add 1 µL of template DNA and 9 µL of the extract to a sterile 1.5-mL tube. Preincubate for 10 min at 30°C (if less than 9 µL of extract is employed, a corresponding amount of NDB buffer is used).
2. Add 2 µL of 10X runoff transcription mix, 0.5 µL of $[^{32}P]$-CTP, 0.8 µL of RNasin (20 U), and 6.7 µL of water. Incubate the reaction at 30°C for 45 min.
3. End transcription by addition of 200 µL of stop mix plus 20 µg of tRNA and 20 µg of proteinase K. Place at 37°C for 15 min.
4. Extract with 1 vol of phenol. Vortex for 30 s, spin for 2 min in microfuge, and recover the upper phase. Extract twice with ether. Precipitate by adding 750 µL of ethanol, placing the tube in dry ice for 15 min and centrifuging in microfuge for 15 min.
5. Resuspend the pellet in formamide dye and load a 6% polyacrylamide/urea gel. Run gel until the bromophenol blue reaches the bottom. Fix the gel for 15 min in 10% acetic acid/10% methanol, dry under vacuum, and expose at –80°C with intensifying screen.

3.3. *In Vitro* Transcription: S1 Analysis

1. In a sterile 1.5-mL tube add: 1 µL of template DNA (50–800 ng/µL), 2 µL 10X S1 transcription mix, 0.8 µL of RNasin, 9 µL of extract, supplemented with NDB buffer if <9 µL are used, and 7–8 µL of sterile water. Incubate at 30°C for 10 min.

2. Add 2 µL of 10X S1 NTP mix to start transcription and incubate at 30°C for a further 45 min.
3. Add a premixed solution of 4 µL 10X of DNase I mix, 2 µL of VRC, 1 µL of RNase-free DNase I, and 13 µL of sterile water. Place the reaction for 5 min at 37°C. This step is necessary to remove template DNA, which might interfere with the hybridization step of S1 analysis.
4. Add 200 µL of stop mix plus 20 µg of tRNA and 20 µg of proteinase K and continue the incubation for a further 15 min at 37°C.
5. The synthesized RNA is then purified by phenol/ether extraction and ethanol precipitation. The pellet is washed once with cold ethanol, dried, and is then ready for S1 analysis (*see* Chapter 28).

4. Notes

1. The extracts can be assayed in two different transcription systems:
   a. The G-less cassette: A run-off transcription system described by Sawadogo and Roeder *(13)*, in which the test promoter is put in front of a synthetic gene devoid of guanine residues. The main advantage of this system is its speed, resulting from quick completion of the experiments (4–5 h from the beginning to gel exposure) and high sensitivity, since the transcribed RNA incorporates radiolabeled nucleotides. A strong promoter like Adenovirus Major Late should give a signal detectable after few hours.

      The main disadvantage is the necessity to have no guanine residues on the transcribed strand since these block chain elongation. This may be a problem for promoters that have regulatory elements (or "Initiators") present downstream of the cap site, within the transcribed region *(14)*. These elements could require the presence of one or more guanine residues.
   b. Transcription of a reporter gene: A plasmid is used where the promoter drives transcription of a reporter gene (for example, β-globin, refs. *15,16*). Unlabeled cold RNA molecules are generated. These are detected by hybridization to an end-labeled single-stranded DNA probe and analyzed by S1 Mapping (*see* Fig. 2). This approach is

Fig. 2. *(opposite page)* Detection of in vitro transcription using extracts from different cell lines by S_1 analysis. Top: The PE3 vector contains the rabbit β-globin reporter gene under the control of the Eα proximal promoter (–215 to +12; *see* ref. *16*). The S1 probe extends on the complementary strand from pos. +60 to position –215, so that hybridization with the transcript RNA and S1 treatment generates a

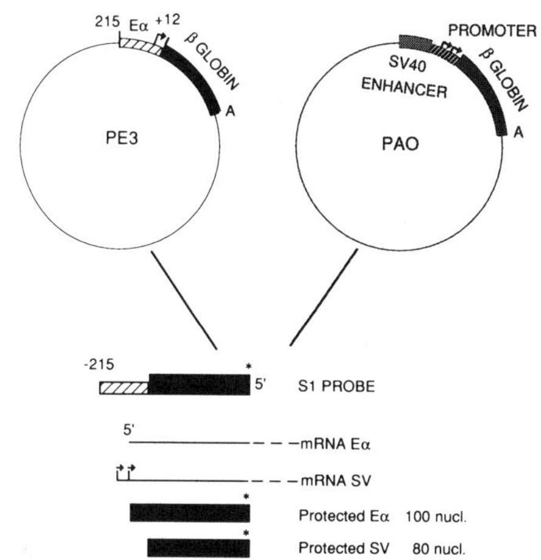

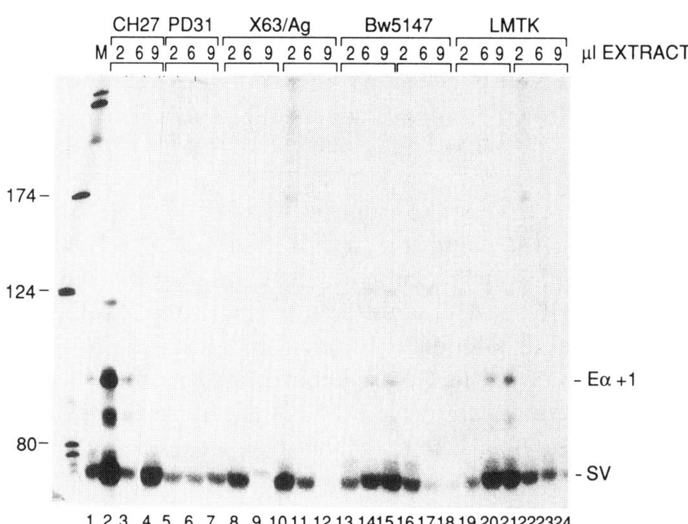

100-bp protected fragment. The presence in the in vitro reaction mixture of the internal control vector PAO that has an identical reporter driven by the SV40 promoter *(15)* will yield a shorter protected fragment of 75 nucleotides. Bottom: The effect of differing amounts of extract (2, 6, and 9 µL respectively). Different extract preparations are also tested from the same cell line in the case of X63/Ag, Bw5147, and LMTK on in vitro transcription from the Eα promoter (100-bp fragment) and SV40 promoter (75-bp fragment). M refers to the DNA molecular weight marker.

more laborious because a single-stranded DNA probe needs to be prepared and purified and takes a minimum of 8 h from transcription to gel exposure. It is less sensitive since only one ^{32}P per RNA molecule is produced. It has no limitation as far as promoter sequences and offers the possibility to test the same construct in other systems, such as by transfection into cells.

2. The plasmid DNA should be prepared by the alkaline lysis method *(12)*, employing either one long CsCl gradient (50,000 rpm 70Ti rotor, 48 h), or two short ones (60,000 rpm VTi 65.1 rotor, 5 h). The DNA is extracted once with butanol, precipitated three times to eliminate CsCl, and finally resuspended in sterile distilled H_2O. Different DNA preps of the same plasmid may have different transcription efficiencies. For accurate comparison, for example when assaying for promoter mutants, it is necessary, therefore, to prepare all samples at the same time. It is also important to test at least a second set of mutant preparations.
3. Addition of the chain terminator GTP analog 3'-O-methylguanosine triphosphate will prevent the formation of spurious transcripts from vector sequences that often cause problems in the interpretation of results in runoff experiments.
4. RNase-free DNase I may be prepared using the following method. Dissolve 5 mg of DNase I in 5 mL of 1 m*M* HCl solution at 4°C and add 1 mL of solution A. To make solution A, add 1.67 g of Iodoacetate (Na salt) to 6 mL of 1*M* acetic acid, adjust to pH 5.3 with NaOH, and bring up to 10 mL with water. Incubate at 50°C for 40 min. Dialyze against 2 L of 2.5 m*M* HCl at 4°C with three buffer changes for a total of 6 h. Centrifuge cloudy solution at 10,000 rpm for 10 min in a Sorvall HB4 rotor at 4°C. Recover the DNase I containing supernatant and add glycerol to a final concentration of 50%. Quick freeze and store at –80°C.
5. The protocol described here has allowed the preparation of active extracts from the following cells:

   PD 31 (Mouse Pre-B)
   70Z/3 (Mouse Pre-B)
   CH27 (Mouse B-Cell)
   X63/Ag (Mouse plasmocytoma)
   BW 5147 (Mouse T-Cell)
   WEHI-3 (Mouse macrophage)
   PD 338 (Mouse macrophage)
   LMTK (Mouse fibroblasts)
   RAJI (Human B-Cell)
   RJ .2.2.5 (Human B Cell)

For any single promoter and cell line it is necessary to standardize the conditions for in vitro transcriptions by carefully titrating the DNA concentration (ranging from 50 to 800 ng/20 µL reaction) and the amount of extract needed (1–9 µL depending on the cell line and the promoter used). An example of extract titration is illustrated in Figs. 1 and 2. As can be seen from Fig. 2, two different extract preps from the same cell line sometimes show a quite different peak of activity (compare, for example, lanes 13–15 to lanes 16–18 in BW5147); therefore it is important to perform this experiment with every new extract preparation.

6. It is very important that the cells are not overgrown; in my experience this is the single most critical point: Extracts prepared from cells at >1.2 × 10^6 density will have a 5- to 10-fold lower transcriptional efficiency.
7. The protocol can be stopped at this point; the pelleted nuclei can be stored in liquid nitrogen or at –80°C for up to 2 mo.
8. Once thawed the extracts can be frozen again without significant loss of activity. If stored at –80°C, however, they progressively lose activity after 2 mo, whereas storage in liquid nitrogen preserves full transcriptional activity for more than 1 yr.

References

1. Mitchell, P. J. and Tjian, R. (1989) Transcriptional regulation in mammalian cells by sequence-specific DNA binding protein. *Science* **245,** 371–378.
2. Johnson, P. F. and McKnight, S. L. (1989) Eukaryotic transcriptional regulatory proteins. *Ann. Rev. Biochem.* **58,** 799–839.
3. Dignam, J. D., Lebovitz, R. M., and Roeder, R. G. (1983) Accurate transcription initiation by RNA polymerase II in a soluble extract from isolated mammalian nuclei. *Nucleic Acids Res.* **11,** 1475–1489.
4. Shapiro, D. J., Sharp, P. A., Wahli, W., and Keller, M. J. (1988) A high efficiency Hela cells nuclear transcription extract. *DNA* **7**, 47–55.
5. Gorski, K., Carneiro, M., and Schibler, U. (1986) Tissue-specific in vitro transcription from the mouse albumin promoter. *Cell* **47,** 767–776.
6. Manley, J. L., Fire, A., Cano, A., Sharp, P. A., and Gefter, M. L. (1980) DNA-dependent transcription of adenovirus genes in a soluble whole-cell extract. *Proc. Acad. Sci. USA* **77,** 3855–3859.
7. Dynan, W. and Tjian, R. (1983) Isolation of transcription factors that discriminate between different promoters recognized by RNA Polymerase II. *Cell* **35,** 79–87.
8. Lichtsteiner, S., Wuarin, J., and Shibler, U. (1987) The interplay of DNA-binding proteins on the promoter of mouse albumin genes. *Cell* **51,** 963–973.
9. Ha, I., Lane, W., and Reinberg, D. (1991) Cloning of a human gene encoding the general transcription initiation factor IIB. *Nature* **352,** 689–695.
10. Bohmann, D. and Tjian, R. (1989) Biochemical analysis of transcriptional activation by Jun: differential activity of c- and v-Jun. *Cell* **59,** 709–717.

11. Workman, J. L. and Roeder, R. G. (1987) Binding of transcription factor TFIID to the Major Late promoter during in vitro nucleosome assembly potentiates subsequent initiation by RNA Polymerase II. *Cell* **51,** 613–622.
12. Birnboim, H. C. (1983) A rapid alkaline extraction procedure for the isolation of recombinant DNA. *Meth. Enzymol.* **100,** 243–255.
13. Sawadogo, M. and Roeder, R. G. (1985) Factors involved in specific transcription by human RNA polymerase II: analysis by a rapid and quantitative in vitro assay. *Proc. Natl. Acad. Sci. USA* **82,** 4394–4398.
14. Smale, T. S. and Baltimore, D. (1989) The "Initiator" as a transcription control element. *Cell* **57,** 103–113.
15. Zenke, M., Gundstrom, T., Matthes, H., Wintzerith, M., Schatz, C., Wildeman, A., and Chambon, P. (1986) Multiple sequence motifs are involved in SV40 enhancer function. *EMBO J.* **5,** 387–397.
16. Viville, S., Jongeneel, V., Koch, W., Mantovani, R., Benoist, C., and Mathis, D. (1991) The Eα promoter: a linker scanning analysis. *J. Immunol.* **146,** 3211–3217.
17. Monaci, P., Nicosia, A., and Cortese, R. (1988) Two different liver specific factors stimulate in vitro transcription from the human α_1-antitrypsin promoter. *EMBO J.* **7,** 2075–2087.

CHAPTER 28

S1 Mapping Using Single-Stranded DNA Probes

Stéphane Viville and Roberto Mantovani

1. Introduction

The S1 nuclease is an endonuclease isolated from *Aspergillus oryzae* that digests single- but not double-stranded nucleic acid. In addition, it digests partially mismatched double-stranded molecules with such sensitivity that even a single base pair mismatch can be cut and hence detected. In practice, a probe of end-labeled double-stranded DNA is denatured and hybridized to complementary RNA molecules. S1 is used to recognize and cut mismatches or unannealed regions and the products are analyzed on a denaturing acrylamide gel. A number of different uses of the S1 nuclease have been developed to analyze mRNA taking advantage of this property *(1,2)*. Both qualitative and quantitative information can be obtained in the same experiment *(3)*.

Qualitatively, it is possible to characterize the start site(s) of mRNA, to establish the exact intron/exon map of a given gene (*see* ref. *4* and Fig. 1), and to map the polyadenylation sites. Qualitatively, it can be used to study gene regulation both in vivo and in vitro, for example, in the study of the Eα gene promoter *(5)*.

In this chapter we describe a S1 mapping method based on the preparation and use of a single-stranded DNA probe (*see* Fig. 2). This offers many advantages:

1. Oligonucleotides allow the exact choice of fragment for a probe.
2. Oligonucleotide labeling is easy and efficient, resulting in a high specific activity probe.

From: *Methods in Molecular Biology, Vol. 31: Protocols for Gene Analysis*
Edited by: A. J. Harwood Copyright ©1994 Humana Press Inc., Totowa, NJ

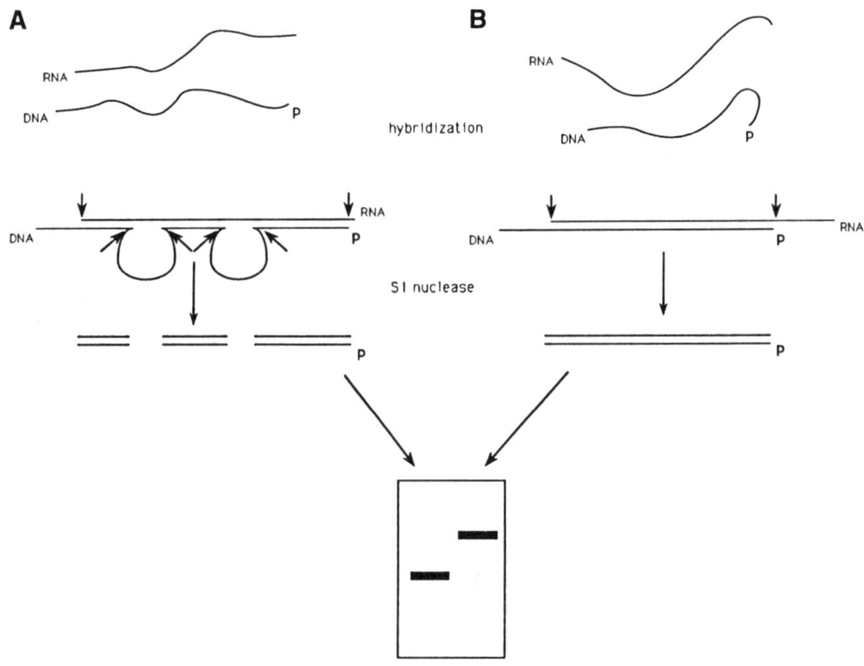

Fig. 1. Illustration of the use of the S1 nuclease. **(A)** Mapping the intron/exon organization of a given gene. The labeled DNA hybridizes to the corresponding mRNA, the introns are cut by the S1 nuclease. **(B)** Mapping of the 5' end of a mRNA. The labeled DNA is used as a probe to map the 5' end of a messenger RNA. In both cases the labeled fragments are visualized on a denaturing acrylamide gel.

3. The probe can be prepared from single-stranded DNA (e.g., M13, BlueScript) as well as a double-stranded template.
4. The single-stranded probe avoids problems often encountered setting up hybridization conditions of double-stranded probes *(6)*.
5. The probe is stable for 3–4 wk.

We illustrate the process with examples from the analysis of the Eα promoter.

2. Materials

2.1. Preparation of Single-Stranded DNA Probe from a Single-Stranded DNA Template

1. 10X Kinase buffer: 400 m*M* Tris-HCl, pH 7.8, 100 m*M* $MgCl_2$, 100 m*M* β-mercaptoethanol, 250 μg/mL bovine serum albumin. Store at –20°C.

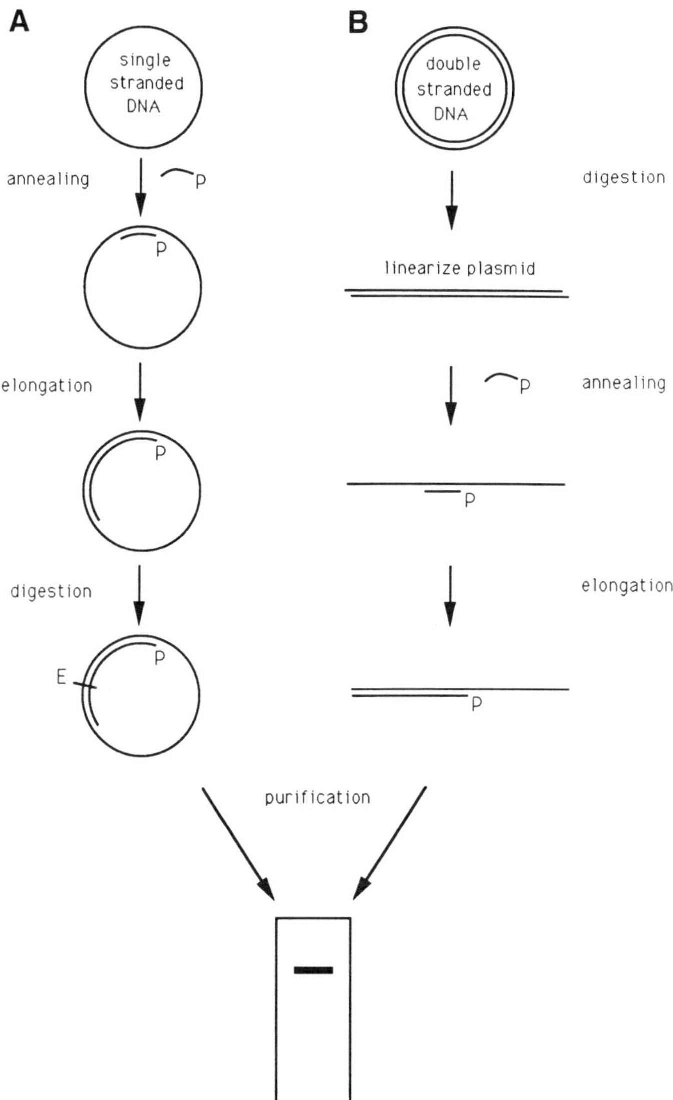

Fig. 2. Schematic illustration of the two methods described in this chapter for the preparation of a single-stranded DNA probe. (**A**) Using a single-stranded plasmid as a template. (**B**) Using a double-stranded DNA plasmid to synthesize the probe.

2. γ-ATP: specific activity 3000 Ci/mmol.
3. Suitable oligonucleotide (*see* Note 1). Prepare a 10 pmol/μL solution in distilled water. Store at –20°C.

4. PNK: Polynucleotide kinase at a stock concentration of 10 U/μL. Store at –20°C.
5. 10X Annealing buffer: 100 m*M* Tris-HCl, pH 7.5, 100 m*M* $MgCl_2$, 500 m*M* NaCl, 100 m*M* dithiothreitol.
6. DNA template: 1 mg/mL (*see* Note 2).
7. 10X dNTP mix: 5 m*M* dATP, 5 m*M* dCTP, 5 m*M* dGTP, and 5 m*M* dTTP.
8. Klenow: The large fragment of *E. coli* DNA Polymerase I at a stock concentration of 10 U/μL. Store at –20°C.
9. Suitable restriction enzyme (*see* Note 3).
10. 6% Polyacrylamide/8*M* urea solution in 0.5X TBE.
11. X-ray film: A film of suitable sensitivity, e.g., Kodak XAR.
12. Elution buffer: 50 m*M* Tris-HCl, pH 7.5, 0.5 m*M* EDTA.
13. Na Acetate: 3*M* sodium acetate, pH 7.4.
14. tRNA: Stock solution 10 mg/mL. Store at –20°C.

2.2. Hybridization and S1 Analysis

15. 4X Hybridization buffer: 1.6*M* NaCl, 40 m*M* PIPES, pH 6.4.
16. Deionized formamide.
17. Paraffin oil.
18. S1 buffer: 300 m*M* NaCl, 30 m*M* Na acetate, pH 4.5, 4.5 m*M* Zn acetate.
19. S1 enzyme: A stock concentration of 10 U/μL. Store at –20° C.
20. S1 stop buffer: 2.5*M* ammonium acetate, 50 m*M* EDTA.
21. Isopropanol.
22. Formamide dye (*see* Chapter 27).

3. Methods

3.1. Preparation of Single-Stranded DNA Probe from a Single-Stranded DNA Template

1. Label the oligonucleotide by incubation of 1 μL of cold oligonucleotide, 10 μL of γ-ATP, 2 μL of 10X kinase buffer, 6 μL of dH_2O and 1 μL of PNK at 37°C for 45 min. Inactivate the PNK at 95°C for 2 min.
2. Add 2 μL of single-stranded DNA template, 4 μL of 10X annealing buffer, 14 μL of H_2O and incubate for 10 min at 65°C, followed 10 min at 55°C, then 10 min at 37°C.
3. Add 4 μL of dNTP mix and 1 μL of Klenow and leave at room temperature for 10 min. Inactivate the Klenow by placing the tube at 65°C for 15 min.
4. Correct the mix for the restriction enzyme digest by either adding a suitable amount of salt or by dilution. Add 20 U of restriction enzyme and incubate at 37°C for 60 min (*see* Note 3).

5. Precipitate the sample by adding 2 µL of tRNA solution, 0.1 vol of Na acetate, 3 vol of ethanol, and placing the sample in dry ice for 10 min. Centrifuge at 12,000*g* for 15 min in a microfuge.
6. Resuspend the pellet in 20 µL of formamide dye, heat at 95°C for 5 min, and load a 6% polyacrylamide/urea gel. Run the gel until the bromophenol blue reaches the bottom.
7. Locate the single-stranded DNA fragment by exposing the gel for 5 min to a X-ray film. Excise the corresponding band and elute it overnight in 500 µL of elution buffer.
8. Add 50 µL of Na acetate, 20 µg of tRNA and 1 mL of ethanol; place at –20°C for 2 h. Centrifuge 20 min at 12,000*g* in a microfuge.
9. Count the pellet and resuspend it in formamide (10^4 cpm/µL). Store at –20°C. The probe is now ready for hybridization and S1 analysis.

3.2. Preparation of Single-Stranded DNA Probe from Double-Stranded DNA Template

Follow the same protocol described in Section 3.1., but for the following exceptions (*see* Fig. 2B).

1. Initially cut 20 µg of double-stranded plasmid with a restriction enzyme of choice. Precipitate and resuspend in H_2O at a concentration of 1 mg/mL.
2. The annealing step (Section 3.1., step 2) must be carried out extremely quickly: After heat inactivation of PNK, add 2–5 µg of restricted plasmid, 4 µL of 10X annealing buffer, 14 µL of H_2O, and incubate for 5 min at 95°C. Immediately place the tube in ice cold water and then proceed to step 3.
3. Omit the restriction enzyme digestion (step 4).

3.3. Hybridization and S1 Analysis

1. In a 750-µL Eppendorf tube put 10 µL of probe (10,000 cpm total), 5 µL of 4X hybridization buffer, and 5 µL of RNA (total poly A^+ or synthesized in vitro). Add 20 µL of paraffin oil to prevent evaporation and incubate 4–16 h at 37°C (*see* Note 4).
2. Add 200 µL of S1 buffer containing 100 U of S1 nuclease, mix well, and place at 37°C for 5–30 min (*see* Note 5). Stop the reaction by adding 50 µL of S1 stop buffer, 40 µg of tRNA, and 300 µL of isopropanol. Place in dry ice for 15 min and centrifuge at 12,000*g* for 15 min.
3. Wash the pellet with 80% ethanol and dry. Resuspend in 3 µL of loading dye, heat for 5 min at 95°C, and load a 6% polyacrylamide/urea gel. Run until the bromophenol blue reaches the bottom. Expose the gel for 8–48 h (*see* Fig. 3).

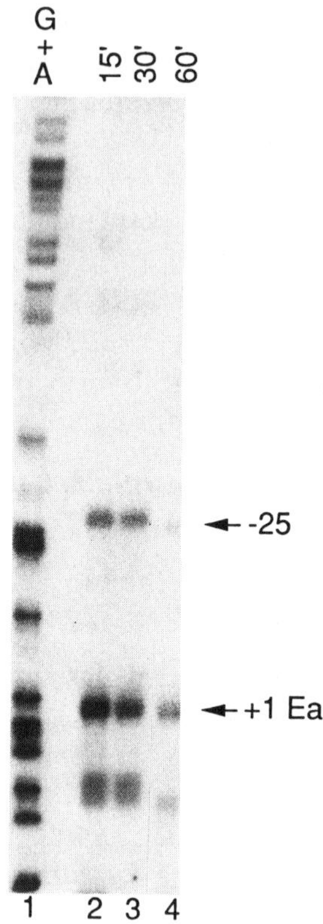

Fig. 3. Time course of S1 analysis. In vitro synthesized RNA hybridized to Eα probe (*see* Chapter 27) is incubated at 37°C for the indicated time with 100 U of S1 nuclease.

4. Notes

1. The oligonucleotide should be 20–25 nucleotides long and should be complementary to the DNA strand to be analyzed.
2. Both the single-stranded and double-stranded DNAs are prepared according to standard protocols. The advantage of using a single-stranded DNA template is basically a higher yield of the probe: 1–1.5 × 10^6 cpm vs the 4–6 × 10^5 cpm expected from a double-stranded plasmid template.

3. The restriction enzyme should be used to cut 200–600 nucleotides 3' of the position to which the oligonucleotide hybridizes to increase fragment recovery. Longer fragments are harder to recover from acrylamide gels. It does not matter if the restriction enzyme cuts the template DNA more than once as long as it is outside the probe sequence.
4. Hybridization time varies with the type of RNA to be analyzed: Using total cytoplasmic or poly A^+ RNA, the samples should be left at 37°C at least 12 h; when hybridizing RNA generated from in vitro transcription 2–4 h are usually sufficient.
5. An S1 time course should be performed as shown in Fig. 3 for in vitro synthesized RNA. This will establish the conditions for which all single-stranded nucleic acids, including the free probe, are completely digested. The optimal cutting time for total cytoplasmic and poly A^+ RNA is usually longer (30 min) than for in vitro synthesized RNA (15 min).

References

1. Berk, A. J. and Sharp, P. A. (1977) Sizing and mapping of early adenovirus mRNAs by gel electrophoresis of S1 endonuclease digested hybrids. *Cell* **12,** 721–732.
2. Favaloro, J., Treisman, R., and Kamen, R. (1980) Transcription maps of polyoma virus-specific RNA: analysis by two-dimensional nuclease S1 gel mapping. *Meth. Enzymol.* **65,** 718–749.
3. Weaver, R. F. and Weissmann, C. (1979) Mapping of RNA by a modification of Berk-Sharp procedure: the 5' termini of 15 S β-globin mRNA precursor and mature 10 S b-globin mRNA have identical map coordinates. *Nucleic Acids Res.* **7,** 1175–1193.
4. Lopata, M. A., Sollner-Webb, B., and Cleveland, D. W. (1985) Surprising S1-resistant trimolecular hybrids: potential complication in interpretation of S1 mapping analysis. *Mol. Cell. Biol.* **5,** 2842–2846.
5. Viville, S., Jongeneel, V., Koch, W., Mantovani, R., Benoist, C., and Mathis, D. (1991) The Eα promoter: a linker-scanning analysis. *J. Immunol.* **146,** 3211–3217.
6. Dean, M. (1987) Determining the hybridization temperature for S1 nuclease mapping. *Nucleic Acids Res.* **15,** 6754.

Chapter 29

Single Primer-Mediated Polymerase Chain Reaction

René L. Myers and Ing-Ming Chiu

1. Introduction

The polymerase chain reaction (PCR) is a powerful method for amplification of DNA segments between two regions of known sequences *(1,2)*. In standard PCR, sequences bound by two unique oligonucleotide primers can be exponentially amplified. However, it is a prerequisite to know the DNA sequence at both ends of the region of interest. To amplify regions of unknown sequence, methods such as inverted PCR *(3)*, *Alu* PCR *(4)*, and rapid amplification of cDNA ends (RACE) *(5)*, also known as anchored PCR *(6)* or one-sided PCR *(7)*, have been developed. These methods require several enzymatic manipulations of DNA that are either tedious or only suitable for certain conditions. While using standard PCR for in vitro amplification of 5' untranslated cDNA sequence of acidic fibroblast growth factor (aFGF) mRNA, we observed amplification of a second specific fragment in addition to the expected fragment. Sequence analysis showed that the additional fragment resulted from binding of the 5' primer to a region that has an imperfect match to the actual primer binding site. Comparing the sequences of the primer (HBGF 1201) and the binding site revealed that only eight out of 32 nucleotides matched (Fig. 1A). Those eight nucleotides cluster in the 19 nucleotides at the 3' end of the primer and therefore provided priming that allowed for a specific amplified product. With this observation, we reasoned that a single primer could be used, binding both to the complementary sequence

From: *Methods in Molecular Biology, Vol. 31: Protocols for Gene Analysis*
Edited by: A. J. Harwood

```
A   HBGF 1201       5'ATCCCCAAGGCTAGGAGGCCAACCTACTAACA 3'
                                      | |  ||   |  |   ||
    Binding Site    5'TGATGGTGAATGGGAACTCCCTTCCTCCTGCA 3'

B   HBGF 1003       5'TCTCAGAGTAGCCCGGCTCTCTGAACCTCCAGGCT 3'
                      |  ||  ||   | | ||||     ||||| ||||
    Binding Site    5'TGCCACCGTCTTCTGTCTCTACCTCCCTCCTGGCT 3'
```

Fig. 1. Similarities of DNA sequences between the primers and the less than perfect binding sites. The identical bases between each pair of oligonucleotides are indicated by vertical bars.

as well as a similar, less than perfect match sequence within several kilobase pairs from the authentic binding site for the PCR. We have explored the possibility of PCR using a single primer. Following reverse transcription (RT) with a gene specific primer, we amplified a specific product using a single internal gene-specific oligonucleotide in a PCR reaction. Cloning and sequence analysis of the amplified products revealed a copy of the primer (HBGF 1003) at both the 5' and the 3' ends of the clone. One site represented the authentic binding site and the second copy of the primer indicated that it had hybridized to a partially homologous sequence (20/35 nucleotides) in the template cDNA (Figs. 1B and 2).

This method takes advantage of the fact that partial complementarity provides sufficient affinity for the oligonucleotide primer to anneal to a secondary, imperfect binding site. Under appropriate conditions, specific PCR products can be generated after amplification of cDNA or genomic sequences in a PCR reaction containing only a single primer. Thus, we have been able to apply single primer-mediated PCR to amplify and clone the 5' untranslated sequences of two different human aFGF mRNA *(8)*. In addition, we have employed this method to amplify end-specific regions of yeast artificial chromosomes (unpublished results). This approach has also been used by others to clone sequences upstream from the coding region of a serotonin receptor gene using genomic DNA as a starting template *(9)*. This method has general applicability and is a fairly simple and straightforward technique. No modification of template DNA is required. Single primer-mediated PCR may not work in every situation;

```
TCTCAGAGTAGCCCGGCTCTCTGAACCTCCAGGCTGGCCAATGGCTCTGTGTTCCTGGGC   60
            HBGF1003
CTGCTGCTGGCTGTCCAGAGTAGGGGTTGCTTAGAGCTGTGTGCATCCCTGCGGGTGGTG  120

TGGGAGTGGGCGGTTGTCTAAAGGCAGGTCCCCTCTACTGATAAACAAGGACCGGAGATA  180

GACCTAGAGGCTGACATTCTTGGCTCCCCCAGCCTACACCCCCCCCACCTCGATTTCCCA  240

CAGAGCCCTAGGGACGGGTAGCCAGCTCTGTGGCATGGTATCTGGAGGCAGGCCAGCAAC  300
            SphI                            PstI
CTGATGTGCATGCCACGGCCCGTCCCTCTCCCCACTCAGAGCTGCAGTAGCCTGGAGGTT  360

CAGAGAGCCGGGCTACTCTGAGAAGAAGACACCAAGTGGATTCTGCTTCCCCTGGGACAG  420
HBGF1003
CACTGAGCGAGTGTGGAGAGAGGTACAGCCCTCGGCCTACAAGCTCTTTAG           471
                            HBGF1105
```

Fig. 2. Nucleotide sequence of the RT-PCR product generated using a single primer. The oligonucleotide HBGF 1105 was used as the primer in the reverse transcription. The single primer HBGF 1003 was used in the PCR. The RT-PCR product was subcloned into pBluescript and sequenced. The primer HBGF 1003 was bound to an upstream region with a less than perfect match and the downstream authentic site to generate the PCR fragment. Primers HBGF 1003 and HBGF 1105 are underlined. The triangle indicates the junction of the upstream nontranslated exon and the first coding exon. The PCR product is 383 bp in length. Nucleotides 384 to 471 are included to show the relative location of oligonucleotide HBGF 1105.

however, given the simplicity of this approach, it is a desirable method for the amplification of regions adjacent to known sequences.

2. Materials

Since this procedure involves the use of RNA, extreme care must be taken to avoid contamination by RNase and degradation of the RNA. Clean gloves should be worn at all times and the work area should be clean. Stock solutions and glassware should be treated with the RNase inhibitor diethyl pyrocarbonate (DEPC) (0.1% at 37°C for at least 12 h). These solutions should then be autoclaved for 15 min at 15 lb/sq. in. on liquid cycle to remove all traces of DEPC prior to use. Remaining DEPC may modify purine residues in RNA by carboxymethylation. Solutions containing Tris-HCl cannot be treated with DEPC since it is unstable in this buffer. Tris solutions should be made with DEPC-treated autoclaved water and then reautoclaved. Glassware should be baked at 180°C for at least 8 h. Commercial sterile plasticware is usually RNase free and can be used without treatment. Separate stocks of reagents and supplies should be maintained for RNA work.

KCl, $MgCl_2$, and EDTA should be DEPC-treated as described above. With the exception of DTT, the other RT buffer components should be made up in DEPC-treated autoclaved water and reautoclaved. Stock solutions of 1*M* DTT should be made up in DEPC-treated 0.01*M* NaOAc (pH 5.2), filter-sterilized, aliquoted, and stored at –20°C. DTT should not be autoclaved. Stocks of dNTPs should be made in DEPC-treated $T_1E_{0.2}$, filter-sterilized, and stored at –20°C. DEPC-treated $T_1E_{0.2}$ is made by combining Tris made in DEPC-treated, autoclaved H_2O with DEPC-treated, autoclaved EDTA and reautoclaving.

Likewise, a major problem in PCR experiments owing to the extremely high level of sensitivity of this technique is DNA contamination. Such contamination leads to false positives and misinterpretation of data. It is necessary to prevent the physical transfer of DNA between amplified samples and between positive and negative experimental controls. There are several precautions that should be taken to avoid such contamination. The PCR preparations and PCR amplified products should be physically isolated. In setting up PCR reactions, separate supplies and pipeting devices should be used. If available, positive displacement pipets should be used to avoid aerosol contamination of DNA. All solutions, including water, should be autoclaved. This treatment degrades DNA to a very low mol wt. Primers, dNTPs, and *Taq* DNA polymerase cannot be autoclaved. All reagents should be aliquoted into multiple tubes and given a lot number. This will minimize the number of repeated samplings and will make it easier to trace the source of contamination if one should arise. The DNA template should be added last and in a separate area, using a clean pipet tip for each sample and capping each tube before moving onto the next one. Run a "no DNA" reagent tube as a negative PCR control as well as a "no reverse transcriptase" reagent control for the RT reaction.

2.1. Reverse Transcription and Polymerase Chain Reaction

1. Oligo$(dT)_{12-18}$: Dissolve in DEPC-treated water at concentration of 0.5 mg/mL.
2. Random hexamers: Dissolve in DEPC-treated water at concentration of 0.5 mg/mL.

3. RNase Inhibitor: Store at –20°C at 20 U/µL.
4. 10X reverse transcription buffer: 500 m*M* Tris-HCl, pH 8.3, 750 m*M* KCl, 30 m*M* $MgCl_2$, 100 m*M* dithiothreitol (DTT). Store at –20°C (*see* Note 1).
5. dNTP mix: A mix of 2m*M* of each dNTP.
6. MMLV-RT: Murine Maloney Leukemia virus reverse transcriptase supplied by US Biochemical, Cleveland, OH. Store at –20°C at 200 U/µL.
7. 5% linear acrylamide: 0.1 g acrylamide, 1.9 mL 5 m*M* Tris-HCl, 0.1 m*M* EDTA, pH 7.4, 80 µL of 25% ammonium persulfate, 1.5 µL TEMED. Rock overnight at room temperature. It will be viscous but will not form a gel in the absence of the crosslinking agent *bis*-acrylamide. Care must be taken handling this material, since acrylamide is a potent neurotoxin!
8. 3*M* Sodium acetate (NaOAc), brought to pH 5.4 with glacial acetic acid, DEPC treated, and autoclaved for 20 min.
9. 10X PCR reaction buffer: 100 m*M* Tris-HCl (pH 8.3), 500 m*M* KCl, 15 m*M* $MgCl_2$, 0.1% Trition X-100, 0.1% Tween 20, 0.1% gelatin.
10. Primer: Synthetic oligonucleotide dissolved in water at concentration of 25 pmol/µL.
11. *Taq* polymerase: Store at –20°C at 5 U/µL.

2.2. Cloning of PCR Amplified Products

12. 10X T4 DNA polymerase reaction buffer: 500 m*M* Tris-HCl, pH 8.0, 50 m*M* $MgCl_2$, 5 m*M* DTT, 0.5 mg/mL BSA (Fraction V, Sigma, St. Louis, MO); divide into small aliquots and store at –20°C.
13. T4 DNA Polymerase: Store at –20°C at 1 U/µL.
14. 0.5*M* EDTA.
15. Phenol/CIA: A mix of phenol, equilibrated with Tris-HCl to pH 7.4, chloroform, and iso-amylalcohol in a 42:42:1 ratio.
16. $T_1E_{0.2}$: 1 m*M* Tris-HCl, pH 8.0, 0.2 m*M* EDTA.
17. 5X ligation buffer: 250 m*M* Tris-HCl, pH 7.6, 50 m*M* $MgCl_2$, 5 m*M* ATP, 5 m*M* DTT, 25% (w/v) polyethylene glycol (PEG 8000). Store buffer at –20°C, shelf life 2 mo.
18. Digested pBluescript vector: Digest pBluescript vector with *Sma*I or *Eco*RV to linearize and leave blunt ends. Phenol/ClA extract the digest reaction, ethanol precipitate, and resuspend in $T_1E_{0.2}$ at 0.2 µg/µL.
19. Polyethylene glycol (PEG 8000): Make a 40% (w/v) stock solution in deionized H_2O, aliquot, and store at –20°C; PEG should be thawed and warmed to room temperature before adding to the ligation mixture.
20. T4 DNA ligase: Store at –20°C at 1 U/µL.
21. Competent *E. coli* cells: Cloning efficiency, e.g., DH5α.

22. SOC: 2% bactotryptone, 0.5% yeast extract, 10 m*M* NaCl, 2.5 m*M* KCl, autoclaved. 10 m*M* $MgCl_2$, 10 m*M* $MgSO_4$, 20 m*M* glucose, filter sterilized.
23. YT plates: 1% bactotryptone, 0.5% yeast extract, 0.17*M* NaCl, 1.1% agar, autoclaved.
24. X-gal: 2% 5-bromo-4-chloro-3-idolyl β-D-galactoside in dimethylformamide. Store at –20°C.
25. IPTG: 100 m*M* isopropyl-β-D-thiogalactosidase in water. Store at –20°C.

3. Methods

3.1. Reverse Transcription and Polymerase Chain Reaction

1. In a 0.5-mL siliconized microfuge tube, mix 1 µL of oligo$(dT)_{12-18}$, 1 µL of random hexamers, 0.5 µL of RNase inhibitor, 1 µg of total cellular RNA, and dH_2O to 22.5 µL. Heat at 70°C for 10 min. Quick chill on ice 2 min; centifuge for 10–20 s.
2. Prepare a master mix containing the following components in the proportions shown (*see* Note 2): 0.5 µL of RNase inhibitor, 5 µL of 10X RT buffer, 20 µL of 2.0 m*M* of each dNTP, and 2 µL of MMLV RT. Dispense 27.5 µL of the master mix into each sample RNA tube. Incubate at 37°C for 1 h.
3. Precipitate the cDNA by adding 1 µL of 5% linear acrylamide (a coprecipitate), 5 µL of 3*M* NaOAc, and 140 µL of absolute ethanol. Precipitate at –20°C for 15 min. Centifuge at 4°C, 13,000*g* for 15 min, and dry pellet. Resuspend in 60 µL of autoclaved dH_2O; store at –70°C.
4. For each sample prepare a PCR master mix containing 10 µL of 10X PCR reaction buffer, 10 µL of dNTP mix, 1 µL of synthetic oligonucleotide primer, and dH_2O to 97 µL/sample. Dispense 97 µL of mastermix into a 0.5-mL siliconized microfuge tube for each sample. Overlay with 100 µL of mineral oil to reduce evaporation. Add 3 µL of cDNA to each tube, pipeting through the oil layer. Microfuge for 20 s.
6. Heat the reaction tube(s) at 95°C for 5 min and cool to 72°C. (*see* Note 3). Add 0.5 µL of *Taq* polymerase.
7. Run 30–40 rounds of temperature cycling as follows: (*see* Note 4): Denature at 94°C for 30 s, anneal at 50°C for 45 s, and extend at 72°C for 2 min.
8. At the end of the run, extend the reaction at 72°C for an additional 15 min.
9. Extract the aqueous sample from the oil using 100 µL of high purity chloroform. The aqueous phase will float on the chloroform/oil mixture. Store at 4°C until further characterization. Analyze 10 µL of the PCR product by electorphoresis on a 3% Nusieve and 1% Seakem agarose gel containing 1 µg/mL ethidium bromide.

3.2. Cloning of PCR Amplified Products

1. For blunt end cloning, isolate the fragment(s) of interest by gel purification (*see* Note 5).
2. Blunt end the fragments by setting up a T4 polymerase fill-in reaction as follows: Combine 0.2–2 µg of DNA, 1 µL of T4 polymerase, 5 µL of 10X T4 polymerase reaction buffer, 3 µL of 2 m*M* dNTPs, and dH_2O to 50 µL. Place tube at 11°C for 20 min.
3. Add 2 µL of 0.5*M* EDTA to inactivate the enzyme. Extract once with 50 µL phenol and once with 50 µL phenol/CIA 1:1 (v/v). Add 5 µL of 3*M* NaOAc and 125 µL of absolute ethanol and precipitate at –20°C for 20 min. Spin at 13,000*g* in microcentifuge at 4°C for 15 min. Dry pellet in a SpeedVac concentrator and resuspend in $T_1E_{0.2}$.
4. Ligate the products with 0.2 µg of vector in a 20 µL volume as follows: Add the fragment at a vector/insert molar ratio of 1:3, add 4 µL of 5X ligation buffer, 1 µL of T4 DNA ligase, 7.5 µL of polyethylene glycol (PEG 8000), and dH_2O to 20 µL. Incubate at 20°C overnight.
5. Add 1–2 µL of ligated DNA to 50 µL of competent *E. coli* DH5α cells. Incubate on ice for 30 min. Heat shock cells at 37°C for 45 s. Incubate on ice 2 min. Add 950 µL SOC and shake at 225 rpm at 37°C for 1 h.
6. Plate 200 µL of the cells with 20 µL of 100 m*M* IPTG and 60 µL of 2% X-gal onto YT plates containing 50 µg/mL ampicillin. Incubate at 37°C overnight. The resulting bacterial colonies can be analyzed by using the alkaline extraction method *(10)* and restriction enzyme digestion; using PCR (*see* Chapter 2) or by hybridization (*see* Note 6).

4. Notes

1. In the RT reaction, we have found that either AMV or MMLV RT performs satisfactorily. However, the RT buffer for the AMV enzyme and the reaction conditions are slightly different. The 10X reaction buffer for AMV RT is 500 m*M* Tris-HCl (pH 8.5 at 25°C), 400 m*M* KCl, and 80 m*M* $MgCl_2$. The reaction is carried out at 42°C for 1 h.
2. For both the reverse transcription and the PCR amplification, master mixes of reagents for all samples can be prepared first and then the appropriate amount aliquoted into reaction tubes. The use of master mixes helps to increase pipeting accuracy by avoiding small volume pipeting, and decrease the number of reagent transfers, which helps to eliminate risk of contamination. Pipet carefully and slowly. The viscosity from the glycerol that is in the buffer of the ribonuclease inhibitor, the RT, and the *Taq* polymerase may lead to pipetting errors.
3. Reverse transcriptase binds to cDNA and is therefore inhibitory to PCR amplification. Incubation at 95°C for 5 min before the addition of the

Taq polymerase in the PCR reaction inactivates any RT that may be remaining and removes the inhibitory effect. Increased concentrations of RT in the synthesis of the cDNA make inactivation more difficult. Therefore, for synthesis of longer RNA transcripts, it is advisable to increase the incubation time of the RT reaction rather than the amount of RT added.

4. The conditons for the PCR reactions reported here were those that we have successfully employed to amplify upstream regions of the aFGF gene using a single primer. However, it is possible that the reaction may have to be optimized for each primer-template pair. In order to improve chances that single-primer mediated PCR will result in amplification, conditions should be adjusted to overall lower stringency. For example, decreasing the annealing temperature will usually result in an increase in "incorrectly" annealed primers. In the PCR cycling protocol, one can use fairly relaxed conditions in early cycles to allow for priming to less than perfect matches, then increase the stringency in later cycles to allow for amplification of specific products from these early priming events. Once priming occurs, the DNA product serves as an ideal template having a perfect match to the primer for the remaining cycles of amplification. Raising the annealing temperature gradually through each cycle would favor the annealing of the perfect match in later cycles and thereby facilitate amplification of specific products. Also, increased annealing times will allow for more hybridization of the primers to a region of only partial complementarity. We have found that annealing of the primer at 50°C for 5 min and extension at 72°C for 40 min prior to the standard cycling protocol increases the amplification of specific products. In addition, higher primer concentrations (>0.1–0.5 µ*M*) may also promote mispriming. Runs of cytosine or guanine residues (more than 3) at 3' ends may promote mispriming at G + C rich sequences. Adding to the versatility of this method, a primer with an inosine or a degeneracy with all four nucleotides at the 3' end of the primer may increase the possibility of locating a secondary binding site within several kilobases of the actual binding site. The effects of such variations on the PCR results can be monitored by examining the intensity and distribution of products on agarose gels.
5. As an alternative to blunt end cloning, we have used T-vector cloning (*see* Chapter 3). This method takes advantage of the template-independent activity of *Taq* polymerase that preferentially adds dA to the 3' end of the PCR duplex products. By constructing vectors that possess dT overhangs the efficiency of cloning is increased by at least 100 times *(11)*.

6. An alternative to screeing of colonies by boiling-PCR *(12)* or standard alkaline extraction of plasmid *(10)*, is a method using microwave colony hybridization that allows for rapid screening of large numbers of colonies. The bacterial colonies are picked onto nylon filters and allowed to grow. The filters are placed on Whatman paper soaked in 2X SSC/0.5% SDS for 2 min and then treated for 2.5 min in a 650 W microwave oven. The colony lysis, DNA denaturation, and fixation occurs in this single step. The treated filters are subjected to standard prehybridization and hybridization without UV fixation.

Further Reading

Kwok, S. and Higuchi, R. (1989) Avoiding false positives with PCR. *Nature* **339,** 237–238.

White, B. A. (ed.) (1993) *PCR Protocols: Current Methods and Applications; Methods in Molecular Biology,* vol. 15, Humana, Totowa, NJ.

References

1. Saiki, R. K., Gelfand, D. H., Stoffel, S., Scharf, S. J., Higuchi, R., Horn, G. T., Mullis, K. B., and Erlich, H. A. (1988) Primer directed enzymatic amplification of DNA with a thermo-stable DNA polymerase. *Science* **239,** 487–491.
2. Erlich, H. A., Gelfand, D., and Sninsky, J. J. (1991) Recent advances in the polymerase chain reaction. *Science* **252,** 1643–1651.
3. Triglia, T., Peterson, M. G., and Kemp, D. J. (1988) A procedure for in vitro amplification of DNA segments that lie outside the boundries of known sequences. *Nucleic Acids Res.* **16,** 8186.
4. Nelson, D. L., Ledbetter, S. A., Corbo, L., Victoria, M. F., Ramirez-Solis, R., Webster, T. D., Ledbetter, D. H., and Caskey, C. T. (1989) *Alu* polymerase chain reaction: a method for rapid isolation of human-specific sequences from complex DNA sources. *Proc. Natl. Acad. Sci. USA* **86,** 6686–6690.
5. Frohman, M. A., Dush, M. K., and Martin, G. R. (1988) Rapid production of full-length cDNA from rare transcripts: amplification using a single gene-specific oligonucleotide primer. *Proc. Natl. Acad. Sci. USA* **85,** 8998–9002.
6. Loh, E. Y., Elliot, J. F., Cwirla, S., Lanier, L. L., and Davis, M. M. (1989) Polymerase chain reaction with single-sided specificity: analysis of T cell receptor δ chain. *Science* **243,** 217–220.
7. Ohara, O., Dorit, R. L., and Gilbert, W. (1989) One-sided polymerase chain reaction: the amplification of cDNA. *Proc. Natl. Acad. Sci. USA* **86,** 5673–5677.
8. Wang, W.-P., Myers, R. L., and Chiu, I.-M. (1991) Single primer-mediated polymerase chain reaction: application in cloning of two different 5'-untranslated sequences of acidic fibroblast growth factor mRNA. *DNA Cell Biol.* **10,** 771–777.
9. Parks, C. L., Chang, L.-S., and Shenk, T. (1991) A polymerase chain reaction mediated by a single primer: cloning of genomic sequences adjacent to a serotonin receptor protein coding region. *Nucleic Acids Res.* **19,** 7155–7160.

10. Birnboim, H. C. and Doly, J. A. (1979) Rapid alkaline extraction procedure for screening recombinant plasmid DNA. *Nucleic Acids Res.* **7,** 1513–1523.
11. Marchuk, D., Drumm, M., Saulino, A., and Collins, F. S. (1991) Construction of T-vectors, a rapid and general system for direct cloning of unmodified PCR products. *Nucleic Acids Res.* **19**, 1154.
12. Gussow, D. and Clackson, T. (1989) Direct clone characterization from plaques and colonies by the polymerase chain reaction. *Nucleic Acids Res.* **17,** 4000.

PART VI

PROTEIN–DNA INTERACTIONS

CHAPTER 30

In Vivo DNA Footprinting by Linear Amplification

Hans Peter Saluz and Jean-Pierre Jost

1. Introduction

Genomic footprinting is an extension of the genomic sequencing technique and is used to study protein/DNA interactions in vivo *(1)* (Fig. 1). It gives information concerning the sequence of protein binding elements and the contact points between a protein and its DNA target extend. Proteins bound to the DNA alter the reactivity of the bases towards UV-light *(2)* or chemicals *(3)*.

In UV-footprinting the extent of photoproduct formed after irradiation differs between DNA that is bound to protein and that which is free of it. These two DNA states will result in DNA fragments of differing sizes since breakage of the DNA backbone only occurs at the sites of photodamage. The differences in the strand-breakage pattern are resolved on a sequencing gel where a "footprint" devoid of bands is produced; an indication of the region where DNA and protein interact *(4)*. This method has been successfully applied in vivo, in studies of lower *(5,6)* and higher eukaryotes *(7)*.

In this chapter we describe genomic footprinting by chemical modification. Dimethyl sulfate (DMS) that modifies the position N7 of guanine residues *(8)* is applied directly to living cells or tissues. Guanine residues in a region bound to protein are generally protected from modification. Some residues, however, may show enhanced reactivity. The DNA is then purified and cleaved at the modified positions with piperidine. Again the resulting footprint is resolved on a sequencing gel.

From: *Methods in Molecular Biology, Vol. 31: Protocols for Gene Analysis*
Edited by: A. J. Harwood

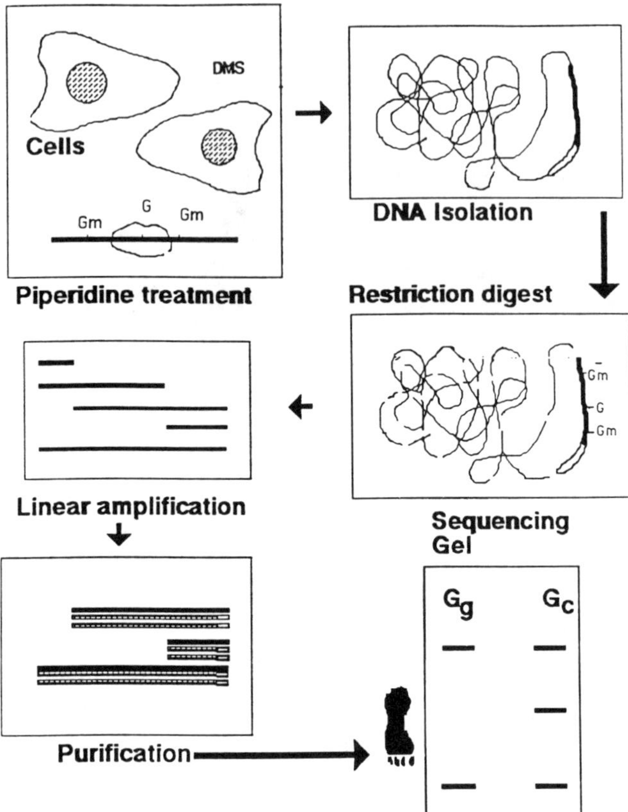

Fig. 1. DNA footprinting in vivo by linear amplification: Cells in suspension are treated with dimethylsulfate (DMS). This results in a modification of unprotected guanosine (Gm) residues. G represents protected and, therefore, unmodified guanosine residues. After DNA isolation, a restriction digestion is performed to reduce the viscosity of the DNA. Piperidine treatment results in cleavage of the DNA at the modified sites. The DNA subfragments (black boxes) are linearly amplified (striped pattern) using a radioactively labeled primer (open box). After purification of the amplification products, the genomic DNA fragments (Gg) are separated on a sequencing gel in parallel with a guanosine control reaction (Gc) allowing the detection of protected sites (footprint).

In the original protocols the DNA cleavage products were visualized by either electroblotting and indirect endlabeling *(1,9)* or via primer extension with labeled primers. The protocol we present here uses linear amplification, with *Taq* or other thermostable DNA polymerases, to allow direct detection of the footprint in the gel, making the technique faster and more sensitive *(10,11)*.

2. Materials

2.1. DMS Treatment of Cells

1. DMS: Use DMS of highest purity. It should be stored under nitrogen. DMS is toxic and suitable handling precautions must be taken.
2. DMS-stop: Phosphate buffered saline (PBS) containing 1% bovine serum albumin and 100 m*M* β-mercaptoethanol. Cool on ice just before use.
3. Nuclei buffer: 0.3*M* sucrose, 60 m*M* KCl, 15 m*M* NaCl, 1 m*M* EDTA, 0.5 m*M* EGTA, 15 m*M* HEPES, pH 7.5, 1 m*M* spermidine, 0.3 m*M* spermine. Add spermine and spermidine just before use.
4. Nonident NP40: A 1% solution (v/v) freshly made in nuclei buffer.
5. Proteinase K buffer: 20 m*M* Tris-HCl, pH 8.0, 20 m*M* EDTA, 20 m*M* NaCl, 1% sodium dodecyl sulfate (SDS).
6. Proteinase K: Stock solution at 20 mg/mL. Store at –20°C.
7. RNase: Pancreatic ribonuclease A. Make a 10 mg/mL stock solution. Store at –20°C.
8. Phenol: Saturated with 1*M* Tris-HCl, pH 8.0, containing 0.1% hydroxyquinoline.
9. EDTA: Prepare 20 L of 0.5*M* EDTA for dialysis.
10. Restriction enzymes and buffers are used as recommended by the manufacturers but without BSA.

2.2. DNA Control Sequence

11. Cloned target DNA: Use a plasmid containing a cloned insert corresponding to the region under investigation. This provides the sequence ladder for the protein-free DNA. Digest 24 µg of plasmid with the same restriction enzyme as the genomic DNA (for a 9 kb) plasmid. Precipitate and resuspend at 2 µg/µL.
12. Sonicated *E. coli* DNA: Sonicate a total of 200 µg of *E. coli* DNA to a size of approx 500 bp. Precipitate and resuspend at 3.7 µg/µL.
13. DMS buffer: 50 m*M* sodium cacodylate, pH 8.0, 1 m*M* EDTA.
14. DMS sequence stop buffer: 1.5*M* sodium acetate, pH 7.0, and 1.0*M* β-mercaptoethanol.
15. Formic acid: Use only highest purity.
16. Hydrazine: Use only highest purity. Hydrazine is toxic and suitable handling precautions must be taken. Store it under nitrogen.
17. Hydrazine (HZ) stop buffer: 0.3*M* sodium acetate, pH 7.5, 0.1 m*M* EDTA.

18. Sodium acetate/EDTA: 0.3*M* sodium acetate and 0.5 m*M* EDTA, pH 5.5.
19. Sample dye: 94% formamide, 10 m*M* Na_2EDTA, pH 7.2, 0.05% xylenecyanole (XCFF) and 0.05% bromophenol blue (BPB).

2.3. Cleavage of the DMS-Treated Genomic DNA with Piperidine

20. Piperidine: Use highest purity (e.g., Fisher Scientific Co., Pittsburgh, PA). Piperidine is toxic and suitable handling precautions must be taken. Store piperidine under nitrogen.

2.4. Synthesis of the Sequencing Primer

21. Oligonucleotides: Two oligonucleotides are required; a 9-mer primer at 0.14 µg/µL and a 33-mer at 0.5 µg/µL (*see* Note 1).
22. 5X Sequenase buffer: 200 m*M* Tris-HCl, pH 7.5, 100 m*M* $MgCl_2$, and 250 m*M* NaCl.
23. dNTP mix: 10 m*M* each of three deoxynucleotide (dNTP) omitting the one corresponding to the radiolabel, in this case dATP. Use sequencing grade quality.
24. α-[^{32}P]dATP: specific activity of at least 3000 Ci/mmol.
25. Sequenase: Use version 2.0 at 13 U/µL, supplied by US Biochemicals, Cleveland, OH.
26. Preparative gel: 15% PAGE. with an acrylamide/*bis*-acrylamide of 29:1 containing 7*M* urea. Make 1 mm thick with slot sizes according to the sample volume.
27. Elution buffer: 0.5*M* ammonium acetate and 0.5 m*M* EDTA, pH 7.5.
28. Ultrafree MC membrane: (0.45 µm: Millipore, Bedford, MA).
29. *E. coli* carrier DNA: Suitable DNA is available from most suppliers but should be of high purity.

2.6. Linear Amplification

30. 10X *Taq* buffer: 166 m*M* $(NH_4)_2SO_4$, 670 m*M* Tris-HCl, pH 8.8, 67 m*M* $MgCl_2$, 100 m*M* β-mercaptoethanol, and 2 mg/mL BSA.
31. *Taq* polymerase: AmpliTaq or Stoffel fragment: Perkin-Elmer/Cetus, Norwalk, CT.

2.7. Purification of the Reaction Products

32. CTAB: 1% *N*-Cetyl-*N,N,N*-trimethyl-ammonium bromide (v/v) in water.
33. 8% Sequence gel: 7*M* urea, 8% PAGE. Acrylamide:*bis*-acrylamide 29:1. Dimensions 60 cm long × 20–40 cm wide × 1 mm thick. Prepare the day before and prerun before loading samples.

34. Footprint loading buffer: 50 m*M* NaOH, 0.5 m*M* EDTA, 4*M* urea, 0.02% bromophenol blue, and 0.02% xylene cyanol. Either freshly prepared or stored in small aliquots at –20°C for some weeks.
35. Fix: 10% methanol/10% acetic acid in water. Make fresh.

3. Methods

3.1. DMS Treatment of Cells

This method describes the treatment and preparation of DNA from cells grown in suspension. This protocol may easily be adapted for cells in monolayer (*see* Note 2). A discussion of the preparation from organs and tissues is beyond the scope of this chapter but we would refer interested readers to ref. *9*.

The concentration of DMS required to produce a footprint varies according to the strength of protein binding and the amount of DNA a cell contains and so in practice it is necessary to treat the cells with a range of DMS concentrations (*see* Note 3).

1. Precool six labeled 15-mL Corex tubes on ice. Add 1 mL of a 10^8 cells/mL cell suspension (in growth medium) to each tube.
2. Add DMS while mixing slowly on a vortex to five of the tubes so that their final concentration ranges from 0.5–0.00005%, decreasing by tenfold at each step. The sixth tube should be left untreated.
3. Incubate for 5 min at 20°C and stop the reaction by adding 10 mL of ice-cold DMS stop. Vortex gently and centrifuge for 5 min at 1000*g* at 4°C.
4. Discard the supernatant. Resuspend the cells in 10 mL of DMS stop and repeat step 3.
5. Resuspend the cells in 1.5 mL nuclei buffer plus 1% Nonident-40. Place the suspension on ice for 5 min and then spin out the nuclei at 3000*g* at 4°C (e.g., in a Sorvall HB-4 rotor).
6. Decant the supernatant and resuspend the nuclei in 1.5 mL of nuclei buffer. Transfer the suspension of nuclei into a sterile 15-mL polypropylene tube (Falcon, Becton Dickinson, Rutherford, NJ). Add 1.5 mL of 2X proteinase K buffer and 45 µL of proteinase K (a final concentration of 300 µg/mL). Seal the lids with parafilm and incubate horizontally in a shaking water bath at 37°C overnight.
7. Add 15 µL of pancreatic RNase A (final concentration 50 µg/mL) and incubate for a further 2 h.
8. Extract seven times with phenol/chloroform as follows: Add 3 mL of phenol into a small Erlenmyer flask on a rotary shaker for 5 min at

room temperature. Add 3 mL of chloroform, mix well, transfer to a Corex centrifuge tube, and centrifuge for 3 min. Extract once with chloroform to remove traces of phenol.

9. Transfer the aqueous phase into dialysis tubing and dialyze against 10 L of 0.5 m*M* EDTA for 2 d. Change the EDTA solution at least once a day.
10. Determine the concentration of the DNA in a photospectrometer (quantity depends on cell type used for the experiment). At this point the DNA may be stored in 1-mL aliquots at –80°C.
11. For each sample, digest 50 µg of DNA in 1 mL of digestion mixture overnight at the appropriate temperature. Six cutter enzymes should be used in a threefold (3 U/µg of DNA) and four cutters in a sixfold excess (*see* Note 4).
12. After digestion extract each sample with phenol/chloroform and precipitate with 100 µL of 3*M* sodium acetate/0.5*M* EDTA, pH 5.5, and 2.5 mL of absolute ethanol precipitation. The digested DNA is then redissolved in 90 µL of water.

3.2. DNA Control Sequence

It is necessary to have a control sequence of purified unmodified DNA against which to compare the footprint. By carrying out a complete sequence we obtain an accurate means of locating the footprint in the DNA sequence. We use cloned unmodified DNA, restriction enzyme digested in the same way as the genomic DNA.

1. G-reaction: Add 2 µL of digested plasmid DNA and 12.5 µL of sonicated *E. coli* DNA to a 1.5-mL Eppendorf tube and lyophilize (in a SpeedVac if available). Dissolve DNA in 6 µL of water and 200 µL of DMS buffer. Chill on ice, add 1 µL of DMS, and mix by tapping the tube and centrifuging for a few seconds in a microfuge at 4°C. Incubate sample in a water bath for 10 min at 20°C. At the end of incubation add 50 µL of DMS stop buffer, mix, and add 750 µL of ethanol precooled to –20°C.
2. (G+A)-reaction: Add 4 µL of digested plasmid DNA and 12.5 µL of sonicated *E. coli* DNA to a 1.5-mL Eppendorf tube and lyophilize. Dissolve DNA in 11 µL of water. Chill on ice, add 25 µL of formic acid, and mix by tapping the tube and centrifuging for a few seconds in a microfuge. Incubate in a water bath at 20°C for 4.5 min. Add 200 µL of HZ stop buffer and then 750 µL of precooled ethanol.
3. (T+C)-reaction: Add 4 µL of digested plasmid DNA and 12.5 µL of sonicated *E. coli* DNA to a 1.5-mL Eppendorf tube and lyophilize. Dis-

solve DNA in 21 μL of water. Chill on ice, add 30 μL of hydrazine, and mix by tapping the tube and centrifuging for a few seconds in a microfuge. Incubate sample in a water bath at 20°C for 10 min. Add 200 μL of HZ stop buffer and then 750 μL of cold ethanol.

4. C-reaction: Add 2 μL of digested plasmid DNA and 12.5 μL of sonicated *E. coli* DNA to a 1.5-mL Eppendorf tube and lyophilize. Dissolve DNA in 5 μL of water and add 15 μL of 5*M* NaCl. Chill on ice, add 30 μL of hydrazine, mix by tapping the tube, and centrifuge a few seconds in a microfuge. Incubate sample at 20°C for 10 min. Add 200 μL of HZ stop buffer and then 750 μL of cold ethanol.
5. Mix all reaction tubes well by inversion and chill for 15 min in a mixture of dry ice and ethanol (–70°C). Centrifuge tubes for 15 min at 15,000*g* at 0°C.
6. Pour out the supernatant very carefully. Resuspend the pellet in 250 μL of sodium acetate/EDTA, pH 5.5. Precipitate with 750 μL of precooled ethanol for 15 min at –70°C and centrifuge as in step 5.
7. Wash the pellet with 1 mL of 70% ethanol/water, centrifuge for 5 min in the microfuge, carefully pour out the supernatant and lyophilize pellet. Resuspend in 90 μL of water.
8. Continue with piperidine treatment as described in Section 3.3.

3.3. Cleavage of the DMS-Treated Genomic DNA with Piperidine

1. Add 10 μL of piperidine to each sample, giving a final piperidine concentration of 1*M*. Mix gently, seal the tubes, and incubate at 90–95°C for 30 min.
2. Freeze samples in dry ice and lyophilize under a high vacuum. Resuspend in 100 μL of water.
3. Repeat the last step at least six times to remove all traces of piperidine (*see* Note 5). After the final wash resuspend the sample in 69 μL of water.

If the DNA is not required immediately store at –70°C.

3.4. Synthesis of the Sequencing Primer

This method produces a highly specific primer that is labeled to high specific activity (*see* Note 1). The primer is generated from two oligonucleotides; a 33-mer with sequence identical to that linearly amplified in Section 3.6. and a 9-mer that is complementary to a region of the 33-mer. Both oligonucleotides are annealed together and then extended using Sequenase enyzme and a mixture of radioactive and

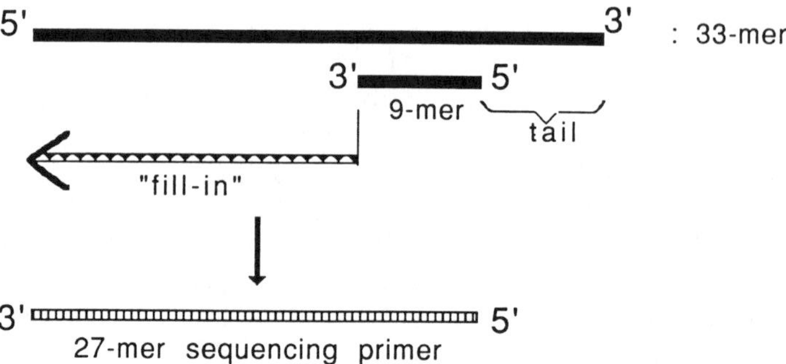

Fig. 2. Synthesis of the radiolabeled sequencing primer by filling-in reaction: A short oligonucleotide (a 9-mer) is annealed to a larger template (a 33-mer). The 9-mer is positioned on the 33-mer such that there is a large single-stranded region at its 5' end and also a short single-stranded tail at 3' end. The 5' single strand is used to incorporate radiolabel by a filling-in reaction. The presence of the 3' tail reduces the size of the labeled primer improving strand separation on a short sequencing gel.

nonradioactive nucleotides (Fig. 2). The labeled primer is then purified on a short sequencing gel and used directly for the linear amplification.

1. To a sterile Eppendorf tube add consecutively 7 µL of 5X Sequenase buffer, 1 µL of the 9-mer, and 1 µL of the 33-mer. Incubate sample at 75°C for 3 min and cool to room temperature over a 10–15 min period. Spin briefly and place on ice.
2. Add 3 µL of dNTP mix, 22 µL of (α-[^{32}P])dATP, and 1 µL of Sequenase. Mix, spin, and incubate at 23°C for 20 min.
3. Add 40 µL of formamide sample dye. Heat denature sample for 3 min at 95°C, quick chill on ice, and load sample on a short 15% sequencing gel. Run the gel until the xylene cyanol dye has migrated 10 cm. Expose the gel for a few seconds and cut out the band of interest.
4. Elute the primer in a 600 µL of elution buffer by vigorous shaking for 2 h.
5. Filter the buffer containing the primer using an Ultrafree MC membrane (0.45 µm: Millipore) into a 1.5-mL Eppendorf tube. Add 2 µL (5 µg) of *E. coli* carrier DNA, 60 µL of 3*M* sodium acetate, 5 m*M* EDTA, pH 5.5. Split into three and add 600 µL of ethanol to each. Mix well and place on dry ice for 20 min. Centrifuge at 12,000*g* for 20 min.
6. Discard the supernatant and dry the pellet. Resuspend in 170 µL of water.

This should be sufficient for 10 reactions (i.e., one complete experiment).

3.5. Linear Amplification and Gel Electrophoresis

1. To a test tube add 69 µL (50 µg) of piperdine-treated genomic DNA, 17 µL (approx 10 ng) of radioactively labeled primer, and 10 µL of 10X *Taq* buffer. Mix gently and spin for a few seconds. Incubate the samples at 95°C for 5 min. This is done in the thermocycler, however do not interrupt its program in the subsequent step.
2. Quick chill the samples in ice/water for 1 min. Spin the samples briefly and place them back in the ice/water.
3. Add 3 µL of each dNTP (dTTP, dGTP, dCTP, and dATP) and 1 µL (5 U) *Taq* polymerase. Mix briefly and replace the samples on ice/water (*see* Note 6).
4. Add 100 µL of mineral oil to each sample and incubate them immediately at 95°C in the thermocycler for 1 min.
5. Amplify using the following program: 1 min at 94°C; 2 min at experimentally determined annealing temperature; 3 min at 72°C; for 30 cycles (*see* Note 7). At the end of the cycles place the samples on ice/water with the minimum delay.
6. By use of a drawn out microcapillary, transfer the sample onto an Ultrafree MC membrane and filter by spinning for 3 min in a bench-top centrifuge. Avoid taking any oil contamination.
7. Add 10 µL of 1% CTAB, mix gently, and put the samples on ice for 20 min. Centrifuge for 15 min at 30,000*g* at 4°C. Remove the supernatants using a drawn out microcapillary and discard. Redissolve the pellet in 100 µL of 0.5*M* ammonium acetate (*see* Note 8).
8. Precipitate with 10 µL of 3*M* sodium acetate 5 m*M* EDTA, pH 5.5, and 300 µL of ethanol at –80°C for 15 min. Centrifuge at 30,000*g* for 15 min at 4°C. Discard the supernatant and in order to remove any residual traces, respin for 1 min and remove supernatant with a drawn out capillary.
9. Redissolve the pellet in 100 µL of 0.5 ammonium acetate and repeat step 4.
10. Carefully redissolve the pellet in 4 µL of footprint loading buffer. Heat for 15 s at 94°C and immediately load onto prerun 8% 1-mm thick sequencing gel (*see* Note 9).
11. Run the sequencing gel until the the xylene cyanol reaches the bottom. This should take about 9 h. After electrophoresis, fix the gel in 1 L of fix, then dry on a gel drier. If necessary, cut the gel laterally into two pieces. Autoradiograph for 12 h on Kodak XAR5 film (*see* Note 10).

Typical results are shown in Fig. 3. The footprint is mapped by comparison of the control sequence to the DMS-treated genomic DNA.

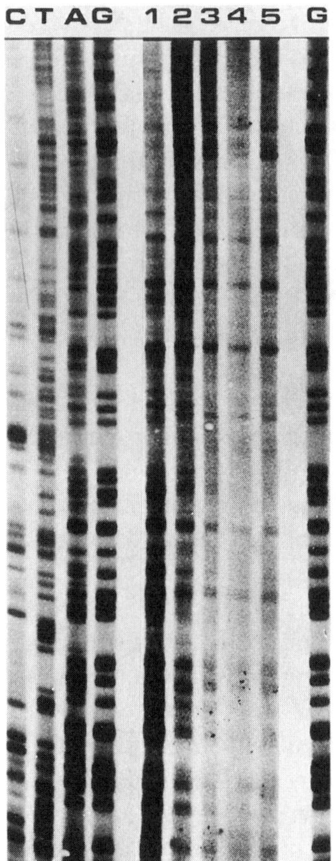

Fig. 3. Example of an in vivo footprint: The left panel shows the four control reactions (C,+T, A+G) performed on cloned target DNA. For genomic footprinting different DMS concentrations were used, whereas the number of cells, incubation time, and temperature were kept constant. Lanes 1–5 demonstrate such a footprint experiment carried out with 0.5%, 0.05%, 0.005%, 0.0005%, and 0.00005% DMS, respectively. This modifies different numbers of guanine residues according to their reactivity, which in turn depends on their interaction with protein. This may give additional information concerning the strength of binding for different DNA/protein complexes.

It may be necessary to footprint both strands to obtain a complete picture of interactions present. The strength of interaction can be estimated from the signal strength relative to the concentration of DMS used on the cells.

4. Notes

1. An oligonuecleotide primer is required for the linear amplification that has a high annealing temperature (at least 60°C) and therefore a high fidelity. This achieved by using a primer of around 30 nucleotides in length. In addition it must be labeled with radioactivity to a high specific activity. A primer with these properties is generated by DNA synthesis in the presence of radiolabeled nucleotides on a primed oligonucleotide template (Fig. 2). In this case we use a short synthetic segment of the sequencing primer (a 9-mer) annealed near to the 3' end of a 33-mer oligo that is complementary to the primer used for linear amplification. This results in a 3'-tail that assists purification of the probe primer on a short sequencing gel (separation from the unlabeled template). Care must be taken with these oligonucleotides taking into consideration the final product. For the gel system described, the oligonucleotide must anneal to the approx 90 nucleotides 5' of the target sequence. In the protocol described we use Sequenase 2.0, which has no 3'–5' exonuclease activity, for the synthesis of the radiolabeled primer.
2. For cells in monolayer steps 2–4 can be performed in tissue culture plates after cooling. The cells are then scraped from the plates with a rubber policeman and pipeted into a suitable corex tube. Steps 5–8 are then carried out as in Section 3.1.
3. For genomic footprinting we obtain good results by using different DMS concentrations while keeping the number of cells, incubation time, and temperature constant. This modifies different numbers of guanine residues according to their reactivity that in turn depends on their interaction with protein. This may give additional information concerning the strength of binding for different DNA/protein complexes.
4. Restriction enzyme digestion is used to reduce the viscosity of the genomic DNA. Care must be taken to avoid choosing an enzyme that cuts within the target sequence since this will reduce the length of the footprint. It is therefore necessary to know the genomic map by either carrying out Southern analysis or having cloned the DNA region of interest. We have had excellent results using regions up to 2 kb. After digestion it is recommended to remove the enzyme by at least one phenol/chloroform extraction and ethanol precipitation.
5. Piperidine must be completely removed by several lyophilization steps from samples before linear amplification since it will interfere with the polymerase. It may also reduce electrophoretic resolution of the footprint. In addition to the method described in Section 3.3, it may also be removed by several rounds of ethanol precipitation. However, lyophilization has the chief advantage that no DNA is lost.

6. Only good quality *Taq* polymerase should be used. We use routinely AmpliTaq (for high G+C content we recommend the Stoffel fragment of *Taq* polymerase) from Perkin-Elmer/Cetus, but other thermostable polymerases may also be used. Failure to use a good quality polymerase may result in artifact bands and low or no signals on the sequencing gels.
7. The annealing temperature is important because all unspecific linearly amplified DNA fragments will be visible on the final autoradiograph. In practice the annealing temperature is determined empirically. It should be 2–4°C below the melting temperature. This point can be established either using a thermoprogrammable spectrophotometer (Perkin-Elmer, λ series with Peltier elements) or by increasing the annealing temperature stepwise by 5°C, starting at approx 60°C. The optimal point resulting in specific signals (sequence clearly visible, no interbands) is used for the linear amplification.
8. Precipitation with CTAB in the presence of ammonium sulfate (already present in the amplication buffer) *(12)* purifies the amplified product away from enzyme, unwanted buffer components, and so on. This avoids smearing on the sequence gel.
9. Since this method results in large amounts of DNA loaded onto the gel it is necessary to run thick sequencing gels to avoid overloading. We use 1-mm thick gels and load the sample into 5-mm slots.

   For good resolution run long gels, at least 60 cm in length. The xylanecyanol marker dye runs at a position equivalent to 90 bp upstream from the start of the primer in these conditions and should be run to the bottom of the gel. Using these conditions it is possible to read 200–300 bp.
10. The control sequence generally gives a stronger signal than the genomic sequence. It is therefore a good idea to develop the autoradiogram after a few hours to obtain good control sequence and to check the quality of separation, fidelity, and so forth. The gel should be re-exposed, usually around 2 d, to visualize the genomic footprint.

Acknowledgments

We would like to thank L. Di Renzo and A. Wallace for proofreading the manuscript.

References

1. Church, G. M. and Gilbert, W. (1984) Genomic sequencing. *Proc. Natl. Acad. Sci. USA* **81,** 1991–1995.
2. Becker, M. M. and Wang, J. C. (1984) Use of light for footprinting DNA in vivo. *Nature* **309,** 682–687.
3. Saluz, H. P. and Jost, J. P. (1989) Genomic sequencing and in vivo footprinting. *Anal. Biochem.* **176,** 201–208.

4. Saluz, H. P., Wiebauer, K., and Wallace, A. (1991) Studying DNA modifications and DNA-protein interactions in vivo. *TIG* **7,** 207–211.
5. Selleck, S. B. and Majors, J. (1987) Photofootprinting in vivo detects transcription dependent changes in yeast TATA boxes. *Nature* **325,** 173–177.
6. Axelrod, J. D. and Majors, J. (1989) An improved method for photofootprinting yeast genes in vivo using Taq polymerase. *Nucleic Acids Res.* **17,** 171–183.
7. Becker, M. M., Wang, Z., Grossmann, G., and Becherer, K. A. (1989) Genomic footprinting in mammalian cells with ultraviolet light. *Proc. Natl. Acad. Sci. USA* **86,** 5315–5319.
8. Maxam, A. M. and Gilbert, W. (1980) Sequencing end-labeled DNA with base-specific chemical cleavages. *Meth. Enzymol.* **65,** 499–560.
9. Saluz, H. P. and Jost, J. P. (1990) *A Laboratory Guide for In Vivo Studies of DNA Methylation and Protein/DNA Interaction.* Birkhaeuser, Basel-Boston, pp. 129–214.
10. Saluz, H. P. and Jost, J. P. (1989) Genomic footprinting with Taq polymerase. *Nature* **338,** 277.
11. Saluz, H. P. and Jost, J. P. (1989) A simple high-resolution procedure to study DNA methylation and in vivo DNA protein interactions on a single-copy gene level. *Proc. Natl. Acad. Sci. USA* **86,** 2602–2606.
12. Jost, J. P., Jiricny, J., and Saluz, H. P. (1989) Quantitative precipitation of short oligonucleotides with low concentrations of cetyl trimethyl ammonium bromide. *Nucleic Acids Res.* **17,** 2143.

CHAPTER 31

DNA Photofootprinting with $Rh(phi)_2bpy^{3+}$

Scott L. Klakamp and Jacqueline K. Barton

1. Introduction

This chapter describes one method for high resolution photofootprinting of proteins or small drugs bound to DNA *(1)*. DNA footprinting is an electrophoretic method that permits the visualization (footprint) of the binding site of a molecule site-specifically bound to DNA. The method entails the nonspecific cleavage of a radioactively end-labeled DNA containing the site-specifically bound molecule. Protection from this sequence neutral cleavage at the binding site reveals the footprint. Photofootprinting differs from other conventional footprinting techniques in that DNA cleavage occurs directly on irradiation of a photocleaving reagent, which binds to the target DNA sequence. Alternative photofootprinting methods produce DNA cleavage indirectly by creating photo-damaged DNA adducts that are labile to base treatment after irradiation of the DNA *(2,3)*. Traditional footprinting procedures utilize either a protein like DNase I for cleavage or small molecule reagents that cleave DNA under specific chemical conditions *(4–8)*.

Here we describe chemical photofootprinting using the metallointercalator $Rh(phi)_2bpy^{3+}$ (phi = phenanthrenequinone diimine, bpy = bipyridyl, *see* Fig. 1), which cleaves DNA with high sequence neutrality on photoactivation *(9)*. The rigid complex binds DNA by intercalation in the major groove. Photoactivation of the rhodium complex bound

From: *Methods in Molecular Biology, Vol. 31: Protocols for Gene Analysis*
Edited by: A. J. Harwood

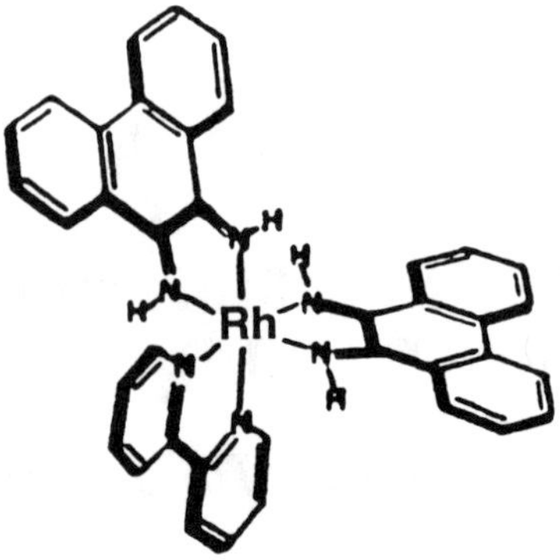

Fig. 1. The structure of $Rh(phi)_2bpy^{3+}$.

to DNA results in the direct abstraction of the C3'-hydrogen atom and DNA strand breakage *(10)*. $Rh(phi)_2bpy^{3+}$ provides remarkably high resolution in its photocleavage pattern because of the rigidity of the complex bound to DNA and because of the lack of a diffusing intermediate in the cleavage chemistry. Other advantages of using $Rh(phi)_2bpy^{3+}$ as a footprinting reagent lie in its sequence neutrality in binding and cleavage, the ability of the intercalator to footprint DNA binding molecules in either the major or minor grooves, the general stability of the complex in the absence of light, and the absence of any additives to induce cleavage other than light activation. DNA cleavage by $Rh(phi)_2bpy^{3+}$ is not inhibited by the common reagents found in protein extract mixtures and can even be conducted under conditions that approximate those that occur intracellularly.

The method of photofootprinting with $Rh(phi)_2bpy^{3+}$ begins with the preparation of a ^{32}P end-labeled DNA restriction fragment containing the target sequence for the molecule to be footprinted using standard molecular biological techniques. $Rh(phi)_2bpy^{3+}$ is then added in the absence or presence of the DNA-binding molecule to be footprinted, and the solution is irradiated for 5–10 min at 313 or 365 nm. The samples are ethanol precipitated twice after irradiation, loaded onto an 8% sequencing gel, and electrophoresed. Autoradiography then reveals the high resolution footprint of the molecule bound to its DNA site. We illustrate this process by mapping the DNA binding site of the antibiotic molecule distamycin A.

2. Materials

In addition to the specific items described below, general molecular biology materials are required. Restriction enzymes, calf intestinal alkaline phosphatase, and T4 polynucleotide kinase should be used with appropriate buffers as described by the manufacturers. The radiolabel, γ-[^{32}P]-ATP is required for labeling of the desired DNA restriction fragment. Phenol:chloroform, 1:1 (v/v) and chloroform are required for DNA purification. Ammonium acetate (6.8*M*) and ethanol (200 proof dehydrated punctilious) are required for precipitation. A native 5% preparatory polyacrylamide gel is required for DNA fragment isolation.

1. Tris-acetate buffer: 50 m*M* Tris base, 20 m*M* sodium acetate, 18 m*M* sodium chloride, pH 7.0. The buffer is initially sterilized by autoclaving and then stored at room temperature and sterilized as needed by filtration through a 0.22-µm filter.
2. Calf thymus DNA: Phenol extracted sonicated calf thymus DNA at 1.3 mg/mL, for use as carrier for the radioactively labeled DNA (*see* Note 1).
3. Distamycin A, used here as the DNA-binding molecule, is prepared fresh for each experiment. Dilute in Tris-acetate buffer (*see* Note 1).
4. $Rh(phi)_2bpyCl_3$: The complex is synthesized as described earlier *(11)*. It is also available from our laboratory on written request. Stock concentration 0.92 m*M* in Tris-acetate buffer (*see* Note 1).
5. UV source: Oriel 1000 Watt Hg/Xe arc lamp focused and filtered with a monochromator (Oriel model 77250) and a 305-nm pass filter placed after the monochromator to eliminate most of the light below 305 nm. Other light sources may be utilized (*see* Note 2).
6. 5X Tris-borate buffer: 0.44*M* Tris-base, 0.44*M* boric acid, 0.01*M* EDTA, pH 8.0.
7. Denaturing gel loading buffer: 80% formamide, 50 m*M* tris-borate buffer, 0.1% xylene cyanol, 0.025% bromophenol blue, 0.01*N* NaOH, and 1 m*M* EDTA.
8. 8% denaturing polyacrylamide gel solution: Mix in the laboratory or buy ready mixed.

3. Methods

Plasmid pJT18-T6 is a derivative of plasmid pUC18 and contains an 18-bp insert (5'-ATATGCAAAAAAGCATAT-3') at the *Sma* I site of pUC18. The 6 AT bp in the middle of the 18-bp insert form a strong distamycin A binding site.

1. Digest 15 μg of pJT18-T6 with *Hind* III for 3 h at 37°C. Purify by ethanol precipitation.
2. Incubate the linearized plasmid with calf intestinal alkaline phosphatase for 1½ h at 37°C . Extract with phenol:chloroform and then chloroform followed by two ethanol precipitations.
3. Label the 5' end of the linearized plasmid with T4 polynucleotide kinase and γ-[^{32}P]-ATP for 1 h at 37°C (*see* Note 3).
4. Digest the 5' end-labeled plasmid with *Pvu* II for 6 h at 37°C and overnight at 4°C. Excise a fragment that contains the binding site(s). Isolate the fragment by electrophoresis at 280 V for 1½ h on a native 5% preparatory polyacrylamide gel. After electrophoresis, the band containing the fragment is excised from the gel and electroeluted for 2 h. Ethanol precipitate and resuspend in Tris-acetate buffer at 6000 cpm/μL.
5. In a 1.7-mL microfuge tube mix 5 μL (30,000 cpm) of 5' end-labeled fragment, 0.65 μL of calf thymus DNA (25 μ*M* bp final concentration), and a range of DNA binding molecule concentrations (in the case illustrated a range of distamycin A concentrations from 0.17–13 μ*M*) in a total volume of 49.2 μL Tris-acetate buffer. Incubate the reaction mixtures for 10 min. Add 0.75 μL of $Rh(phi)_2bpy^{3+}$ (to make 13.8 μ*M* final) and incubate for a further 15 min (*see* Notes 4 and 5). Also set up light and dark control samples that contain all the components of the cleavage and footprinting reactions but lack either the rhodium complex or the binding molecule (distamycin A), respectively.
6. Place the open tubes in front of the 1000 W Hg/Xe lamp oriented such that the 50 μL reaction solution adhering to the bottom of the tube is directly at the focal point. Irradiate for 7 min.
7. Add 1 μL of 5 μg/μL calf thymus DNA and 17 μL of ammonium acetate to each reaction mixture. Heat at 90°C for 2 min and then precipitate by the addition of 136 μL of EtOH (*see* Note 6). Vortex the samples for 5 s and place on dry ice for 15 min. Spin at 13,000*g* for 10 min. Discard the supernatant and add 50 μL of deionized water to each sample. Resuspend the samples by vortexing 2 min followed by a quick spin.
8. Reprecipitate by the addition of 15 μL of ammonium acetate and 136 μL of ethanol, placing in dry ice again for 15 min and spinning at 13,000*g* for 10 min. Discard the supernatant and rinse the pellets with 70% ethanol and dry *in vacuo* (*see* Note 7).
9. Dissolve the DNA pellets in 3 μL of loading buffer. Load on a standard 40 cm, 8% denaturing polyacrylamide gel and electrophorese for 1½ h at 1500 V (bromophenol blue is at the bottom of the gel) with 1X TBE as a running buffer. Dry gel and audioradiograph.

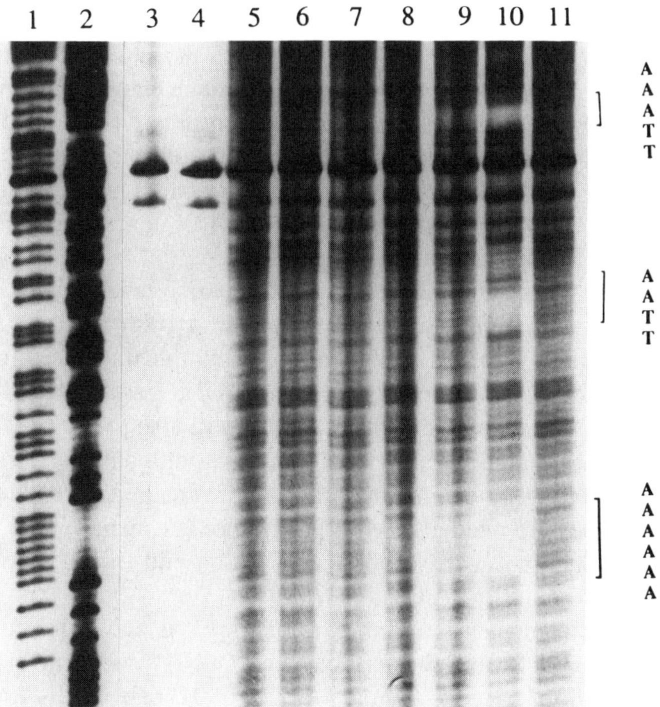

Fig. 2. $Rh(phi)_2bpy^{3+}$ photofootprinting of distamycin A. Lanes 1 and 2: Maxam–Gilbert A + G and T + C reactions, respectively; lanes 3 and 4: light and dark controls, respectively, the 5'-[^{32}P]-end labeled fragment in the absence of metal complex with and without irradiation (7 min at 313 nm) with light; Lanes 5 and 11: DNA irradiated in the presence of 13.8 μM metal complex but in the absence of distamycin; Lanes 6–10: in the presence of 0.17, 0.30, 1.4, 2.5, and 13 μM distamycin, respectively. Lanes showing fragments cleaved by the metal complex without the DNA-binding molecule (5 and 11) are essential to show the extent of sequence neutrality and for comparison with lanes containing the DNA-binding molecule to indicate the footprint. The bars indicate the A-T rich footprints.

A typical result is shown in Fig. 2; the distamycin A footprint is clearly seen in Lane 10.

4. Notes

1. The concentration of the $Rh(phi)_2bpy^{3+}$ and distamycin A stock solutions for the cleavage reactions is determined spectrophotometrically with molar absorptivities at 350 nm of 23,600M^{-1}cm^{-1} (in Tris-acetate

buffer) for the rhodium complex, and 34,000$M^{-1}cm^{-1}$ at 302 nm for distamycin A (in deionized water). The calf thymus DNA solution concentration is also determined optically in Tris-acetate buffer with a molar absorptivity of 6600$M^{-1}cm^{-1}$. The calf thymus DNA, $Rh(phi)_2bpy^{3+}$, distamycin A, and 5' end-labeled fragment solutions used for preparation of the final cleavage reactions are all made up by dilution with Tris-acetate buffer so as not to dilute the Tris-acetate concentration in the final reaction solution.

2. Photofootprinting with $Rh(phi)_2bpy^{3+}$ can also be achieved by using a transilluminator light source *(1)* instead of a 1000 W Hg/Xe lamp. As expected, owing to the lower intensity of the transilluminator light source, irradiation times must be increased by at least threefold to achieve sufficient DNA cleavage. $Rh(phi)_2bpy^{3+}$ will also promote DNA cleavage when irradiated with 365 nm light, although cleavage is much less efficient at longer wavelengths and therefore significantly longer irradiation times are needed for footprinting experiments.
3. DNA fragments may also be labeled by an end-filling reaction using Klenow and α-[^{32}P]-dATP.
4. The photocleavage reaction of $Rh(phi)_2bpy^{3+}$ is not inhibited by conditions necessary for in vivo applications *(1)*. High concentrations of salt, such as 100 m*M* NaCl, 10 m*M* $MgCl_2$, or 10 m*M* $CaCl_2$ may also be employed. The cleavage reaction is also unaffected by 1 mM EDTA, 1 m*M* dithiothretitol, 10 m*M* 3-mercaptopropionic acid, bovine serum albumin (60 mg/mL), and 10% glycerol.
5. The relative concentrations of rhodium and DNA-binding molecule being footprinted to the target DNA sequence and the irradiation times utilized for the photocleavage reaction must be determined empirically for every system studied. The rhodium complex binds DNA with an affinity of $10^7 M^{-1}$. Hence, higher concentrations of molecules that bind DNA with lower affinity are required to achieve footprints. Molecules bound either in the major or minor groove may be footprinted with the rhodium intercalator.
6. It is important to heat the samples at 90°C for 2 min after irradiation, since it has been shown previously *(1)* that this heating step helps $Rh(phi)_2bpy^{3+}$ to dissociate from the DNA and to produce clearer sequencing gels.
7. After the second precipitation of the irradiated samples and drying of the resulting DNA pellets *in vacuo*, it is helpful to add 50 μL deionized water, followed by 20 s of vortexing and a quick spin prior to removal of the water *in vacuo*. If ammonium acetate is used in the final ethanol

precipitation, the excess salt is removed as acetic acid and ammonia, which will also produce clearer sequencing gels.

Acknowledgments

We are grateful to the National Institutes of Health (GM33309 to J. K. Barton; NRSA CA08891 to S. L. Klakamp) for their financial support of this research.

References

1. Uchida, K., Pyle, A. M., Morii, T., and Barton, J. K. (1989) High resolution footprinting of *Eco* RI and distamycin with $Rh(phi)_2bpy^{3+}$, a new photofootprinting reagent. *Nucleic Acids Res.* **17**, 10,259–10,279.
2. Selleck, S. B. and Majors, J. (1987) Photofootprinting *in vivo* detects transcription-dependent changes in yeast TATA boxes. *Nature* **325**, 173–177.
3. Becker, M. M. and Wang, J. C. (1984) Use of light for footprinting DNA *in vivo*. *Nature* **309**, 682–687.
4. Galas, D. J. and Schmitz, A. (1978) DNase I footprinting: a simple method for the detection of protein-DNA binding specificity. *Nucleic Acids Res.* **5**, 3157–3170.
5. Van Dyke, M. W., Hertzberg, R. P., and Dervan, P. B. (1982) Map of distamycin, netropsin, and actinomycin binding sites on heterogeneous DNA: DNA cleavage-inhibition patterns with methidiumpropyl-EDTA· Fe(II). *Proc. Natl. Acad. Sci. USA* **79**, 5470–5474.
6. Hertzberg, R. P. and Dervan, P. B. (1984) Cleavage of DNA with Methidiumpropyl-EDTA-iron(II): reaction conditions and product analyses. *Biochemistry* **23**, 3934–3945.
7. Sigman, D. S. (1986) Nuclease activity of 1, 10-phenanthroline-copper ion. *Acc. Chem. Res.* **19**, 180–186.
8. Tullius, T. D., Dombroski, B. A., Churchill, M. E. A., and Kam, L. (1987) Hydroxyl radical footprinting: a high resolution method for mapping protein-DNA contacts. *Meth. Enzymol.* **155**, 537–558.
9. Pyle, A. M., Long, E. C., and Barton, J. K. (1989) Shape-selective targeting of DNA by phenanthrenequinone diimine rhodium(III) photocleaving agents. *J. Am. Chem. Soc.* **111**, 4520–4522.
10. Sitlani, A., Long, E. C., Pyle, A. M., and Barton, J. K. (1992) DNA photocleavage by phenanthrenequinone diimine complexes of rhodium(III): shape-selective recognition and reaction. *J. Am. Chem. Soc.* **114,** 2303–2312.
11. Pyle, A. M., Chiang, M. Y., and Barton, J. K. (1990) Synthesis and characterization of physical, electronic, and photochemical aspects of 9, 10 phenanthrenequinone diimine complexes of ruthenium(II) and rhodium(III). *Inorg. Chem.* **29**, 4487–4495.

CHAPTER 32

The Gel Retardation Assay

Valerie Scott, Andrew R. Clark, and Kevin Docherty

1. Introduction

The gel retardation or electrophoretic mobility shift assay (EMSA) is a sensitive technique for studying protein–DNA interactions. Originally devised by Fried and Crothers *(1)* to study the kinetics of protein–DNA interactions, the method relies on the stability of protein DNA complexes when subject to nondenaturing polyacrylamide gel electrophoresis. The DNA is radiolabeled to enable rapid detection and in its native state it migrates quickly through the gel matrix. Protein binding generates slower mobility protein–DNA complexes that resolve as discrete bands.

Although the original method used purified proteins, the gel retardation assay is now widely used to investigate the formation of protein–DNA complexes in a crude nuclear protein extract. A number of modifications have subsequently been introduced to increase the specificity of the assay.

The detection of nonspecific protein–DNA complexes, which may mask sequence-specific interactions, can be reduced by the inclusion of nonspecific competitor DNA. Heterologous competitor DNA, such as *E. coli* DNA, may be used. The use of the synthetic copolymer poly(dI · dC) · poly(dI · dC) by Singh et al. *(2)*, however, aims to increase the sensitivity of the assay by minimizing the binding of sequence-specific DNA-binding proteins to related sequences within the heterologous competitor DNA.

From: *Methods in Molecular Biology, Vol. 31: Protocols for Gene Analysis*
Edited by: A. J. Harwood Copyright ©1994 Humana Press Inc., Totowa, NJ

DNA restriction fragments of 300 bp or less may be used in the assay although the results may be complex if the fragment comprises several protein binding sites. In order to focus on a particular protein–DNA interaction, synthetic oligonucleotides that span a single protein binding site are more commonly used. We routinely use 30 base pair (bp) oligonucleotides in our studies, however the limiting factor is the size of the binding site. Oligonucleotides as short as 16 bp have been used, although in one case the shorter oligonucleotide appeared to have a lower affinity for its binding protein than the equivalent 30 bp oligonucleotide (Scott, unpublished).

In the first instance, it is necessary to have a defined region of DNA with which to investigate the binding protein(s) of potential interest. The gel retardation assay can then be used to map the binding site in further detail. The use of competition assays with homologous and heterologous oligonucleotides can provide information on the sequence-specificity of any protein–DNA interactions, and mutagenesis of the binding site DNA sequence may aid in the identification of particular residues that play a key role in determining protein binding.

2. Materials

2.1. Preparation of Nuclear Protein Extracts

1. PBS: 10 m*M* phosphate buffer, pH 7.4, 0.15*M* NaCl.
2. Buffer A: 10 m*M* HEPES/KOH, pH 7.9, 10 m*M* KCl, 1.5 m*M* $MgCl_2$, 0.5 m*M* dithiothreitol (DTT). Store at –20°C. Place at 4°C just before use.
3. Buffer C: 20 m*M* HEPES/KOH, pH 7.9, 10% (v/v) glycerol, 0.42*M* NaCl, 1.5 m*M* $MgCl_2$, 0.5 m*M* DTT, 0.2 m*M* EDTA, 1 m*M* PMSF. Make up a 100-m*M* stock of PMSF (phenylmethylsulfonyl-fluoride) in propan-2-ol and store at –20°C. **Caution:** PMSF is very toxic by inhalation and in contact with skin. Add DTT and PMSF just before use and cool the solution to 4°C.
4. Buffer D: 20 m*M* HEPES/KOH, pH 7.9, 10% (v/v) glycerol, 0.5 m*M* DTT, 0.2 m*M* EDTA, 1 m*M* PMSF. DTT and PMSF should be added just prior to use and the solution cooled to 4°C. **Note:** In addition to PMSF a cocktail of protease inhibitors may be added to buffers C and D (2.5 µg/mL of aprotinin, and 1 m*M* each of leupeptin, pepstatin, and antipain).
5. Ammonium sulfate.
6. 10% (v/v) Nonidet NP-40.

2.2. Radiolabeling of Oligonucleotide Probes

1. Synthetic oligonucleotides are usually obtained as a lyophilized powder. Resuspend in 1 mL H_2O and extract three times with 400 µL butan-1-ol (not water-saturated). The upper organic phase is discarded and the aqueous phase evaporated to dryness using a SpeedVac centrifuge. The oligonucleotides are resuspended in 200 µL H_2O and the A_{260} recorded.
2. 10X kinase buffer: 0.5 m*M* Tris-HCl, pH 7.6, 0.1*M* $MgCl_2$, 50 m*M* DTT, 1 m*M* spermidine, 1 m*M* EDTA. Store at –20°C.
3. Adenosine 5'-γ-[^{32}P] triphosphate, triethylammonium salt, specific activity 3000 Ci/mmol (Amersham, Arlington Heights, IL).
4. 0.5*M* EDTA, pH 8.0.
5. 20X oligo annealing buffer: 200 m*M* Tris-HCl, pH 7.9, 40 m*M* $MgCl_2$, 1*M* NaCl, 20 m*M* EDTA. Store at –20°C.
6. TE buffer, pH 7.6: 10 m*M* Tris-HCl, pH 7.6, 1 m*M* EDTA.
7. Sephadex G-50 equilibrated in TE, pH 7.6.

2.3. Gel Retardation Assay

1. 40% (w/v) acrylamide stock solution: Prepare a 40% (w/v) solution of acrylamide:*N,N'*-methylene *bis*-acrylamide (19:1, w/w) using Ultrapure grade chemicals and store at 4°C in the dark. **Caution:** Acrylamide is a neurotoxin—wear gloves.
2. TEMED (*N,N,N',N'*-tetramethylethylenediamine). Store at 4°C.
3. 10% (w/v) ammonium persulfate (APS): usually made fresh as required but can be stored at 4°C for at least a week.
4. 10X TBE: 890 m*M* Tris, 890 m*M* boric acid, 20 m*M* EDTA. Dilute stock to 0.5X TBE for use.
5. 10X gel retardation assay buffer, 50 m*M* salt (GRAB50): 100 m*M* Tris-HCl, pH 7.6, 500 m*M* KCl, 10 m*M* EDTA, 50% (v/v) glycerol. Store at –20°C (*see* Note 1).
6. Poly(dI · dC) · poly(dI · dC) (Pharmacia, Piscataway, NJ): 1 mg/mL in distilled H_2O. Store at –20°C (*See* Note 2).
7. 20 m*M* DTT. Store at –20°C.
8. 10X gel loading buffer: 50% (v/v) glycerol, 0.25% (w/v) xylene cyanol, 0.25% (w/v) bromophenol blue.

3. Method

3.1. Preparation of Nuclear Protein Extracts (3)

1. Harvest 1–5 × 10^9 cells and wash twice with 5 vol of PBS by centrifugation at 400*g* for 10 min. Adherent cell lines are harvested in 5–10 mL of PBS by gently scraping the cells off the plate using a plastic spatula. All subsequent steps are performed at 4°C or on ice.

2. Resuspend the cells in 5 packed cell volumes of cold buffer A and allow to stand on ice for 10 min. Pellet the cells by centrifugation (400*g* for 10 min at 4°C) and resuspend in 2 packed cell volumes of buffer A.
3. Lyse the cells on ice using a Dounce homogenizer with the B pestle (50–100 strokes are usually sufficient). The cells can be examined microscopically for lysis by staining a small aliquot of the homogenate with trypan blue—nuclei take up the stain immediately.
4. Pellet the nuclei (400*g*, 20 min). Take care when removing the supernatant since the pellet may be loose. The supernatant may be retained to prepare a cytoplasmic S100 extract.
5. Resuspend the nuclei in 3 mL of ice-cold buffer C/10^9 cells and homogenize with 20 strokes of the B pestle in a Dounce homogenizer.
6. Stir the homogenate gently with a magnetic stirrer for 30 min at 4°C, then centrifuge at 100,000*g* for 20 min (e.g., 32,700 rpm in a Beckman SW50.1 rotor).
7. Collect the nuclear supernatant and precipitate the proteins by the gradual addition of ammonium sulfate to 0.33 g/mL, with gentle stirring as in step 6.
8. Harvest the nuclear precipitate by centrifugation at 25,000*g* for 20 min (e.g., 16,400 rpm in a Beckman SW50.1 rotor), and resuspend the pellet in 0.5–1.0 mL buffer D/10^9 cells.
9. Dialyse the resuspended nuclear protein extract against 100 vol of buffer D at 4°C for 12–18 h.
10. Centrifuge the dialysate in a microcentrifuge for 10 min. Store the supernatant at –70°C in 0.1–0.25 mL aliquots in order to minimize repeated freezing and thawing.

3.2. Preparation of "Mini-Extracts"

We find this protocol more simple and rapid than the Dignam method (Section 3.1.), taking little more than an hour and providing adequate nuclear extract for several assays *(4)*. Retardation patterns do not appear to vary substantially between the two types of extract. Protease inhibitors as in Section 2.1. step 4 may be added to buffer C. Note that the final extract is in 400 m*M* NaCl; this must be taken into account when making up the binding mixes.

1. Harvest 0.5–1 × 10^6 cells and wash twice with 5 vol of PBS. Resuspend the cell pellet in 400 μL of cold buffer A, and allow the cell suspension to stand for 15 min on ice.
2. Add 25 μL of a 10% (v/v) solution of Nonidet NP-40 and vortex the tube vigorously for 10 s. Centrifuge the homogenate for 30 s in a microcentrifuge.

3. Resuspend the nuclear pellet in 50 µL of ice-cold buffer C and rock the tube vigorously at 4°C on a shaking platform.
4. Centrifuge for 5 min in a microcentrifuge at 4°C. Collect the supernatant and store in aliquots at –70°C.

3.3. Preparation of Radiolabeled Double-Stranded Oligonucleotide Probes

1. Set up the following reaction mixture in a 1.5 mL microcentrifuge tube: 1 µL of 10X kinase buffer, 20 pmol sense oligonucleotide, 5 µL of (50 µCi) γ-[^{32}P]ATP, and H_2O to a final volume of 10 µL. Then add 1 µL of T4 polynucleotide kinase (10 U/µL), and incubate at 37°C for 30–60 min.
2. Heat inactivate the enzyme at 70°C for 10 min. Add a 1.5 molar excess (30 pmol) of unlabeled complementary oligonucleotide, 1 µL of 20X oligo annealing buffer, and H_2O to a final volume of 20 µL.
3. Heat to 90°C for 5 min in a dri-block. Remove the block from the dri-block unit and allow to cool slowly to room temperature to promote annealing.
4. Make the volume up to 100 µL with TE, pH 7.6, and pass through a Sephadex G-50 spun column to separate the radiolabeled oligonucleotide from unincorporated nucleotides.
5. Collect the eluate (100 µL) and determine the level of radioactivity. We generally obtain 0.5–1 × 10^6 dpm/µL (*see* Note 3).

3.4. Gel Retardation Assay

1. For an 8% native polyacrylamide gel, mix the following: 10 mL of 40% acrylamide/*bis* (19:1), 2.5 mL of 10X TBE, and H_2O to a final volume of 50 mL. Then add 150 µL of 20% APS and 75 µL of TEMED, and pour into a 15 cm × 15 cm × 1.5 mm gel assembly. Allow to polymerize (15–30 min).
2. Prerun for 20–30 min at 10–15 V/cm in 0.5X TBE, then replace the contents of the top reservoir with fresh 0.5X TBE and continue to prerun for a further 20–30 min before loading samples.
3. Set up the protein–DNA binding reaction as follows: 2 µL of 10X GRAB50, 1 µL of 20 m*M* DTT, 1 µg of Poly(dI · dC) · poly(dI · dC), 10 µg of nuclear protein extract, and H_2O to a final volume of 20 µL. Each reagent is applied to the side of a microcentrifuge tube that is centrifuged for 1–2 s. Finally add 0.5–1 × 10^5 dpm of radiolabeled double-stranded oligonucleotide, and incubate for 30 min at room temperature (*see* Notes 1 and 2).
4. Add 2 µL of 10X gel loading buffer and apply samples to the gel in order to resolve protein-bound DNA and free probe (*see* Note 4). Continue the electrophoresis until the free probe has traversed the length of

the gel (for a 30-bp oligonucleotide and 8% acrylamide gel run until the bromophenol blue is 1–2 cm from the end of the gel).

5. Transfer the gel onto Whatman 3MM paper and dry under vacuum using a slab gel dryer.
6. Autoradiograph overnight at –70°C with intensifying screens (*see* Fig. 1).

4. Notes

1. Tris or HEPES-based buffers may be used for the gel retardation assay. The buffer composition may require modification in order to achieve optimal protein–DNA complex formation. Factors to consider include:
   a. Salt concentration—initially it is advisable to carry out a KCl/NaCl titration over the range 50–500 m*M*;
   b. Concentration of divalent cations such as magnesium; and
   c. Cofactor requirements—for example, the detection of zinc finger protein complexes may require the omission of metal chelators from the buffer and supplementation with zinc.
2. The concentration of nonspecific competitor DNA such as poly(dI · dC) · poly(dI · dC) may need to be optimized for each particular probe-nuclear extract combination. In general we find 1 μg competitor DNA and 10 μg nuclear protein usually allows detection of specific protein–DNA complexes. Figure 1 shows the effect of poly(dI · dC) · poly(dI · dC) concentration. In the absence of competitor DNA, specific interactions are masked by a smear of nonspecifically retarded probe and some nonspecific complexes fail to enter the gel. Specific protein–DNA interactions are detected even in the presence of relatively high concentrations of poly(dI · dC) · poly(dI · dC).
3. Although we find this is generally not necessary, gel purification of the probe may sometimes be used to achieve a cleaner result. The radiolabeled double-stranded oligonucleotide or restriction fragment may be separated from contaminants by resolution on a native polyacrylamide gel, excised, and eluted from the gel. This method (plus an additional anion-exchange purification step) is described in detail in an earlier volume in this series *(5)*.
4. It may be found that gel loading buffer interferes with protein–DNA complex formation, although we have only rarely observed this phenomenon. If no retarded bands appear, electrophoresis can be performed in the absence of loading buffer (glycerol is already present in the sample), and its progress monitored by running 2 μL of loading buffer in an adjacent track.
5. Gel retardation may be used as a sensitive assay to monitor the purification of DNA-binding proteins. When using semipurified or purified

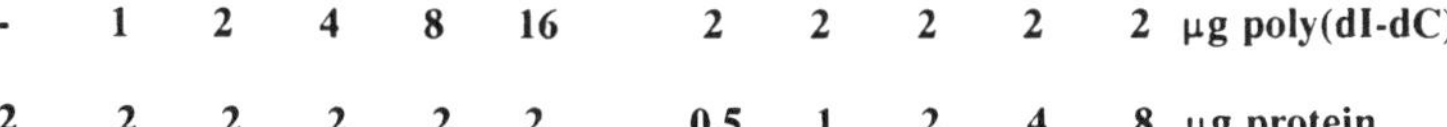

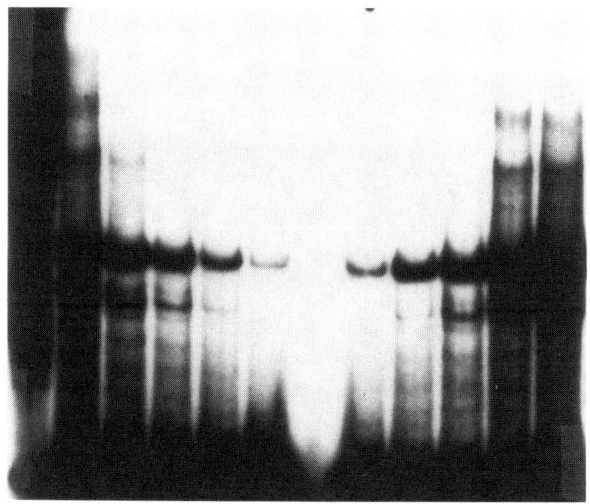

Fig. 1. The effect of poly(dI · dC) · poly(dI · dC) and quantity of protein on the detection of sequence-specific protein–DNA interactions. The gel retardation assay was set up as in Section 3.4. using 10^5dpm of a 16-bp oligonucleotide, 0.5–8 μg nuclear protein extract from pancreatic β cells, and in the presence of increasing amounts of poly(dI · dC) · poly(dI · dC).

proteins, Triton X-100 is included in the reaction buffer at a final concentration of 0.1% (v/v). It is also necessary to decrease or omit the nonspecific competitor DNA.

6. The competition gel retardation assay may be used to investigate the specificity of a particular protein–DNA interaction. Unrelated and mutant oligonucleotides can be used to further delineate the DNA binding site. Figure 2 shows an example of a competition assay. Interactions that are not competed by cold homologous oligonucleotides are unlikely to be sequence-specific. In contrast, complexes that are abolished by self-competition but unaffected by unrelated competitor oligonucleotides may be attributed to a sequence-specific protein–DNA interaction. The molar excess required for competition by cold oligonucleotide varies between protein–DNA complexes, and in some cases may be very high. In order to determine whether a complex is specific it may be necessary to investigate self-competition and competition by

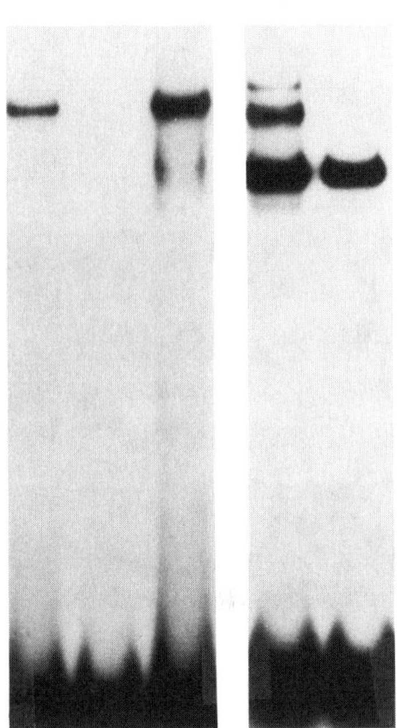

Fig. 2. Competition gel retardation assay. Nuclear protein extract (10 µg) from pancreatic β cells (lanes 1–3) and HeLa cells (lanes 4 and 5) were incubated with 10^5 dpm of a ^{32}P-labeled 30-bp oligonucleotide. Lanes 2, 3, and 5 represent extracts preincubated with an approx 200-fold molar excess of cold homologous or heterologous competitor oligonucleotide. Lanes 1 and 4, no competitor DNA; lanes 2 and 5, homologous competitor; lane 3, unrelated heterologous competitor.

unrelated oligonucleotides across a range of molar excesses. In most cases, a 100-fold excess of cold oligonucleotide will be sufficient for cold competition. The competition assay is set up as in Section 3.4. except that radiolabeled probe is omitted and the protein extract is preincubated with a 50–500 molar excess of unlabeled competitor oligonucleotide. After a 10 min incubation at room temperature, the radiolabeled oligonucleotide is added and the incubation continued for a further 20 min.

7. In order to investigate the potential involvement of a particular protein in a specific protein–DNA interaction, antibody studies may be used. Protein extracts that have been preincubated with antibody will gener-

ate the normal mobility shift profile only if the antibody does not bind the DNA-binding protein. If an interaction occurs between the antibody and the DNA-binding protein, the DNA-bound complex will have a slower mobility on gel retardation resulting in a super-shift. Alternatively, antibody binding may mask the DNA-binding site and preclude formation of the protein–DNA complex.

References

1. Fried, M. and Crothers, D. M. (1981) Equilibria and kinetics of lac repressor-operator interactions by polyacrylamide gel electrophoresis. *Nucleic Acids Res.* **9,** 6505–6525.
2. Singh, H., Sen, R., Baltimore, D., and Sharp, P. A. (1986) A nuclear factor that binds to a consensus sequence motif in transcriptional control elements of immunoglobulin genes. *Nature* **319,** 154–158.
3. Dignam, J. D., Debovitz, R. M. I., and Roeder, R. G. (1983) Accurate transcription initiation by RNA polymerase II in a soluble extract from isolated mammalian nuclei. *Nucleic Acids Res.* **11,** 1475–1489.
4. Schreiber, E., Matthias, P., Muller, M. M., and Schaffner, W. (1989) Rapid detection of octamer binding proteins with "miniextracts" prepared from a small number of cells. *Nucleic Acids Res.* **17,** 6419.
5. McGookin, R. (1988) Purification of synthetic oligonucleotides by preparative gel electrophoresis, in *Methods in Molecular Biology, vol. 4: New Nucleic Acid Techniques* (Walker, J. M., ed.), Humana, Clifton, NJ, pp. 215–220.

CHAPTER 33

The Southwestern Assay

Jacques Philippe

1. Introduction

Determination of cellular phenotypes results from the expression of a limited number of genes whose products interact to establish a unique environment. The mechanisms by which individual cells can selectively express only a few of all the genes in a specific cell have been the focus of intense research during the last 10 yr. It has become apparent that developmental, tissue-specific, and hormone-regulated gene expression is, for the most part, controlled at the level of transcriptional initiation. This involves the interaction of specific DNA binding proteins with control elements present in the gene promoters. To better understand the process of gene transcription, characterization of these DNA binding proteins is a mandatory step.

Detection of sequence-specific DNA binding proteins has been achieved by a variety of techniques, including nitrocellulose filter binding *(1)*, DNase I footprinting *(2)*, and the electrophoretic mobility shift assay (gel retardation; ref. *3*). These techniques, however, are limited since they do not inform on the proteins themselves. To this end alternative methodologies have been developed to directly characterize DNA-binding proteins. Ultraviolet light can be used to crosslink binding proteins to labeled DNA elements *(4)*. The molecular weight of the protein can then be estimated by SDS polyacrylamide gel electrophoresis (SDS-PAGE). In this chapter I describe a second approach that avoids the need to crosslink the protein–DNA complex, and hence any inaccuracies of size determination resulting from

From: *Methods in Molecular Biology, Vol. 31: Protocols for Gene Analysis*
Edited by: A. J. Harwood

this process. This procedure was first developed by Bowden et al. *(5)* and since it closely follows the principles of the Western assay, it has been referred to as the Southwestern assay. In this technique, crude protein extracts are first separated by SDS-PAGE and blotted onto a nitrocellulose membrane. Specific DNA binding proteins are detected by incubating the membrane with a labeled DNA probe that is derived from the protein recognition site. This technique has now been refined into a simple, reliable, and versatile assay, and is commonly used not only to characterize nuclear proteins from cellular extracts, but also as a preliminary step to set up conditions for *in situ* detection of DNA-binding proteins expressed by recombinant bacteriophages *(6)* or to select DNA sequences recognized by a specific DNA-binding protein. By analogy, other procedures derived from the same principles, but based on protein–protein or protein–RNA interactions, have been established (*5,8*; *see* Notes).

2. Materials

All solutions should be made with distilled, deionized water.

1. Running gel solution: 33.5% acrylamide/0.3% *bis*-acrylamide (*see* Note 1). Bring 33.5 g of acrylamide and 0.3 g of *N,N'* methylene*bis*-acrylamide to 100 mL. Filter and store at 4°C. Protect from light. The solution can be kept refrigerated for at least 3 mo. Wear a mask and gloves to weigh and handle acrylamide and *bis*-acrylamide since they are potent neurotoxic agents that may be absorbed through the skin.
2. Running gel buffer: 1*M* Tris-HCl, pH 9.1. Dissolve 12.1 g of Tris-base in 80 mL. Adjust pH to 9.1 with concentrated HCl and make the volume up to 100 mL. Filter and refrigerate.
3. 10% SDS: Wear a mask to handle SDS powder.
4. 10% ammonium persulfate: 0.5 g of ammonium persulfate to 5 mL. Store at 4°C for up to 3 wk; for longer storage aliquot and freeze at –20°C.
5. TEMED.
6. Stacking gel solution: 30% acrylamide/0.44% *bis*-acrylamide. Bring 3 g of acrylamide and 0.44 g of *N,N'* methylene*bis*-acrylamide to 10 mL. Filter and store at 4°C. Protect from light.
7. Stacking gel buffer: 0.5*M* Tris-HCl, pH 6.2. Dissolve 1.5 g of Tris-base, pH 6.8, with concentrated HCl.
8. Tank buffer: 25 m*M* Tris-base, 250 m*M* glycine, 0.1% SDS. Dissolve 12 g of Tris-base, 57.6 g of glycine, and 4 g of SDS in 4 L. It is not necessary to pH this solution.

9. 2X sample buffer: 125 m*M* Tris-HCl, pH 6.8, 4% SDS, 20% glycerol, 1.44*M* β-mercaptoethanol, 0.1% bromophenol blue. For 10 mL, use 2.5 mL of 0.5M Tris-HCl, pH 6.8, 4 mL of 10% SDS, 2 mL of glycerol, 1 mL of β-mercaptoethanol, and a few crystals of bromophenol blue; make up to volume with water. Store at –20°C.
10. Suitable electrophoresis apparatus: A vertical electrophoresis apparatus. Gels are poured between two clamped plates. These are available from many manufacturers, e.g., Hoefer (San Francisco, CA) and Bio-Rad (Richmond, CA) (*see* Note 2).

2.2. Transfer of Proteins

11. Nitrocellulose membrane: Good quality nitrocellulose, e.g., BA 85, Schleicher and Schuell (*see* Note 3)
12. Transfer tank buffer: 25 m*M* Tris-base, 192 m*M* glycine, 20% v/v methanol. Dissolve 12.1 g of Tris-base, 57.6 g of glycine in 3 L of water. Add 800 mL of methanol and bring to 4 L with water.
13. Transfer apparatus: A rectangular tank with a large network of platinum electrodes on each side to generate an even field. The gel/nitrocellulose sandwich is held vertically in place between two sponge pads by tight-fitting cassettes (*see* Fig. 1). These are available from many manufacturers, e.g., Hoefer and Bio-Rad.

2.3. Probe Labeling

14. Probe DNA: Two complementary synthetic oligonucleotides that anneal to provide the recognition site for the protein under investigation (*see* Note 4). Dilute to 2 ng/μL and store at –20°C.
15. γ-[^{32}P]ATP: specific activity >3000 Ci/mmol.
16. 10X T4 kinase buffer *(8)*: 200 m*M* Tris-HCl, pH 7.6, 50 m*M* $MgCl_2$, 50 m*M* DTT, and 500 μg/mL BSA (DNase-free). Store in small aliquots at –20°C.
17. T4 polynucleotide kinase: 10 U/μL. Store at –20°C.
18. 10X Ligase buffer: 500 m*M* Tris-HCl, pH 7.6, 100 m*M* $MgCl_2$, 100 m*M* DTT, 10 m*M* ATP, and 500 μg/mL BSA (DNase-free). Store in small aliquots at –20°C.
19. T4 DNA ligase: 1 U/mL. Store at –20°C.
20. G50 Sephadex spun column: (*see* Note 5).

2.4. Probe Binding

21. 10X Binding buffer: 250 m*M* HEPES, pH 7.9, 30 m*M* $MgCl_2$, 500 m*M* KCl, 1 m*M* DTT. To make a 10X stock solution, use 25 mL of 1*M* HEPES (titrated to pH 7.9 with 1*M* NaOH), 3 mL of 1*M* $MgCl_2$, and 50 mL

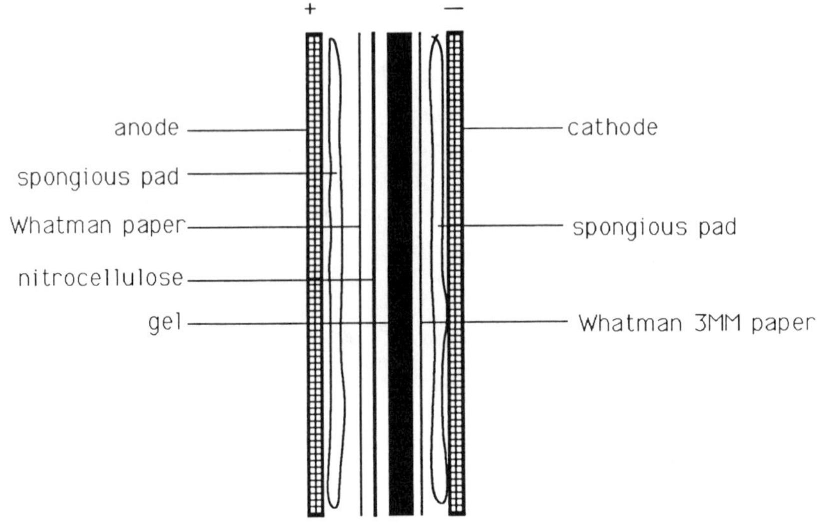

Fig. 1. Arrangement for electrophoretic transfer: + and – designate the polarity of the plastic supports containing the gel/nitrocellulose sandwich. These supports are oriented toward the anode and cathode of the electrophoretic transfer tank, respectively.

of 1*M* KCl; adjust to 100 mL with water. Store at 4°C. Add DTT to the diluted solution just before use.

22. Denaturation buffer: 6*M* guanidine hydrochloride in 1X binding buffer. For a 100-mL solution, dissolve 57.3 g in 1X binding buffer, but add DTT just before use.
23. Blocking buffer: 5% nonfat dry milk in 1X binding buffer.
24. Poly (dI · dC): A competitor used to block nonspecific DNA binding. Store at –20°C as a 1 mg/mL stock (*see* Note 6).
25. Specific and nonspecific oligonucleotide competitors: Synthetic oligonucleotides made as for the probe (*see* Section 3.3.) but with unlabeled ATP and at 50-fold the concentration, i.e., 100 ng/µL (*see* Note 7).
26. Probe solution: 0.25% nonfat milk and 10 µg/mL of poly(dI · dC) in 1X binding buffer.

3. Methods

Proteins from many sources can be analyzed by the Southwestern assay (*see* Note 8). Loadings should be adjusted according to the source. I routinely load 75 µg of crude nuclear extract per electrophoretic lane, but this may vary depending on the source of the extract, the abundance of the binding protein, and the affinity of the protein for the DNA sequence. If very high concentrations of extract are used, preliminary gels should be run to ensure that overloading of the gel does not occur.

To confirm the specificity of the proteins detected using this assay, it is necessary to demonstrate that binding is abolished by specific competitors but unaffected by those that are known to be nonspecific. For each probe it is therefore necessary to run at least three lanes of protein extract. The nitrocellulose membrane is cut into strips for each lane; the first lane is incubated with probe alone, the second with probe and specific competitor, and the third with probe and a nonspecific competitor.

3.1. SDS Gel Electrophoresis

1. Mix 12 mL of running gel solution, 15.2 mL of running gel buffer, 11.2 mL of water, 0.4 mL of 10% SDS, and 0.33 mL of 10% ammonium persulfate. Just before pouring, add 20 µL of TEMED and swirl the components. Pour between the glass plates of the gel apparatus, but leave the last 4 cm for the stacking gel. Overlay the mix with water to ensure even polymerization. After approx 30 min the gel should have polymerized. Pour off the overlaying water and dry with an absorbent paper towel.
2. Mix 1.3 mL of stacking gel solution, 2.5 mL of stacking gel buffer, 6.1 mL of water, 0.1 mL of 10% SDS, and 33 µL of 10% ammonium persulfate. Just before pouring, add 10 µL of TEMED. Fill the remaining space above the running gel and insert the comb. Once the gel has polymerized, remove the comb, wash out the wells with water, and assemble the gel apparatus. Ensure that no bubbles are trapped at the bottom of the gel when the tank buffer is added. Do not prerun the gel before loading.
3. Typically each sample should contain 75 µg of protein. Dilute each nuclear extract to 10 µL with the appropriate extraction buffer and mix with 10 µL of 2X sample buffer. These samples can be scaled up for the number of gel loadings, e.g., for a triple loading 60 µL may be prepared. Prepare two samples containing marker proteins of known molecular weights.
4. Boil the samples for 3 min to dissociate proteins and inactivate proteases (*see* Note 9). Insoluble material should be removed by centrifugation at 12,000*g* for 5 min to avoid streaking during electrophoresis.
5. Load 20 µL for each sample onto the gel (*see* Note 10). Apply a voltage of 8 V/cm for the run through the stacking gel and increase to 15 V/cm

for the running gel. Run at room temperature. The run should take 4 h and is complete when the marker dyes reach the bottom.

6. Remove the glass plates and mark the orientation of the gel by cutting off one corner.

3.2. Transfer of Proteins

Wear gloves through the rest of the procedure.

1. Equilibrate the gel in transfer buffer for 30 min. The size of the gel may change slightly owing to the methanol in the buffer.
2. While the gel is equilibrating, cut the nitrocellulose membrane and two pieces of Whatman 3MM paper to the size of the gel. Wet these and the sponge pads of the transfer apparatus in transfer buffer.
3. Assemble the transfer apparatus as follows (*see* Fig. 1). Lay the cathode flat and onto it pace one of the sponge pads. On top place a sheet of Whatman 3MM paper, followed by the gel, the nitrocellulose membrane, the second sheet of Whatman 3MM paper, the second sponge pad, and finally the anode. Using a glass pipet as a roller, squeeze out any trapped air bubbles. Immerse the sandwich into the transfer chamber filled with buffer. Ensure the nitrocellulose membrane is between the gel and the anode.
4. Although voltage and current readings depend on many variables, for a standard size gel, transfer with a constant voltage of 30 V (initial current of approx 180 mA) for 14 h at 4°C. Cool and stir constantly (*see* Note 11).
5. After transfer rinse the filter in water and dry at room temperature before DNA binding (*see* Note 12).

3.3. DNA Labeling

The probe should ideally contain multiple copies of the control element to enhance sensitivity. The multiple sites allow a single molecule to interact with more than one protein and alleviate problems of rapid probe dissociation. The probe is generated by ligating labeled oligonucleotides to create a concatemer (*see* Note 13).

Since the probe is labeled with ^{32}P, necessary precautions should be taken when handling it.

1. Add 5 µL (10 ng) of each complementary oligonucleotide to 3 µL of 10X kinase buffer, 6 µL of γ-[^{32}P]ATP, and 10 µL of water. Add 1 µL of T4 polynucleotide kinase and incubate at 37°C for 1 h.

2. Place the reaction in a beaker containing 500 mL of water at 90°C and then place the beaker in ice. Let it stand for 2–3 h until the water temperature reaches 20°C.
3. Add 3.3 µL of 10X ligation buffer and 1 µL of T4 DNA ligase. Incubate overnight at 15°C.
4. Purify the catenated DNA from unincorporated label using a G50 Sephadex spun column (*see* Note 5).

This should produce a concatenated probe suitable for high affinity binding to the immobilized DNA binding protein. (*see* Note 14).

3.4. DNA Binding

The proteins blotted onto nitrocellulose are denatured and then renatured. This is thought to help appropriate folding or multimerization of the denatured proteins to occur and reduce nonspecific DNA interactions (*see* Fig. 2A and Notes 15–17).

1. Cut the membrane into strips corresponding to each electrophoretic lane and immerse in denaturation solution. Choose a small glass or plastic dish to spare solutions; 20–40 mL is adequate. Agitate gently for 10 min at 4°C.
2. Dilute with the same volume of 1X binding buffer and shake for 5 min at 4°C. This dilutes the solution to 3*M* guanadine HCl.
3. Repeat step 2 four times to give 1.5*M*, 0.75*M*, 0.38*M*, and 0.185*M* guanadine HCl sequential dilutions. Agitate each time for 5 min at 4°C. Finally wash in 1X binding buffer.
4. Place the membrane in 40 mL of blocking buffer and agitate gently for 1 h at 4°C.
5. Add the labeled probe to sufficient volume of probe solution for binding. This should be carried out in the smallest volume possible (0.1–0.2 mL/cm^2) By using heat sealable bags for each strip of membrane, the volume of solution can be minimized. A 50-fold excess of specific or nonspecific competitors should be included in the solutions of the relevant strips. Agitate gently for 2 h at 4°C.
6. Wash membrane strips for 5 min in 100 mL of 1X binding buffer at 4°C. Repeat twice. Dry strips on Whatman paper for 1–2 min and autoradiograph for at least 12 h.

Figure 2 shows a typical result using this technique. To maximize the sensitivity of the assay it is important to carry out each step efficiently. Problems with probe synthesis, protein transfer, or renatur-

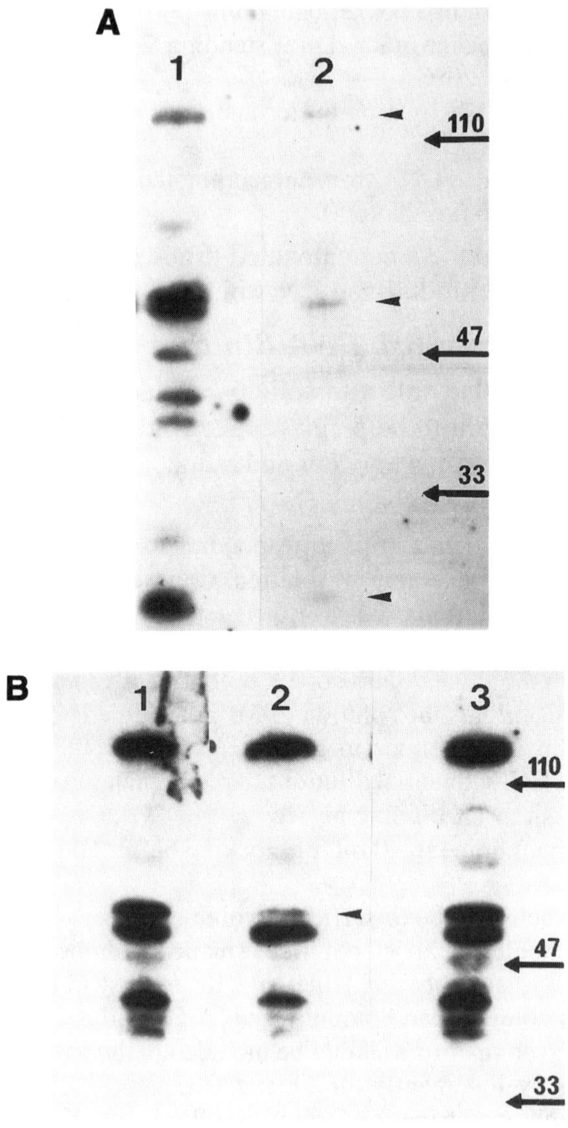

Fig. 2. Southwestern assay. Denaturation/renaturation: 75 μg nuclear extracts were separated by SDS-PAGE, transferred, and probed with a concatenated 42-bp element from the rat glucagon gene promoter. The mol wt of the protein markers are indicated **(A)** Lane 1; the membrane was not subject to denaturation/renaturation. Lane 2; the membrane underwent denaturation/renaturation with guanidine hydrochloride, as described in Section 3.4. The denaturation/renaturation treatment reduced the ten bands detected in lane 1 to the three specific bands in lane 2 (marked by arrow point) **(B)** Competition: To illustrate the effectiveness of competition, the

ation of the protein will certainly reduce the sensitivity of the assay. Some problems, however, may be inherent to the protein under investigation (*see* Notes 9–11). A major problem encountered is if the protein is a heterodimer and therefore unable to reform the DNA binding domain after separation by SDS-PAGE (*see* Note 18). There are a number of other applications of the technique discussed in Note 19.

4. Notes

1. Both the percentage of polyacrylamide and the ratio of acrylamide to *bis*-acrylamide need to be determined to optimize separation and transfer of proteins, particularly for high molecular weight proteins. The lower the total monomer the higher the porosity of the gel. In this protocol I use a 10% gel with an acrylamide:*bis*-acrylamide ratio of 110:1; this resolves proteins of molecular weights from 10–200 kDa *(10)*.
2. Polyacrylamide gels for protein electrophoresis can be run on 1.5-mm thick standard (16 × 18 cm) or mini (8 × 10 cm) gels. The recipe is given for a standard size gel; the volume should be scaled down according to the gel size.
3. Although nylon membranes have a higher protein binding capacity than nitrocellulose, they give unacceptably high background even with increasing the nonfat milk concentration of the blocking buffer to 5%. If the background is unacceptable with nonfat milk (at 0.25–5% concentrations), it may be necessary to optimize the signal-to-noise ratio by using lipid-free BSA as a blocking agent *(1)*.
4. The sequence of the oligonucleotides used for the probe must first have been established by other means, e.g., deletion analysis and footprinting studies. They should be synthesized so that after annealing there is a complementary sequence overhang (or "sticky end"). This aids the concatemerization process described in Section 3.3.
5. G50 Sephadex spun columns are used to purify the radiolabeled DNA. This can be prepared in a 1-mL syringe or bought ready prepared (e.g., Boehringer Mannheim, Indianapolis, IN). The elution limit for double-stranded DNA is less than 70-bp and recovery is greater than 90%. Alternatives are to use ion exchange columns (e.g., Elutip, Schleicher and Shuell) or to precipitate the DNA using ethanol.

---

membranes were not subject to denaturation/renaturation treatment but probed with the same probe as in (A). A 50-fold excess of multimerized specific competitor (unlabeled wildtype DNA) was added to probe for lane 2 and 50-fold excess of multimerized nonspecific competitor (a mutant element) was added to probe for lane 3. A single band is systematically and reproducibly competed (arrow head).

6. DNA from many sources, e.g., Salmon sperm DNA, may also be used as a nonspecific block. Again, use at a final concentration of 10 μg/mL.
7. To confirm the identity of the signals detected in the assay, it is necessary to determine their specificity for the probe. Both specific and nonspecific competition must be carried out (*see* Fig. 2B). At its simplest the specific competitor can be unlabeled probe. Additional specific competitors, derived from related recognition sites, should be used if available. Nonspecific competitors at their best should correspond to mutants of the recognition site that are known not to bind the protein. If not available, oligonucleotides unrelated to the recognition site should be used.

   The competitors should be synthesized in exactly the same manner as the probe. In this protocol, competitors are used at a 50-fold excess. It is possible that this may vary for different binding proteins.
8. One of the major advantages of the Southwestern assay resides in its power to discriminate a specific DNA-binding protein among many thousands. Whole cell, nuclear and, in the case of the glucocorticoid receptor, cytosolic extracts have been used as the starting source of the proteins. Other sources, such as fusion proteins that contain DNA binding domains expressed by recombinant bacteriophage or truncated proteins synthesized from cloned cDNA in vitro or in bacteria, are also perfectly suitable.
9. During sample preparation, the boiling step might cause some proteins to precipitate and be lost from the assay. This step therefore may be omitted if the Southwestern assay is unsuccessful.
10. Some nuclear nonhistone proteins have a tendency to form aggregates and remain at the top of the gel. Addition of 8*M* urea before loading may sometimes allow them to enter the gel.
11. Efficient protein transfer is critical for good results. Different factors can be modified to optimize transfer efficiency.
    a. Gel composition (*see* Note 1).
    b. Transfer time and current. These variables may have to be adapted for complete transfer. Overnight applications with high voltages should only be attempted with caution and heating should be minimized. High currents can be used satisfactorily for short transfer times (e.g., 0.5 A for 2–3 h). Reducing the buffer strength may allow the use of higher voltages.
    c. Buffer type. The standard buffer can adversely affect transfer of basic proteins. These may require pH 9.5–10 to elute from the gel. In this case, methanol may have to be omitted, but this can diminish binding to the nitrocellulose membrane. Similarly, addition of 0.1% SDS to the standard buffer may help elution of proteins from the gel but

may reduce binding efficiency to the membrane and also increase the relative current and cause heating.

12. The transfer efficiency may be checked by staining the gel and a strip on the membrane containing one of the marker lanes with Coomassie blue. Do this by fixing the gel for 30 min in 100 mL of gel fix (50% methanol, 10% acetic acid), stain for 30 min in 100 mL of stain solution (0.2% Coomassie blue, 50% methanol, 10% acetic acid), and destain in 100 mL of 10% methanol, 10% acetic acid for three changes of 10 min each. The gel should be dried and the filter rinsed in distilled water.
13. In this protocol the concatenated DNA is synthesized by simple ligation of monomers. It also may be made by subcloning the monomers into a vector. Probes generated in this second way must be labeled by nick translation, but should be of a higher specific activity.
14. To check the efficiency of the ligation reaction, run an aliquot of the labeling reaction on a nondenaturing polyacrylamide gel. Dry the gel down and autoradiograph. This will indicate the degree of concatemerization. Multimers of between 3 and 5 oligonucleotides should be sufficient to give a good probe.
15. The renaturation procedure described in the protocol is based on conditions established for *in situ* detection of sequence specific DNA binding fusion proteins from λgt11 expression libraries *(6)*. Alternative protocols to renature proteins have been described in the literature, but it is unclear to what extent these procedures allow renaturation and whether they may result in better protein DNA interactions.
16. Protein renaturation can also be carried out directly in the gel before transfer. After electrophoresis, the gel is incubated in 200 mL of renaturation buffer (50 m*M* NaCl, 10 m*M* Tris-HCl, pH 7.2, 0.2 m*M* EDTA, 0.1 m*M* DTT, and 4*M* urea) with gentle agitation for 2 h at room temperature. This incubation will remove SDS from the proteins, but the extent of renaturation is unknown *(12)*. The gel is then equilibrated in transfer buffer for 30 min. The proteins are electrophoretically transferred onto a nitrocellulose membrane and directly placed in blocking buffer. After urea treatment, it has been observed that there is a decrease in transfer efficiency (personal observation; *13*). This may be owing to the reduced solubility of the protein samples when SDS is replaced by urea resulting in proteins becoming trapped in the gel. Compared to the guanidine-HCl treatment, more nonspecific protein–DNA binding is observed after urea incubation.
17. For some binding proteins, a specific renaturation step may not be necessary since removal of SDS from proteins during transfer may be sufficient to allow the proteins to renature *(14)*.

18. Failure to detect a specific DNA binding protein may indicate that the protein is a heterodimer. Dissociation of the subunits in the process of SDS-PAGE may destroy the binding activity. This possibility may be tested by addition of nuclear or cellular extracts during the binding step. If a subunit or cofactor is required for high affinity binding the addition of extract may restore the binding *(15)*.
19. Variants to the Southwestern assay:
    a. Two-dimensional Southwestern assay *(10):* Characterization of DNA-binding proteins can also be attempted after two-dimensional electrophoresis. Distinct isoelectric point variants can thus be occasionally revealed. Two-dimensional electrophoresis separates proteins by charge in the first dimension and by size in the second, allowing discrimination of proteins of similar size but varying charge. Protein transfer, renaturation, and DNA binding are conducted as in the standard protocol.
    b. Double replica Southwestern: Although double replica from a single gel is not recommended in initial attempts to detect a nuclear protein, once conditions are established for a specific protein it may be possible to transfer proteins onto two different filters. One way to do this is to sandwich the gel between two nitrocellulose membranes and blot the proteins at 250 mA in Tris/glycine buffer but with frequent polarity changes *(15)*. A simple alternative is to place two membranes on top of the gel before transfer. Sufficient protein may travel through the first membrane and be retained on the second.
    c. Variants on the Western blot: Multiple variants of the Western blot based on the same principles but utillizing different probes have been developed. High affinity interactions may be detected between proteins and RNA (Northwestern) and other proteins (Farwestern). (*See* refs. *5* and *6* and Chapter 36).

References

1. Lin, S. Y. and Riggs, A. D. (1975) The general affinity of lac repressor for *E. Coli* DNA: implication for gene regulation in prokaryotes and eukaryotes. *Cell* **4,** 107–111.
2. Galas, D. and Schmidt, A. (1978) DNase footprinting, a simple method for the detection of protein-DNA binding specificity. *Nucleic Acids Res.* **5,** 3157–3170.
3. Garner, M. M. and Revzin, A. (1981) A gel electrophoresis method for quantifying binding of proteins to specific DNA regions: applications to components of the *E. coli* lactose operon regulatory system. *Nucleic Acids Res.* **9,** 3047–3059.
4. Chodosh, L. A., Carthew, R. W., and Sharp, P. A. (1986) A single polypeptide possesses the binding and transcription activities of the adenovirus major late transcription factor. *Mol. Cell. Biol.* **6,** 4723–4733.

5. Bowen, B., Steinberg, J., Laemmli, U. K., and Weintraub, H. (1980) The detection of DNA-binding proteins by protein blotting. *Nucleic Acids Res.* **8,** 1–20.
6. Vinson, C. R., LaMarco, K. L., Johnson, P. F., Landschulz, W. H., and McKnight, S. L. (1988) In situ detection of sequence-specific DNA binding activity specified by a recombinant bacteriophage. *Genes Dev.* **2,** 801–806.
7. Keller, A. D. and Maniatis, T. (1991) Selection of sequences recognized by a DNA binding protein using a preparative Southwestern blot. *Nucleic Acids Res.* **19,** 4675–4680.
8. Hoeffler, J. P., Lustbader, J. W., and Chen, C. Y. (1991) Identification of multiple nuclear factors that interact with cAMP-response element-binding protein and activating transcription factor-2 by protein-protein interactions. *Mol. Endo.* **5,** 256–266.
9. Sambrook, J., Fritsch, E. F., and Maniatis, T. (1989) *Molecular Cloning. A Laboratory Manual.* Cold Spring Harbor Laboratory, Cold Spring Harbor, NY.
10. Dreyfuss, G., Adam, S. A., and Choi, Y. D. (1984) Physical change in cytoplasmic messenger ribonucleoproteins in cells treated with inhibitors of mRNA transcription. *Mol. Cell. Biol.* **4,** 415–423.
11. Papavassiliou, A. G. and Bohmann, D. (1992) Optimization of the signal-to-noise ratio in Southwestern assays by using lipid-free BSA as blocking reagent. *Nucleic Acids Res.* **20,** 4365–4366.
12. Silva, C. M., Tully, D. B., Petch, L. A., Jewell, C. M., and Cidlowski, J. A. (1987) Application of a protein blotting procedure to the study of human glucocorticoid receptor interactions with DNA. *Proc. Natl. Acad. Sci. USA* **84,** 1744–1748.
13. Jack, R. S., Gehring, W. J., and Brack, C. (1981) Protein component from Drosophila larval nuclei showing sequence specificity for a short region near a major heat shock protein gene. *Cell* **24,** 321–331.
14. Miskimins, W. K., Roberts, M. P., McClelland, A., and Ruddle, H. (1985) Use of a protein blotting procedure and a specific DNA probe to identify nuclear proteins that recognizes the region of the transferrin receptor gene. *Proc. Natl. Acad. Sci. USA* **82,** 6741–6744.
15. Matsuno, K., Suzuki, T., Takiya, S., and Suzuki, Y. (1989) Complex formation with the fibroin gene enhancer through a protein–protein interaction analyzed by a modified DNA-binding assay. *J. Biol. Chem.* **264,** 4599–4604.
16. Hübscher, U. (1987) Double replica Southwestern. *Nucleic Acids Res.* **15,** 5486.

CHAPTER 34

Cloning DNA Binding Proteins from cDNA Expression Libraries Using Oligonucleotide Binding Site Probes

Ian G. Cowell and Helen C. Hurst

1. Introduction

Central to the regulation of transcription of eukaryotic genes is the interaction of specific DNA binding proteins with promoter and enhancer elements. The molecular cloning of such DNA binding factors is an important step toward understanding this process and a number of strategies have been devised to meet this aim. Oligonucleotide probes derived from peptide sequence data may be used to screen cDNA libraries by hybridization or, alternatively, cDNA expression libraries may be screened immunologically, using antibodies raised against the factor of interest. The protocol described here, which is based on a method described by Singh et al. *(1)*, is analogous to immunological screening but uses a labeled DNA binding site oligonucleotide probe to screen a cDNA expression library; DNA binding factors are thus cloned by virtue of their binding activity.

This approach has been successfully applied to the cloning of a number of bZIP proteins *(2–4)* and to other classes of DNA binding protein *(5,6)*. A note of caution, however, concerns the posttranslational modification of many DNA binding proteins. If a modification such as phosphorylation is required for efficient DNA binding, then the cloned factor expressed in *E. coli* may not bind sufficiently well

From: *Methods in Molecular Biology, Vol. 31: Protocols for Gene Analysis*
Edited by: A. J. Harwood Copyright ©1994 Humana Press Inc., Totowa, NJ

to be detected by this method. However, using this procedure we have successfully cloned a bZIP factor that in its native form normally requires phosphorylation for maximum binding efficiency.

2. Materials

2.1. Library Plating and Replica Lifts

1. L-broth, agar, and maltose for bacterial growth and λ library manipulation.
2. SM buffer for diluting λ phage: 100 m*M* NaCl, 10 m*M* Mg_2SO_4, 50 m*M* Tris-HCl, pH 7.5, 0.01% gelatin. Autoclave.
3. Isopropyl β-D-thiogalactopyranoside (IPTG). 1*M* stock solution in water. Store at –20°C.
4. Nitrocellulose filters (BA85, Schleicher and Schuell, Keene, NH). 132 and 82 mm circles.
5. Southwestern block (SW block) buffer: 2.5% dried milk powder, 50 m*M* HEPES, pH 8.0, 10% glycerol, 50 m*M* NaCl, 0.05% LDAO (lauryl dimethylamideoxide, Calbiochem, La Jolla, CA), 1 m*M* EDTA, 1 m*M* dithiothreitol (DTT). Add DTT fresh. Store at 4°C.
6. TNE-50: 50 m*M* NaCl, 10 m*M* Tris-HCl, pH 7.5, 1 m*M* EDTA; 1 m*M* DTT. Add fresh DTT.
7. Wash buffer: TNE-50 plus 0.05% LDAO.

2.2. Probe Labeling

1. T4 Polynucleotide kinase (Life Technologies, Gaithersburg, MD) and 10X polynucleotide kinase (PNK) buffer: 500 m*M* Tris-HCl, pH 7.5, 100 m*M* $MgCl_2$, 5 m*M* DTT, 1 m*M* spermidine. Make up fresh.
2. γ–[^{32}P]-ATP at 5000 Ci/mmol (Amersham, Arlington Heights, IL).
3. Sephadex G-50-M (Pharmacia, Piscataway, NJ) autoclaved in water.
4. T4 DNA ligase and 10X ligase buffer: 500 m*M* Tris-HCl pH 7.5, 100 m*M* $MgCl_2$, 10 m*M* DTT, 10 m*M* ATP, 10 m*M* spermidine.

2.3. Screening and Selection of Positive Plaques

Poly(dA)/poly(dT) (Pharmacia). Stock of 1 mg/mL made up in water and stored at –20°C.

3. Method

Before trying to clone a DNA binding factor it is necessary to characterize its optimum binding requirements in terms of specific oligonucleotide sequence, nonspecific competitor, and salt concentration. This may already be known from gel retardation experiments with purified protein. Also, it might be worth trying a Southwestern experiment (*see* Note 1) to test if the factor of interest will bind DNA when attached to nitrocellulose. Described below is a method we have found to work well when cloning factors containing a bZIP binding and dimerization domain. In the notes we discuss possible modifications that are worth testing when optimizing the system to clone other types of factors.

3.1. Library Plating and Replica Lifts

The method described here assumes the use of a λgt11 library and *E. coli* Y1090 *(7)*. However, library selection is an important feature of this method and is discussed further in Note 2.

1. Having determined the phage titer of the library, plate out at least 10^6 phage at a density of up to 10^5 plaque forming units (pfu) per 150-mm Petri dish using fresh Y1090 plating cells, top LB agarose (0.7%), and dry LB plates (i.e., 1–2 d old). Once the agarose has set, incubate the plates at 42°C until the plaques are just visible on the bacterial lawn—this usually takes 3–4 h.
2. Soak one nitrocellulose filter for each plate in 10 m*M* IPTG and pat dry. Quickly place a damp, numbered filter on each plate and transfer to a 37°C incubator. It is best to handle one plate at a time since the temperature of the agar should not drop below 37°C. Incubate for a further 1–2 h at 37°C.
3. Mark the position of each filter relative to the plate using a syringe needle dipped in India ink. Carefully remove each filter with a pair of forceps into a lunch box containing a large volume of wash buffer. Place a second IPTG-soaked filter on each plate and incubate for a further 2–3 h at 37°C, again marking the filters before removal. This time a permanent marker pen can be used to make dots on the filters over the syringe needle holes that should be visible in the agar. Use a light box to make this easier.
4. After the second set of lifts have been taken, wrap the plates in SaranWrap and store at 4°C. Meanwhile, rinse all of the filters for 5–10 min in Wash buffer to remove any loose fragments of top agarose. This reduces background when the filters are probed. Transfer the filters to SW block (approx 500 mL for 10–20 filters) and leave at 4°C overnight. It is best to probe filters as soon as possible, but they may be left 2–3 d in SW block at 4°C. On no account should they be allowed to dry out.

3.2. Probe Labeling

The selection of a suitable probe is another important feature that is discussed in Note 3. In general, the specific factor binding sequence should be represented by a 20–25-mer double-stranded oligonucleotide with 3' hydroxyl groups and compatible "sticky" ends to allow easy ligation.

1. Form the double stranded probe by annealing the single-stranded oligonucleotides in a 100 µL incubation containing 10 µL of 10X PNK buffer and up to 40 µg of each oligonucleotide resuspended in water. Incubate for 2 min at 90°C, 10 min at 65°C, 10 min at 37°C, and 5 min at room temperature. Ethanol precipitate and resuspend in TE at 0.1 µg/µL.
2. To 4 µL of double-stranded oligonucleotide from above, add 2 µL of 10X PNK buffer, 6 µL (60 µCi) of γ-[^{32}P]-ATP, 2 µL (20 U) of T4 PNK, and water to 20 µL final volume. Mix and incubate at 37°C for 1 h.
3. Add 80 µL of TE (10 m*M* Tris, pH 8.0, 0.1 mm EDTA) to the incubation and separate the labeled double-strand oligonucleotide from the unincorporated label by spun column chromatography using Sephadex G50-M. Radioisotope incorporation can be checked by liquid scintillation counting of 1 µL of the labeled oligo probe. This method of labeling should give 1–2 × 10^8 cpm in total.
4. To the labeled oligonucleotide (100 µL) add 11 µL of 10X ligation buffer, 2 µL of 10 m*M* ATP, and 2 µL (20 U) of T4 DNA ligase. Ligate overnight at 15°C to produce a concatemerized probe. This can be stored at –20°C, but is best used immediately for primary screens.

3.3. Screening and Selection of Positive Plaques

1. Remove the filters from the SW block (this may be reused several times) and rinse in TNE-50.
2. Make up the probe mixture by adding the concatemerized probe at 5 × 10^5–10^6 cpm/mL to TNE-50 together with poly dA/dT (*see* Notes 4 and 5) to 10 µg/mL.
3. Probing is most conveniently performed in 150-mm Petri dishes containing 50 mL of the probe mixture. Add filters one at a time, protein side up, and swirl to cover with buffer before adding the next. Put 5–6 filters in 50 mL of probe mixture and shake gently on an orbital shaker for 1 h at room temperature (*see* Note 6), checking that each filter is moving freely to allow even distribution of the probe. Filters can be processed in batches (i.e., reusing the probe mixture), meanwhile keeping unprobed filters in the SW block or TNE-50 at 4°C.
4. Remove probed filters to a large tray or lunch box and wash 3–4 times for 5–10 min each in large volumes of Wash buffer (*see* Note 6) again checking that each filter moves freely for efficient rinsing. The probe mixture can be retained, stored at –20°C for up to a week, and used for second round screens adding fresh DTT before use.

5. After washing, blot the filters dry and expose to X-ray film. Include fluorescent markers to orient the film to the filters. An overnight exposure is usually sufficient, but first round screens should also be left down longer to check for weaker signals (*see* Note 7).
6. Use the alignment markers to orient the film to the filters and hence the film to the plates to check that any positive signals appear on duplicate filters. Examples of such signals are shown in Fig. 1. Pick plugs of agar corresponding to these areas of the plate into 1 mL SM buffer using the wide end of a Pasteur pipet.
7. Using the methods described in Section 3.1., replate the phage from these plugs on 90-mm plates at a range of densities aiming for densely packed but clearly isolated plaques, i.e., about 200 plaques/plate. Take lifts off the best looking plates as described above and probe. Repeat the process (3–4 rounds) until phage purification is achieved. After the first round it is not usually necessary to do duplicate filters; the presence of multiple positive signals (*see* Fig. 1) on the second round plates is a good indication that the first round signals were "real." Once purified, the phage can be used to generate lysogens *(7)* to produce protein for further binding studies (*see* Note 8).

4. Notes

1. The conditions employed to screen the library can be tested with a Southwestern blot. Briefly, the purified protein of interest is run on SDS-PAGE and blotted, as in a Western blot, onto nitrocellulose. The filter is then treated as described in Section 3.3. using a probe made according to the method in Section 3.2. (*see* Chapter 33).
2. The protocol as written here assumes the use of a λgt11 library. Another commonly used expression vector for cDNA library construction is λZAP (Stratagene, La Jolla, CA). With the exception of details concerning the growth of the phage, which are given by the supplier, λZAP libraries may be handled identically to λgt11. Libraries may be made or purchased commercially in either λgt11 or λZAP. In either case, the complexity of the library should be at least 10^6 (only 50% of the recombinant phage will contain inserts in the correct orientation for expression and only one third of these will be in frame to generate a β-galactosidase fusion protein, although some proteins may be synthesized from internal translation initiations).

   Clearly the cells or tissue from which a library is generated should be known to contain the factor of interest. We have used a human placental library; placenta, being a heterogenous tissue, contains a large variety of DNA binding proteins.

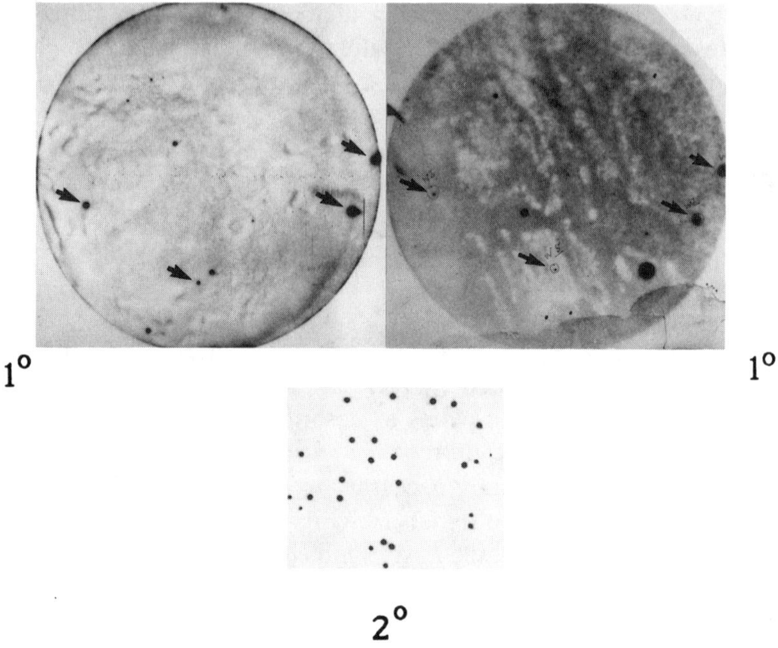

Fig. 1. A λgt11 cDNA library was screened as described in the protocol, using a double-stranded ATF site oligonucleotide probe. A pair of duplicate lifts from one primary screen plate are shown (marked 1°); typically, there are several radioactive "spots" on each filter, only some of which are genuine signals and thus appear on both lifts. Duplicate signals have been marked here with an arrow; on the original autoradiographs, duplicate spots were easily identified by overlaying the developed films on a light box and looking for coinciding "spots." The result of a typical secondary screen is shown below (marked 2°). The background in secondary and subsequent screens is less than that in the first screen because of the lower plating density resulting in larger and more well defined plaques.

3. A concatemerized probe is described in this protocol, sincc we have found that this gives a better signal than that obtained with a monomeric binding site oligonucleotide. Concatemerization, however, generates new sequences at the points of ligation. Therefore, as far as is possible, sequences should be avoided that create in this way the binding site for another DNA binding protein.

   Another point to consider is the selectivity of the probe, since it is frequently the case that multiple DNA binding factors bind the same or closely related sites. Some groups check the specificity of first round signals by probing duplicate lifts with oligonucleotide probes containing point mutations known to be detrimental to the binding of the factor of interest.

4. TNE-50 is a low salt (i.e., low stringency) binding buffer and thus encourages binding of bacterially synthesized proteins that may bind poorly for a number of reasons, such as lack of posttranslational modification. It is advisable to check the binding activity of the factor in question in TNE-50 using a gel retardation assay before embarking on screening.
5. The gel retardation assay may also be used to determine empirically the best nonspecific competitor to use. We have found poly(dA)/poly(dT) to be suitable, but poly(dI)/poly(dC) or herring sperm DNA may also be used.
6. The hybridization conditions we describe are given as a guide; longer incubation times at a lower temperature may improve the signal for a poorly binding factor. Similarly, the washing conditions described are those we find to be the most effective. Longer washes at room temperature are detrimental owing to dissociation of the probe from the filters, but longer washing times with cold wash buffer can reduce the background without significant loss of bound probe.
7. Some DNA binding proteins may be inactive when expressed as fusion proteins in *E. coli*, but binding activity can sometimes be recovered by denaturation in guanidine hydrochloride followed by slow renaturation. Immerse the filters one at a time in 6*M* guanidine hydrochloride, allowing 40 mL of solution/150 mm filter. Shake the filters at room temperature on a slowly moving orbital shaker, ensuring that the filters do not stick together. After 10 min dilute the guanidine hydrochloride 1:1 with TNE-50 and shake for a further 10 min. Repeat this dilution another seven times and finally discard the solution and replace with TNE-50. Block the filters with SW block and probe as described in Section 3.3. Some laboratories have found these denaturation/renaturation steps essential for the detection of their factor of interest *(8,9)*. A sensible approach would be to initially probe the filters directly as we have described and only to denature, renature, and reprobe the filters after autoradiography if no positive signals are obtained.
8. Once a positive signal has been plaque purified, it is essential that the bacterially made protein (usually generated via a λ lysogen; *see* ref. *7*) is tested for binding to the monomer binding site in a standard gel retardation assay (*see* Chapter 32). One can then be reasonably confident that the cloned factor is indeed recognizing the binding sequence of interest.

References

1. Singh, H., LeBowitz, J. H., Baldwin, A. S., and Sharp, P. A. (1988) Molecular cloning of an enhancer binding protein: isolation by screening of an expression library with a recognition site DNA. *Cell* **52,** 415–423.
2. Maekawa, T., Sakura, H., Kanei-Ishii, C., Sudo, T., Yoshimura, T., Fujisawa, J.-I., Yoshida, M., and Ishii, S. (1989) Leucine zipper structure of the protein CRE-BP1 binding to the cAMP response element in brain. *EMBO J.* **8,** 2023–2028.
3. Hai, T., Liu, F., Coukos, W. J., and Green, M. R. (1989) Transcription factor ATF cDNA clones: an extensive family of leucine zipper proteins able to selectively form DNA binding heterodimers. *Genes Devel.* **3,** 2083–2090.
4. Poli, V., Mancini, F. P., and Cortese, R. (1990) IL-6DBP, a nuclear protein involved in interleukin-6 signal transduction, defines a new family of leucine zipper proteins related to C/EBP. *Cell* **63,** 643–653.
5. Klemsz, M. J., McKercher, S. R., Celada, A., Van Beveren, C., and Maki, R. A. (1990) The macrophage and B cell specific transcription factor PU. 1 is related to the ets oncogene. *Cell* **61,** 113–124.
6. Xiao, J. H., Davidson, I., Matthes, H., Garnier, J.-M., and Chambon, P. (1991) Cloning, expression and transcriptional properties of the human enhancer factor TEF-l. *Cell* **65,** 551–568.
7. Huynn, T. V., Young, R. A., and Davis, R. W. (1985) Construction and screening of cDNA libraries in λgt11 and gt10, in *DNA Cloning—A Practical Approach,* vol. 1, (Glover, D. M., ed.) IRL, Oxford, pp. 49–78.
8. Vinson, C. R., LaMarco, K. L., Johnson, P. F., Landshulz, W. H., and McKnight, S. L. (1988) In situ detection of sequence-specific DNA binding activity specified by a recombinant bacteriophage. *Genes Dev.* **2,** 801–806.
9. Lum, L. S. Y., Sultman, L. A., Kaufman, R. J., Linzer, D. I., and Wu, B. J. (1990) A cloned human CCAAT-box-binding factor stimulates transcription from the human hsp70 promoter. *Mol. Cell. Biol.* **10,** 6709–6717.

PART VII

Protein Function

CHAPTER 35

6xHis-Ni-NTA Chromatography as a Superior Technique in Recombinant Protein Expression/Purification

Joanne Crowe, Heinz Döbeli, Reiner Gentz, Erich Hochuli, Dietrich Stüber, and Karsten Henco

1. Introduction

The 6xHis/Ni-NTA system is a fast and versatile tool for the affinity purification of recombinant proteins and antigenic peptides. It is based on the high-affinity binding of six consecutive histidine residues (the 6xHis tag) to immobilized nickel ions, giving a highly selective interaction that allows purification of tagged proteins or protein complexes from <1% to >95% homogeneity in just one step *(1,2)*. The tight association between the tag and the resin allows contaminants to be easily washed away under stringent conditions, yet the bound proteins can be gently eluted by competition with imidazole, or a slight reduction in pH. Moreover, because the interaction is independent of the tertiary structure of the tag, 6xHis labeled proteins can be purified even under the strongly denaturing conditions required to solubilize inclusion bodies.

The six histidine residues that comprise the 6xHis tag can be attached at either end of the recombinant protein, are uncharged at physiological pH, and are very poorly immunogenic in all species except monkeys. Consequently, the 6xHis tag very rarely affects the

From: *Methods in Molecular Biology, Vol. 31: Protocols for Gene Analysis*
Edited by: A. J. Harwood

structure or function of the tagged protein, and need not be removed after purification *(3)*. Its small size makes it ideal for incorporation into any expression system.

We have combined the advantages of 6xHis/Ni-NTA purification with a high level bacterial expression system to create an elegant yet simple strategy, allowing protein purification whether it is expressed at low or high levels; denatured; or associated with other proteins, DNA, or RNA. It is currently used in a wide variety of applications, ranging from the large-scale purification of proteins for antibody production, to the isolation of subunits and substrates through their interactions with the tagged proteins.

1.1. Expression of Proteins Using the pQE Expression Vectors

The pQE expression vectors provide high level expression in *E. coli* of proteins or peptides containing a 6xHis affinity tag. The tag may be placed at the N-terminus of the protein to create a Type IV construct; at the C-terminus of the protein to create a Type III construct; or at the C-terminus of a protein utilizing its original ATG start codon to create a Type ATG construct (pQE-60) (Fig. 1). If small peptides are being synthesized, they can be fused to mouse DHFR to create a Type II construct, where the poorly immunogenic DHFR stabilizes the peptide during expression, and enhances its antigenicity.

The pQE plasmids were derived from plasmids pDS56/RBSII and pDS781/RBSII-DHFRS *(1)*. They contain the following elements as shown for two typical vectors pQE-30 (Type IV) and pQE-40 (Type II) (Fig. 2):

1. An optimized, regulatable promoter/operator element N250PSN250P29, consisting of the *E. coli* phage T5 promoter containing two *lac* operator sequences for tight regulation;
2. A synthetic ribosome binding site, RBSII, designed for optimal recognition and binding;
3. Optimized 6xHis affinity tag coding sequence;
4. The mouse DHFR coding sequence (in some vectors only);
5. A multicloning site in three reading frames;
6. Translation stop codons in all reading frames;
7. The transcriptional terminator "to" from phage lambda;
8. The promoter-free gene for chloramphenicol acetyltransferase with its genuine translational signals, no direct function for expression;

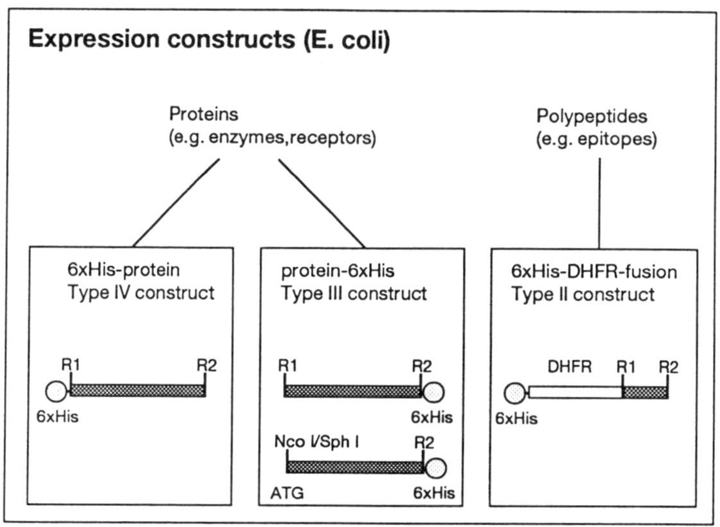

Fig. 1. Expression constructs available with pQE expression vectors.

9. The transcriptional terminator T1 of the *E. coli* rrnB operon; and
10. The replication region and the gene for β-lactamase of plasmid pBR322.

The *E. coli* host cells contain multiple copies of the plasmid pREP4. This carries the gene for neomycin phosphotransferase (NEO) conferring kanamycin resistance, and the *lac I* gene encoding the *lac* repressor. The strains are called M15 [pREP4] and SG13009[pREP4]. The multiple copies of pREP4 present in the host cells ensure high levels of *lac* repressor and tight regulation of protein expression. The plasmid is maintained in *E. coli* cells in the presence of kanamycin at a concentration of 25 µg/mL *(1)*. Expression from pQE vectors is rapidly induced by the addition of IPTG, which inactivates the repressor. The level of IPTG used for induction can be varied to control the level of expression.

We use and recommend the *E. coli* host strains M15[pREP4] or SG 13009[pREP4] carrying the repressor (pREP4) plasmids, for the production of recombinant proteins. *E. coli* strains that contain the *lac* I^q gene, such as JM109 and TG1, are suitable for storing and propagating the pQE plasmids, since they produce enough *lac* repressor to block expression without carrying the pREP4 plasmid. They may also be used as expression hosts.

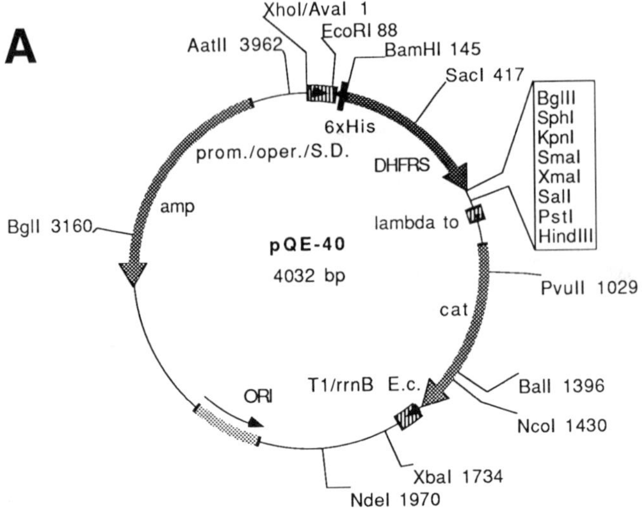

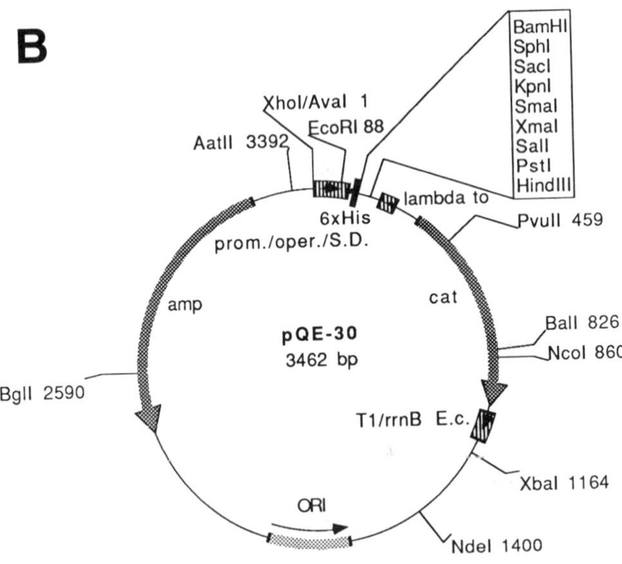

Fig. 2. Typical pQE expression vectors. **(A)** pQE-40 (Type II construct) contains a DHFRS sequence between the 6xHis tag and the polycloning region. DHFRS stabilizes short protein sequences. **(B)** pQE-30 (Type IV construct). The polycloning region is directly 3' to the 6xHis tag sequence.

The affinity tag for purification on the Ni-NTA resin consists of just six consecutive histidine residues. This small size means that there is minimal addition of extra amino acids to the recombinant protein. It is nonimmunogenic, or at most very poorly immunogenic, in all species except some monkeys, and being uncharged at pH 8.0, rarely affects the secretion or the folding of the protein to which it is attached. In over 150 proteins purified using this system, the 6xHis tag has never been found to interfere with the structure or function of the purified protein. There is consequently rarely any need to go through time-consuming and inefficient protease cleavage reactions in order to remove the 6xHis tag from the protein after purification. If it is desirable to remove the tag, a protease cleavage site (IEGR [Factor Xa], DDDDK [Enterokinase]) can be inserted into the construct by PCR.

The small size of the 6xHis tag makes it ideally suited for inclusion in a variety of other expression systems. It works well in prokaryotic, mammalian, yeast, and other eukaryotic systems. The six histidine residues can be easily inserted into the expression construct, at the C- or N-terminus of the protein, by PCR, mutagenesis, or ligation of a small synthetic fragment.

1.2. The Use of Ni-NTA Resin to Purify the Expressed Proteins

Immobilized metal chelate affinity chromatography was first used to purify proteins in 1975 *(4)* and has become a relatively widely used technique. Until the development of NTA (nitrilo-tri-acetic acid), the chelating ligand iminodiacetic acid (IDA) was charged with metal ions such as Zn^{2+} and Ni^{2+}, and then used to purify a variety of different proteins and peptides *(5)*. However, IDA, which has only three chelating groups, does not bind metal ions tightly. As a consequence, the ions may be washed out of the resin on loading with strong chelating proteins and peptides, or during the washing of the bound proteins, resulting in low yields and impure products.

NTA is a novel chelating adsorbent that was developed in order to overcome these problems (Fig. 3). The NTA occupies four of the ligand binding sites in the coordination sphere of the Ni^{2+} ion (leaving two sites free to interact with the 6xHis tag), and consequently binds the metal ions more stably *(6)*. As a result this resin binds proteins about

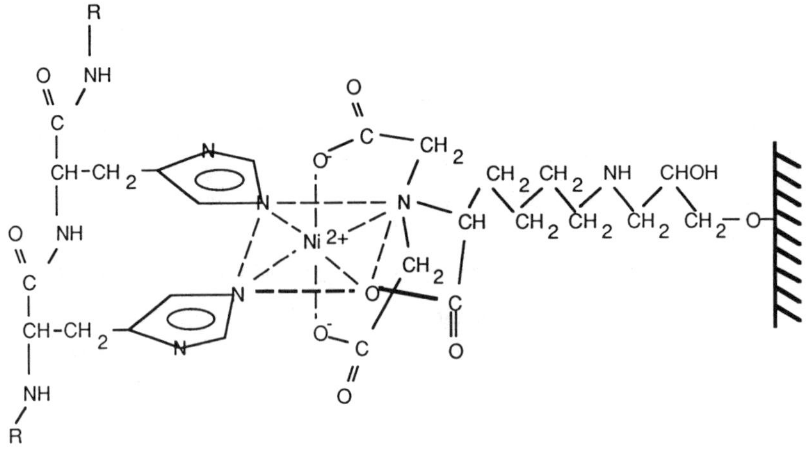

Fig. 3. Model for the binding of neighboring 6xHis residues to Ni-NTA resin.

100–1000× more tightly than IDA, allowing the purification of proteins from <1% to >95% homogeneity in just one step *(2)*.

The Ni-NTA resin is composed of a high surface concentration of NTA ligand attached to Sepharose CL-6B, and is sufficient for the binding of approx 5–10 mg of 6xHis tagged protein/mL of resin. The Ni-NTA resin can be reused 3–5 times for purification of the same protein, and is very stable and easy to handle.

Proteins containing 6xHis tags, located at either the amino or carboxyl terminus of the protein, bind to the Ni-NTA resin with an affinity far greater than the affinity between most antibodies and antigens. As a consequence, the background of proteins that bind to the resin owing to the presence of neighboring histidine residues, can be easily washed away under relatively stringent conditions without affecting the binding of the tagged proteins. The high binding constant also allows proteins in very dilute solutions, such as those expressed at low levels or secreted into the media, to be efficiently bound to the resin.

The binding of tagged proteins to the resin does not require any functional protein structure and is thus unaffected by strong denaturants such as 6*M* guanidine hydrochloride or 8*M* urea. This means that, unlike purification systems that rely on antigen/antibody or enzyme/substrate reactions, Ni-NTA can be used to purify almost all proteins—

even those that are insoluble under nondenaturing conditions. In addition, *E. coli* proteins that could copurify owing to the formation of disulfide bonds can be easily removed by the addition of low levels of β-mercaptoethanol to the loading buffer.

The presence of low levels of detergents, such as Triton X-100 and Tween-20, or high salt concentrations, also has no effect on the binding, allowing the complete removal of proteins that would normally copurify because of nonspecific hydrophobic interactions. Nucleic acids that might associate with certain DNA and RNA binding proteins can also be efficiently removed without affecting the recovery of the tagged protein.

Elution of the tagged proteins from the column can be achieved by several methods. Reducing the pH will cause the histidine residues to become protonated, which allows them to dissociate from the Ni-NTA. Monomers are generally eluted at around pH 5.9, whereas aggregates and proteins that contain more than one tag elute at around pH 4.5. Elution can also be achieved by competition with imidazole buffer, which binds to the Ni-NTA and displaces the tagged protein. Low levels of imidazole can also be used to selectively elute contaminants that bind less strongly to the resin *(2)*. A typical result of a Ni-NTA-purification is shown in Fig. 4.

1.3. General Considerations

In this chapter we describe a mini-prep procedure to enable the investigator to confirm correct protein expression. We also provide protocols for bulk purification of proteins from *E. coli* under both denaturing and nondenaturing conditions. Although each procedure works very well for most proteins, some modifications may be necessary if host systems other than *E. coli* are used. The purification power of the 6xHis Ni-NTA system will be enhanced if the conditions are optimized for each individual protein. Possible modifications are considered in Notes 1–5.

2. Materials

The pQE-vectors, *E. coli* host strains, and Ni-NTA-resin are available exclusively from Diagen GmbH, Dusseldorf, Germany or Qiagen Inc. (Chatsworth, CA).

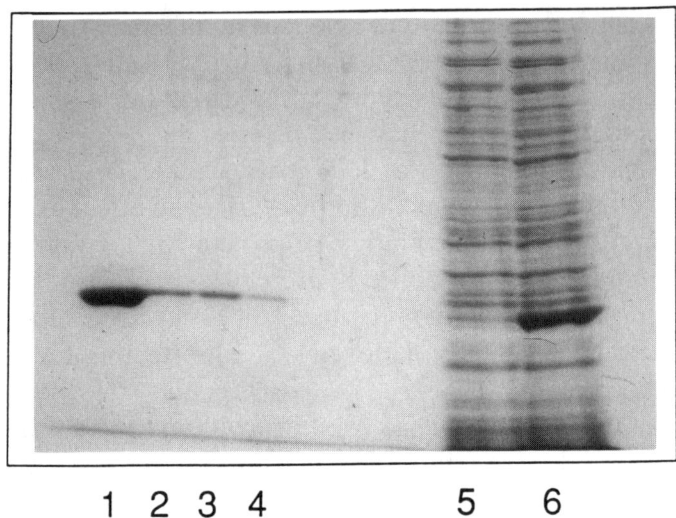

Fig. 4. The typical profile of NTA purified proteins from *E. coli* (HIV-antigen) on SDS-PAGE is shown. The 6xHis labeled protein (Type III) was bound to Ni-NTA resin from a crude lysate in 6*M* guanidinium/HCl, 10 m*M* β-ME, pH 8.0, washed and eluted at pH 4.6. (1–4) Eluates at pH 4.6 (4 fractions); (5) uninduced control; (6) induced control.

2.1. Rapid Screening of Mini Expression Cultures

1. Culture media: Use LB-medium and its modifications, 2X YT or Super Broth, containing 100 µg/mL ampicillin and 25 µg/mL kanamycin for growth of M15 cells containing pQE expression and pREP4 repressor plasmids (*see* Note 6): LB medium: 10 g of bacto-tryptone, 5 g of bacto-yeast extract, and 5 g of NaCl/L. 2X TY medium: 16 g of bacto-tryptone, 10 g of bacto-yeast extract, and 5 g of NaCl/L. Super medium: 25 g of bacto-tryptone, 15 g of bacto-yeast extract, and 5 g of NaCl/L.
2. IPTG: Stock concentration 1*M*.
3. Buffer B: 8*M* urea, 0.1*M* NaH_2PO_4, 0.01*M* Tris, pH adjusted to 8.0 with NaOH. Owing to the dissociation of urea, the pH has to be adjusted immediately before use.
4. Buffer C: Same composition as buffer B, but pH adjusted to 6.3 with HCl. Owing to the dissociation of urea, the pH has to be adjusted immediately before use.
5. EDTA solution: 100 m*M* EDTA dissolved in buffer C.
6. 5X PAGE sample buffer: 15% β-mercaptoethanol, 15% SDS, 1.5% bromophenol blue, and 50% glycerol.
7. 12.5% polyacrylamide gels containing 0.2% SDS *(7)*.

2.2. Native Purification Protocol

8. Sonication buffer: 50 m*M* Na-phosphate, pH 8.0 (adjust with NaOH), 300 m*M* NaCl.
9. Lysozyme: Stock concentration 10 mg/mL.
10. RNase: Stock concentration 200 mg/mL.
11. DNase: Stock concentration 60 mg/mL.
12. Wash buffer: 50 m*M* Na-phosphate, 300 m*M* NaCl, 10% glycerol, pH 6.0.

2.3. Denaturing Purification of Insoluble Proteins

13. Buffer A: 6*M* guanidine hydrochloride, 0.1*M* NaH_2PO_4, 0.01*M* Tris, pH adjusted to 8.0 with NaOH.
14. Buffer D: Same composition as buffer B, but pH adjusted to 5.9 with HCl. Owing to the dissociation of urea, the pH has to be adjusted immediately before use.
15. Buffer E: Same composition as buffer B, but pH adjusted to 4.5 with HCl.
16. Buffer F: 6*M* guanidinium-HCl, 0.2*M* acetic acid.

3. Methods

3.1. Rapid Screening of Mini Expression Cultures

The following is a basic protocol for the expression and screening of small cultures. It contains no stringent washing steps, and therefore cannot be expected to give very pure product, but is useful to show whether there is a protein being expressed with an affinity tag, and the level of expression. It is performed under denaturing conditions, which will lead to the isolation of any tagged protein, independent of its location within the cell.

When analyzing a time course of the expression characteristics, it is best to begin with a 10-mL culture, and to take 1-mL samples at t = 0, 1, 2, 3, and 4 h after induction (*see* Note 7). All culture media should contain ampicillin at 100 µg/mL and kanamycin at 25 µg/mL.

1. Pick single colonies of transformants into 1.5 mL of culture media containing both ampicillin and kanamycin. Also inoculate one 1.5-mL culture with a colony transformed with the control plasmid, pQE 16, which expresses DHFR. Inoculate one extra culture to serve as an uninduced control. Grow the cultures overnight.
2. Inoculate 1.5 mL of prewarmed media (+ antibiotics) with 500 µL of the overnight cultures (check for approx 0.4 A_{600}), and grow at 37°C for 30 min, with vigorous shaking, until 0.7–0.9 A_{600} is reached (*see* Note 8). Induce expression by addition of IPTG to a final concentration of 2 m*M*. Grow the cultures for an additional 3–5 h. Do not add IPTG to the culture that will serve as an uninduced control.

3. Transfer 1 mL of culture to a microcentrifuge tube. Harvest the cells by centrifugation for 3 min at 15,000*g*, and discard supernatants. Resuspend cells in 200 µL of buffer B and lyse cells by gently vortexing, taking care to avoid frothing (*see* Note 9).
4. Centrifuge the lysate at 15,000*g* for 10 min (in a microcentrifuge) to remove the cellular debris, and transfer the supernatant to a fresh tube.
5. Add 30–50 µL of a 50% slurry of Ni-NTA resin to each tube, and mix gently for 30 min at room temperature.
6. Centrifuge 10 s at 15,000*g* to pellet the resin, transfer 10 µL of the supernatant to a fresh tube, and discard the remaining supernatant. Store the supernatant samples on ice (*see* Note 10).
7. Wash the resin three times with 1 mL of buffer C. Then add 20 µL of EDTA solution to each tube. The EDTA chelates the Ni^{2+} ions from the NTA resin and elutes the protein. Incubate at room temperature for 2 min with gentle mixing. Centrifuge for 10 s at 15,000*g* and carefully remove 20 µL supernatant to a fresh tube.
8. Add 5 µL of 5X PAGE sample buffer to all samples, including the unbound fractions, and boil for 7 min at 95°C. Analyze samples on a 12.5% polyacrylamide gel and visualize proteins by staining with Coomassie blue.

3.2. Native Bulk Purification Protocol

1. Grow and induce a 1-L culture as described in Section 3.1.
2. Harvest the cells by centrifugation at 4,000*g* for 20 min. Resuspend the pellet in sonication buffer at 2–5 vol/g of wet weight (*see* Note 11). Freeze sample in dry ice/ethanol (or overnight at –20°C), and thaw in cold water. Alternatively, add lysozyme to 1 mg/mL and incubate on ice for 30 min.
3. Sonicate on ice (1 min bursts/1 min cooling/200–300 W) and monitor cell breakage by measuring the release of nucleic acids at A_{260} until it reaches a maximum.
4. If the lysate is very viscous, add RNase A to 10 µg/mL and DNase I to 5 µg/mL, and incubate on ice for 10–15 min. Alternatively, draw the lysate through a narrow gage syringe needle several times. Centrifuge at >10,000*g* for 20 min, and collect the supernatant.
5. Add 8 mL of a 50% slurry of Ni-NTA resin, previously equilibrated in sonication buffer, and stir on ice for 60 min (*see* Notes 2 and 12).
6. Load resin into 1.6-cm diameter column, and wash with sonication buffer (flow rate 0.5 mL/min), until the A_{280} of the flow through is <0.01 (approx 40–80 mL).

7. Wash the resin with wash buffer until the flow through A_{280} is <0.01 (approx 80 mL).
8. Elute the protein with a 30-mL gradient of 0–0.5*M* imidazole in wash buffer (*see* Note 13). Collect 1-mL fractions, and analyze 5 µL samples on SDS-PAGE (*see* Note 14).

3.3. Denaturing Bulk Purification of Insoluble Proteins

1. Grow and induce a 500-mL culture as described in Section 3.1. Harvest the cells by centrifugation at 4,000*g* for 20 min. Store at –70°C if desired.
2. Thaw cells for 15 min and resuspend in buffer A at 5 mL/g wet weight (*see* Note 15). Stir cells for 1 h at room temperature. Centrifuge lysate at 10,000*g* for 15 min at 4°C. Collect supernatant.
3. Add 8 mL of a 50% slurry of Ni-NTA resin, previously equilibrated in buffer A. Stir at room temperature for 45 min, then load resin carefully into a 1.6-cm diameter column (*see* Note 16).
4. Wash with 10 column vol of buffer A, and 5 column vol of buffer B. If necessary, wash further until a flow through of <0.01 A_{280} is achieved.
5. Wash with buffer C until the A_{280} is below 0.01.
6. Elute the recombinant protein with 10–20 mL of buffer D, followed by 10–20 mL of buffer E (*see* Note 17). Collect 3-mL fractions from each elution, and analyze by SDS-PAGE (*see* Notes 18 and 19).
7. Elute further with 20 mL of buffer F. Collect 3-mL fractions and analyze by SDS-PAGE (*see* Note 20).

Problems that may be encountered during purification are discussed in Notes 21–26.

4. Notes

1. Many proteins remain soluble during expression and can be purified in their native form under nondenaturing conditions on Ni-NTA resin; others, however, form insoluble precipitates. Since almost all of these proteins are soluble in 6*M* guanidinium hydrochloride, Ni-NTA chromatography and the 6xHis tag provide a universal system for the purification of recombinant proteins. The decision whether to purify the tagged proteins under denaturing or nondenaturing conditions depends on both the location of the protein and the accessibility of the 6xHis. Proteins that remain soluble in the cytoplasm, or are secreted into the periplasmic space, can generally be purified under nondenaturing conditions (but note the exception below). If the protein is insoluble, or located in inclusion bodies, then it must generally be solubilized by denaturation before it can be purified. Some proteins, however, may be

solubilized by the addition of detergents, and it is worth experimenting with different solubilization techniques if it is important to retain the native configuration of the protein. Many proteins that form inclusion bodies are also present at some level in the cytoplasm, and may be efficiently purified in their native form, even at very low levels, on Ni-NTA resin.

In rare cases the 6xHis tag is hidden by the tertiary structure of the native protein, so that soluble proteins require denaturation before they can be purified on Ni-NTA resin. If denaturation of the protein is undesirable, the problem is usually solved by moving the tag to the opposite terminus of the protein. Proteins that have been purified under denaturing conditions can either be used directly or refolded in dilute solution by dialyzing away the denaturants. It is also possible to renature proteins on the Ni-NTA column.

2. Proteins may be purified on Ni-NTA resin in either a batch or a column procedure. The batch procedure entails binding the protein to the Ni-NTA resin in solution, and then packing the protein/resin complex into a column for the washing and elution steps. This may promote more efficient binding and reduce the amount of debris that is loaded onto the column. In the column procedure, the Ni-NTA column is packed and washed, and the cell lysate is applied slowly to the column.
3. Background contamination arises from proteins that contain neighboring histidines, and thus have some affinity for the resin; proteins that copurify because they are linked to the 6xHis-tagged protein by disulfide bonds; proteins that associate nonspecifically with the tagged protein; and nucleic acids that associate with the tagged protein. All of these contaminants can be easily removed by washing the resin under the appropriate conditions. Proteins that contain neighboring histidines are not common in bacteria, but are quite abundant in mammalian cells. These proteins bind to the resin much more weakly than proteins with a 6xHis tag, and can be easily washed away, even when they are much more abundant than the tagged protein *(2)*. The addition of 10–20 m*M* β-mercaptoethanol to the loading buffer will reduce background owing to crosslinked proteins. Do not use DTT, which will reduce Ni^{2+} ions.

   Proteins that are associated with the tagged protein or the resin owing to nonspecific interactions and nucleic acids that copurify can be removed by washing with low levels of detergent (up to 2% Triton X-100 or 0.5% Sarcosyl); increasing the salt concentration up to 1*M* NaCl; or including low levels of ethanol or 30% glycerol to reduce hydrophobic interactions. The optimum levels of any of these reagents should be determined empirically for different proteins.

4. Removal of background proteins and elution of tagged proteins from the column may be achieved by either lowering the pH in order to protonate the histidine residues or by the addition of imidazole, which competes with tagged proteins for binding sites on the Ni-NTA resin. Although both methods are equally effective, the imidazole is somewhat milder, and is recommended in cases where the protein would be damaged by a reduction in pH (e.g., tetrameric aldolase) *(3)*.

   In bacterial expression systems, it is seldom necessary to wash the bound protein under very stringent conditions, since proteins are expressed to high levels and the background is low. In mammalian systems, however, or under native conditions where many more neighboring histidine residues will be exposed to the resin, it may be necessary to increase the stringency of the washing considerably. This can be done by gradually decreasing the pH of the wash buffer, or by slowly increasing the concentration of imidazole. The pH or imidazole concentration that can be tolerated before elution begins will vary slightly for each protein.

   In situations where the tagged protein is very dilute and the background is likely to be high, it is also useful to bind the 6xHis tagged protein to the resin under conditions in which the background proteins do not compete for the binding sites, i.e., at a slightly lower pH or in the presence of low levels of imidazole. Likewise, the purification process will be optimized if the amount of tagged protein is closely matched to the capacity of the of the resin used, i.e., if the amount of resin is minimized (H. Stunnenberg, personal communication). Since the 6xHis tagged protein has a higher affinity for the Ni-NTA resin than do the background proteins, it can fill all the available binding sites and very few background proteins will be retained on the resin.
5. Do not use strong reducing agents such as DTT or DTE on the column, because they will reduce the Ni^{2+} ions and cause them to elute from the resin. In most situations, ß-mercaptoethanol can be used at levels up to 10 m*M*, but even these low levels might cause problems occasionally when the protein itself has a strongly reducing nature. Use any reducing agent with care, and if in doubt, test it out on a small amount of Ni-NTA resin first. Strong chelating agents will chelate the Ni, and also cause it to elute from the NTA resin. Do not use EDTA, or EGTA, or any other chelating agents.
6. We suggest that expression should be tried in all three media in parallel, and a time course taken of expression after induction. There are often striking differences noted between the level of expression in different media at different times.

7. If 10-mL cultures are being grown to prepare a time course of expression, inoculate 8.75 mL of medium with 1.25 mL of saturated culture, and grow for 1 h before IPTG induction. A sample taken at $t = 0$ sample can serve as the uninduced control. Take 1 mL samples at hourly intervals, collect the cell pellets, and store at –20°C until all the samples are ready for processing.
8. The short second growth from an aliquot of the saturated culture is to ensure that all cultures are grown to a similar cell density before induction. If expression levels are being closely compared, it is advisable to measure the A_{600} to ensure uniform cell density.
9. The solution should become translucent when lysis is complete. Most proteins are soluble in buffer B, which contains 8*M* urea.
10. The supernatant samples will contain any proteins that have not bound to the resin.
11. The composition of the sonication, wash, and elution buffers can be modified to suit the particular application, e.g., by adding 1–2% Tween, 5–10 m*M* β-mercaptoethanol, 1 m*M* PMSF, or increased NaCl or glycerol concentrations.
12. An alternative is to directly load the lysate at 2–3 column vol/h onto a 4-mL Ni-NTA column, previously equilibrated with sonication buffer.
13. The protein may be eluted by a number of different means. If preferred, elution can be performed using pH, either as a continuous or step gradient decreasing from pH 6.0 to 4.5. Most proteins will be efficiently eluted by wash buffer at pH 4.5, although many (particularly monomers) can be eluted at a higher pH. If elution at a higher pH is desired, most proteins can be efficiently removed from the resin with 250 m*M* imidazole in either wash or sonication buffer. If it is necessary to keep the pH above 7.0 at all times, the column can be washed with sonication buffer containing 0.8–40 m*M* imidazole, and eluted with a 50 mL gradient of 0–0.5*M* imidazole in sonication buffer.
14. Do not boil a sample that contains imidazole before SDS-PAGE, since it will hydrolyze acid labile bonds. Heat the sample for a few minutes at 37°C immediately before loading the gel.
15. The purification is performed in 8*M* urea, since 6*M* guanidinium hydrochloride precipitates in the presence of SDS, making SDS-PAGE analysis of samples difficult. Otherwise, both urea and GuHCl or combinations thereof can be used throughout the whole purification procedure.
16. An alternative is to load the lysate, at a flow rate of 10–15 mL/h, onto a 4-mL Ni-NTA column pre-equilibrated in buffer A.
17. Alternative elution procedures may be used. The protein can be eluted with 50 mL of buffer C containing 250 m*M* imidazole. Elution can also

be carried out using a pH 6.5–4.0 gradient in 8*M* urea, 0.1*M* Na-phosphate, 0.01*M* Tris-HCl.

18. Where possible, follow the chromatography by A_{280} and collect pools rather than fractions. Note that discolored or impure reagents may affect optical density readings, and that imidazole will also absorb light at 280 nm.
19. On minigels it is usually sufficient to analyze 5 μL samples of each fraction in an equal volume of SDS-PAGE loading buffer, with or without 3% β-mercaptoethanol. Since the fractions that contain GuHCl will precipitate with SDS, they must either be very dilute (1:6), dialyzed before analysis, or separated from the guanidinium hydrochloride by TCA precipitation: Dilute samples to 100 μL; add equal vol of 10% TCA; leave on ice 20 min; spin 15 min in a microfuge; wash pellet with 100 μL of ice cold ethanol; dry; and resuspend in sample buffer. If there is any guanidinium hydrochloride present, samples must be loaded immediately after boiling for 7 min at 95°C.
20. Monomers usually elute in buffer D, whereas multimers, aggregates, and proteins with two 6xHis tags will generally elute in buffer E.
21. The expression level may be too high. If so, one or more of the following may help. Use less IPTG (0.5 m*M*, 0.1 m*M*) or reduce time of induction and/or lower temperature; try the induction at higher cell densities (0.8 A_{600}); start with the nondenaturing protocol, followed by the denaturing protocol for the residual, nondissolved cellular debris and inclusion bodies.
22. In case of precipitation during purification, check for aggregates of purified proteins and try Tween or Triton additives (up to 2%) or adjust to 10 m*M* β-mercaptoethanol (max. 20 m*M* β-mercaptoethanol) and check for stabilizing cofactor requirements (e.g., Mg^{2+}). Make sure that the salt concentration is ≥150 m*M* NaCl (e.g., 300 m*M*). Check room temperature (>20°C) for the denaturing protocol.
23. In case of an insufficient binding of 6xHis to Ni-NTA resin, check for the absence of chelating agents (EDTA/EGTA) and repeat the binding step. If the 6xHis-tag is hidden in the native protein structure, improve the exposure by adding small concentrations of urea to the nondenaturing sample preparation buffer. Slower binding kinetics can be compensated by longer contact times with NTA, preferably under batch binding conditions. Alternatively, try 6xHis at the opposite terminus, or use completely denaturing conditions (6*M* Guanidinium-HCl buffer with 10 m*M* β-mercaptoethanol). Avoid any Ni^{2+} complexing reagents.
24. Background binding can be suppressed by adjusting the Ni-NTA column size to the 6xHis-protein expression level. The binding capacity

should not exceed the amount of tagged protein by more than a factor of 2.

25. Partially hidden 6xHis tags in native NTA-bound proteins can cause earlier elution characteristics.
26. Do not determine the size of the recombinant product by SDS gel. 6xHis tags often seem to shift protein bands suggesting a mol wt several kDa higher than expected. Therefore, this specific sequence effect cannot be sized this way.

Further Reading

Reviews

Hochuli, E. (1990) Purification of recombinant proteins with metal chelate adsorbent. *Gene. Engineer.*, vol. 12 (Setlow, J. K., ed.), Plenum, New York, pp. 87–98.

Selected Examples

Abate, C., Luk, D., Gentz, R., Rauscher III, F. J., and Curran, T. (1990) Expression and purification of the leucine zipper and DNA-binding domains of Fos and Jun: both Fos and Jun contact DNA directly. *Proc. Natl. Acad. Sci. USA* **87,** 1032–1036.

Bush, G. L., Tassin, A., Friden, H., and Meyer, D. I. (1991) Purification of a translocation-competent secretory protein precursor using nickel ion affinity chromatography. *J. Biol. Chem.* **266,** 13,811–13,814.

Gentz, R., Certa, U., Takacs, B. J., Matile, H., Döbeli, H., Pink, R., Mackay, M., Bone, N., and Scaife, J. G. (1988) Major surface antigen p190 of Plasmodium falciparum: detection of common epitopes present in a variety of plasmodia isolates. *EMBO J.* **7,** 225–230.

Gentz, R., Chen, C., and Rosen, C. A. (1989) Bioassay for trans-activation using purified immunodeficiency virus tat-encoded protein: trans-activation requires mRNA synthesis. *Proc. Natl. Acad. Sci. USA* **86**, 821–824.

Hochuli, E., Bannwarth, W., Döbeli, H., Gentz, R., and Stüber, D. (1988) Genetic approach to facilitate purification of recombinant proteins with a novel metal chelate adsorbent. *Bio/Technology* **6,** 1321–1325.

Le Grice, S. F. J. and Grueninger-Leitch, F. (1990) Rapid purification of homodimer HIV-1 reverse transcriptase by metal chelate affinity chromatography. *Eur. J. Biochem.* **187,** 307–314.

Stüber, D., Bannwarth, W., Pink, J. R. L., Meloen, R. H., and Matile, H. (1990) New B cell epitopes in the plasmodium falciparum malaria circumsporozoite protein. *Eur. J. Immunol.* **20,** 819–824.

Takacs, B. J. and Girard, M.-F. (1991) Preparation of clinical grade proteins produced by recombinant DNA technologies. *J. Immunol. Meth.* **143,** 231–240.

References

1. Stüber, D., Matile, H., and Garotta, G. (1990) System for high-level production in Escherichia coli and rapid purification of recombinant proteins: application to epitope mapping, preparation of antibodies, and structure-function analysis, in *Immunological Methods* (Lefkovits, I. and Pernis, B., eds.), vol. IV, Academic, New York, pp. 121–152.
2. Janknecht, R., de Martynoff, G., Lou, J., Hipskind, R. A., Nordheim, A., and Stunnenberg, H. G. (1991) Rapid and efficient purification of native histidine-tagged protein expressed by recombinant vaccinia virus. *Proc. Natl. Acad. Sci. USA* **88,** 8972–8976.
3. Döbeli, H.,Trecziak, A., Gillessen, D., Matile, H., Srivastava, I. K., Perrin, L. H., Jakob, P. E., and Certa, U. (1990) Expression, purification, biochemical characterization and inhibition of recombinant Plasmodium falciparum aldolase. *Molec. Biochem. Parasitol.* **41,** 259–268.
4. Porath, J., Carlsson, J., Olsson, I., and Belfrage, G. (1975) Metal chelate affinity chromatography, a new approach to protein fractionation. *Nature* **258,** 598–599.
5. Sulkowski, E. (1985) Purification of proteins by IMAC. *Trends Biotechnol.* **3,** 1–7.
6. Hochuli, E., Döbeli, H., and Schacher, A. (1987) New metal chelate adsorbent selective for proteins and peptides containing neighbouring histidine residues. *J. Chromatog.* **411,** 177–184.
7. Takacs, B. J. (1979) in *Immunological Methods* (Lefkovits, I. and Pernis, B., eds.), vol. 1, Academic, New York, p. 81.

CHAPTER 36

Production of ^{35}S-Labeled Proteins in *E. coli* and Their Use as Molecular Probes

Michael A. Lydan and Danton H. O'Day

1. Introduction

The ability to obtain essentially unrestricted quantities of a cloned protein via expression in a bacterial, fungal, or eukaryotic cell expression system facilitates the structural and functional characterization of that protein. The rapidly increasing availability of diverse expression vectors and complementary cellular systems will eventually allow for the heterologous expression of any protein *(1–3)*. Heterologous protein expression with concomitant radiolabeling provides an attractive means to produce a readily detectable, biochemically active form of the protein without the potential chemical modifications that may arise when a radioactive or other detectable marker is covalently linked to it.

In this chapter we describe a protocol to ^{35}S-label proteins by expression in *E. coli* in the presence of ^{35}S methionine and their subsequent use as molecular probes. The techniques presented should be readily applicable to any protein that can be functionally expressed in *E. coli*. As an example we focus on the in vivo labeling of calmodulin (CaM) and the use of [^{35}S]-CaM as a probe to identify intracellular CaM binding proteins (CaMBPs) *(4)*. A synthetic CaM gene (VU-1 CaM) was cloned into the pKK233-2 expression vector to produce pVUC-1 *(5,6)* and expressed in *E. coli* by utillizing the hybrid trp/*lac* (tac) promoter *(6,7)*. With this expression system, VU-1

From: *Methods in Molecular Biology, Vol. 31: Protocols for Gene Analysis*
Edited by: A. J. Harwood

CaM is both efficiently labeled with ^{35}S and easily purified (Fig. 1). The use of ^{35}S as a radiolabel provides the greatest sensitivity when used as a probe and can detect nanogram quantities of CaMBPs in a calcium-dependent manner (Fig. 2).

2. Materials

2.1. Production and Purification of ^{35}S-Labeled Proteins

1. *E. coli* strain that expresses the protein of interest: In this case the strain UT481 is used, which has been transformed with the plasmid pVUC-1 that expresses the VU-1 CaM protein *(6)*.
2. Luria Bertani (LB) agar and media containing, respectively, 100 and 50 µg/mL of ampicillin *(8)*.
3. Modified M9 medium: For 1 L mix 6 g of Na_2HPO_4, 3 g of KH_2PO_4, 0.5 g of NaCl, 1 g of NH_4Cl, pH 7.4; autoclave, cool, and add from sterile stocks 2 mL of 1*M* $MgCl_2$, 50 mL of 20% glucose, 0.1 mL of 1*M* $CaCl_2$, and ampicillin to 50 µg/mL *(8)*. The $MgSO_4$ component of standard M9 medium has been replaced by $MgCl_2$ (*see* Note 1).
4. [^{35}S]-methionine: 5 mCi of cell labeling grade. Store at –70°C This is a low level β emitter.
5. IPTG: 1*M* of isopropyl-β-D thiogalactopyranoside. Store at –20°C.
6. Buffer A: 50 m*M* Tris-HCl, pH 7.6.
7. Buffer B: 50 m*M* Tris-HCl, pH 7.6, 2 m*M* ethylenediaminetetraacetic acid (EDTA), 1 m*M* dithiotreitol (DTT), 1 mg/mL lysozyme. Add lysozyme just before use.
8. Buffer C: 50 m*M* Tris-HCl, pH 7.6, 2 m*M* EDTA, 1 m*M* DTT, 75 m*M* $MgCl_2$, 300 U/mL DNase. Add DNase just before use.
9. Buffers and solutions for phenyl-Sepharose column: Store phenyl-Sepharose in 20% ethanol at 4°C. To make column, add 1.0 mL of phenyl-Sepharose to a 2 mL disposable polypropylene column (Bio-Rad, Richmond, CA) and equilibriate with wash 1. Wash 1 contains 50 m*M* Tris-HCl, pH 7.5, and 1 m*M* $CaCl_2$. Wash 2 contains 50 m*M* Tris, pH 7.5, 500 m*M* NaCl, and 1 m*M* $CaCl_2$. Elution buffer contains 50 m*M* Tris-HCl, pH 7.5, and 2 m*M* EDTA.
10. Dry Sephadex G15: Store at room temperature.

2.2. Use of ^{35}S-Labeled Proteins as Molecular Probes

11. Buffers and reagents required for discontinuous SDS-PAGE, mini SDS-PAGE apparatus (*see* vol. 1 of this series).
12. 40% Methanol/7% acetic acid: Make as required.

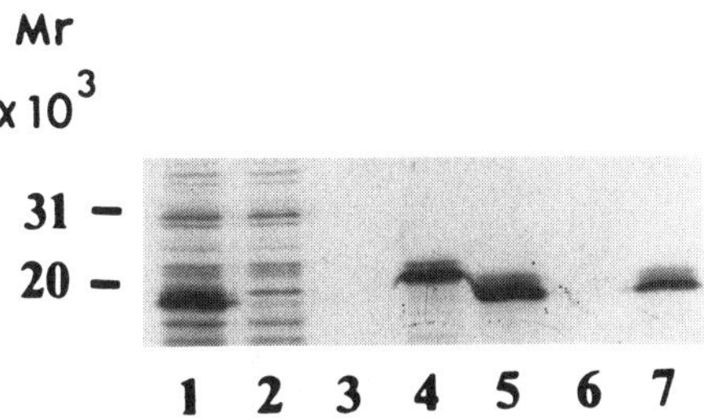

Fig. 1. Expression and purification of [^{35}S]-VU-1 calmodulin. A major 18,000-Mr soluble protein ([^{35}S]-VU-1 CaM) was present following the labeling and lysis procedure (lane 1) and was retained by the phenyl-Sepharose column (lane 2). After extensive washing of the column (lane 3), [^{35}S]-VU-1 CaM was eluted with EDTA and exhibited a Mr of about 20,000 (lane 4). When [^{35}S]-VU-1 CaM was made 5 m*M* $CaCl_2$, the protein migrated with a Mr of about 18,000 (lane 5). This calcium dependent electrophoretic mobility shift is diagnostic for CaM *(3)*. [^{35}S]-VU-1 CaM is stable during its application as a probe for CaM-binding proteins. All of the [^{35}S]-VU-1 CaM present in the probe buffer bound to the phenyl-Sepharose column since it was absent in the flow through (lane 6). When eluted with EDTA, it exhibited an electrophoretic mobility consistent with freshly prepared [^{35}S]-VU-1 CaM (lane 7).

13. 10% Ethanol: Make as required.
14. Buffer D: 100 m*M* imidazole, pH 7.0. Make as required.
15. Buffer E: 20 m*M* imidazole, pH 7.0, 200 m*M* KCl, 1 m*M* $CaCl_2$, or 1 m*M* ethylene glycol-*bis*-(B-amino-ethyl ether) *N,N,N',N'*-tetra-acetic acid (EGTA), 0.1% BSA (fraction 5). Make as required.
16. 0.1% Coomassie blue R250 in 40% methanol/7% acetic acid, stable at room temperature.
17. Autoradiographic film: For example, Beta-max (Amersham Corp., Arlington Heights, IL).

3. Methods

3.1. Production and Purification of ^{35}S-Labeled Proteins

This protocol involves the use of radioactive materials, so all necessary precautions must be made. The labeling steps must be done in a fume hood because of volatile [^{35}S]-methionine degradation products.

1. Inoculate a single colony from an overnight plate of *E. coli* UT481 containing pVUC-1 into 25 mL LB media in a 125-mL Erlenmeyer flask. Incubate at 37°C with shaking overnight.

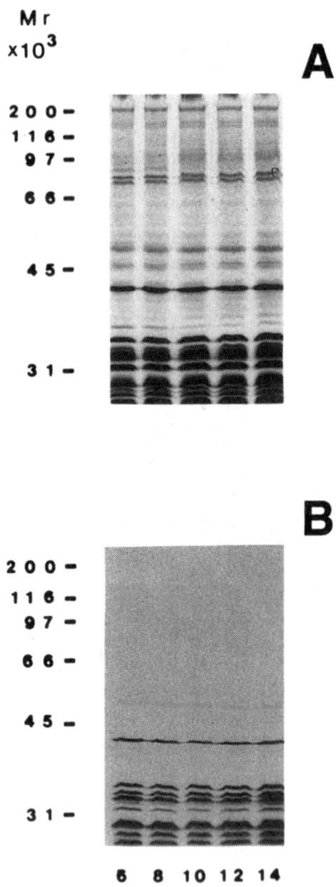

Fig. 2. Calmodulin-binding proteins during sexual development in *Dictyostelium discoideum*. Samples of cells from sexual cultures of *D. discoideum* at 6–14 h of development were electrophoresed in a 10% mini SDS-PAGE at 175 V and probed with [^{35}S]-VU-1 CaM in the presence of either 1 m*M* $CaCl_2$ (**A**) or 1 m*M* EGTA (**B**).

2. Inoculate 4 mL of the overnight culture into 50 mL of LB in a 250-mL Erlenmeyer flask; incubate at 37°C with shaking until 1.0 A_{550} is reached. This should take approx 2 h.
3. Harvest cells by centrifugation in a capped 50-mL disposable polypropylene tube at 1500*g* (e.g., 3500 rpm for 5 min in a HN-S centrifuge with #816 rotor). Wash cell pellet once with modified M9 media. Resuspend cells in 50 mL M9 media in a 250-mL Erlenmeyer flask and incubate at 37°C with shaking for 15 min (*see* Note 2).
4. Add 5 mCi [^{35}S]-methionine and 100 µL of 1*M* IPTG and incubate at 37°C with shaking for 4 h (*see* Notes 3 and 4).

5. Harvest cells as in step 3 and wash once in buffer A. Resuspend in 5 mL of buffer B and incubate at 0°C for 30 min. Add 500 μL of buffer C and incubate at 0°C for a further 15 min. During the lysozyme and DNase treatments, the cells are gently shaken in an ice bucket on a platform shaker to ensure complete mixing.
6. Sonicate 3× for 10 s. Centrifuge at 27,000*g* for 30 min at 4°C in a 15-mL Corex tube. If the protein is heat-stable, heat the supernatant to 90°C for 3 min, cool on ice, then centrifuge at 27,000*g* for a further 30 min at 4°C in a 15-mL Corex tube and add 25 μL of 1*M* $CaCl_2$ (*see* Note 5).
7. To purify [^{35}S]-VU-1 CaM, apply the lysate to the phenyl-Sepharose column. Wash sequentially with 20 bed vol of wash 1, 20 bed vol of wash 2, and then 10 bed vol of wash 1 again. Elute the [^{35}S]-VU-1 CaM with 5 bed vol of elution buffer.
8. To desalt, add approx 600 μL of the eluate to a microcentrifuge tube that has a small pinhole in the bottom and add sufficient dry Sephadex G15 to soak up the liquid. Place the microcentrifuge tube in a disposable plastic test tube, wait about 5 min, and spin at 200*g* for 2 min. This will also concentrate the [^{35}S]-VU-1 CaM approx fourfold (*see* Notes 6–9).

Synthesized proteins are stored at –70°C in a buffer that is known to maintain the biochemical activity of the specific protein.

3.2. Use of ^{35}S-Labeled Proteins as Molecular Probes

The ^{35}S-labeled protein may be used as a molecular probe in a number of situations. We describe here a technique to identify other proteins that bind to the labeled protein after SDS-PAGE. This has been used successfully to identify CaMBPs (*see* Note 10 and Fig. 2) and is a modification of that described by Slaughter and Means *(9)*.

1. Run the protein samples under investigation on a discontinuous mini SDS-PAGE gel. Fix the gel for 30 min in 100 mL of 40% methanol/7% acetic acid.
2. Treat the gel with 200 mL of 10% ethanol overnight. Change to a fresh 200 mL of 10% ethanol for a further hour. All treatments are carried out at room temperature. The volumes should be increased proportionally with the number of gels treated (*see* Note 11).
3. The gel is washed with 100 mL of buffer D for 1 h at room temperature, followed by 100 mL of buffer E for 30 min at room temperature. Probe with 5 mL of buffer E containing 2 μCi/mL of [^{35}S]-VU-1 CaM for 20 h at 4°C.

4. The gel is washed 3 times in 250 mL of buffer E for 2 h each at 4°C. Wash 5 times in 200 mL of dH_2O for 5 min each at room temperature and then in 200 mL of 40% methanol/7% acetic acid for 1–12 h at room temperature.
5. Stain with Coomassie blue and destain with 40% methanol/7% acetic acid. Dry and autoradiograph with Beta-max autoradiography film according to the the manufacturer's instructions (*see* Notes 12–15).

4. Notes

1. The modified M9 medium used lacks sulfate ions enhancing the incorporation of [^{35}S]-methionine during the synthesis of CaM.
2. The culturing protocol separates the periods of rapid cell division from the labeling period. Initial growth in LB medium produces a large number of viable cells. When these cells were transfered to the modified M9 medium, growth continues at a linear rather than an exponential rate, allowing the exogenously induced CaM to be preferentially expressed over proteins required for division or growth.
3. Radiolabeling of CaM using $^{35}SO_4^{2-}$ as the source of ^{35}S has been reported, but [^{35}S]-methionine increases the yield by approx 200% *(10)*. When $^{35}SO_4^{2-}$ is used, [^{35}S]-cysteine is synthesized by the cell in addition to [^{35}S]-methionine, decreasing the proportion of the total ^{35}S activity available for incorporation into CaM. In general, [^{35}S]-methionine is the labeled precursor of choice when labeling proteins in *E. coli* with ^{35}S. Cysteine-rich proteins should be avoided since disulfide bonds do not form very well in the protoplasm of *E. coli.* Their expression would be better suited to a different expression system *(3)*.
4. Five milliCuries of [^{35}S]-methionine has been empirically determined to be optimal. The yield remains the same with 10 mCi, but is reduced when 2.5 mCi is used.
5. Sonication is used to ensure that *E. coli* rupture is complete and to denature any DNA unaffected by the DNase. Treatment with SDS should be avoided since it can precipitate proteins and interferes with phenyl-Sepharose chromatography. Heating the lysate to 90°C simplifies the affinity chromatography step since many heat-labile proteins precipitate, whereas the heat-stable [^{35}S]-VU-1 CaM remains soluble. Heating the lysate must be done in the fume hood since there is typically at least 500 mCi of ^{35}S actvity present at this stage of the purification, some of which can volatilize.

   $CaCl_2$ is added to the lysate and is also present in both phenyl-Sepharose and CaMBPs. The calcium-dependent hydrophobic binding is a central biochemical characteristic of CaM.
6. Several methods work well for desalting and concentrating [^{35}S]-VU-1 CaM. We describe the easiest *(11)*. If the [^{35}S]-VU-1 CaM is sufficently

concentrated, it is not always necessary to desalt since the probe buffer for CaMBP detection is essentially identical to the column elution buffer. Here, the eluate is made to 1 m*M* $CaCl_2$ if the probing is to be performed in the presence of calcium or is added directly to the probe buffer if the probing is to be performed in the absence of calcium.

7. [^{35}S]-VU-1 CaM is the predominant labeled protein and demonstrates the electrophoretic characteristics of homologously expressed CaM (*3*; Fig. 1). The yield of pure [^{35}S]-VU-1 CaM is approx 90–100 μCi at 0.5 μCi/μg, about 2% of the total ^{35}S activity added. In addition, about 20% of the total ^{35}S activity incorporated into all soluble proteins was present in [^{35}S]-VU-1 CaM. This level of VU-1 CaM expression is typical for genes expressed under the control of the tac promoter *(7)*.
8. The use of [^{35}S]-methionine does not preclude the incorporation of either ^{3}H or ^{14}C into an expressed protein since the same labeling protocol should be effective with any radiolabeled amino acid. However, the labeling time should be empirically determined for other radiolabeling applications. This can be done by taking samples at sequential time points and determining, through techniques such as SDS-PAGE, when continued accumulation of the labeled protein stops.
9. [^{35}S]-VU-1 CaM was determined to be biochemically identical to homologously expressed CaM by a functional assay for the activation of CaM-dependent 3':5' cyclic AMP phosphodiesterase employing a novel micro assay technique.
10. The detection of CaMBPs, in a calcium-dependent manner, in sexually developing *Dictyostelium discoideum* cells (Fig. 2) demonstrates the effectiveness of [^{35}S]-VU-1 CaM as a general probe for CaMBPs *(12)*.
11. The ethanol washes remove the SDS and allow the proteins to reestablish their CaM-recognition abilities. An isopropyl alcohol wash followed by guanidine hydrochloride denaturation and slow renaturation has also been used to restore the CaM-binding and -activated characteristics of the multifunctional CaM-dependent protein kinase following SDS-PAGE *(13)*.
12. The washing times for the probed gels are significantly shorter than techniques using [^{125}I]-CaM, eliminating about 12 h from the total time required to perform the procedure.
13. If greater sensitivity is required, the fixed gels can be treated with a fluorography reagent such as Amplify (Amersham) without changing the pattern or relative intensities of the CaMBPs detected.
14. The free [^{35}S]-VU-1 CaM present in the probe buffer after probing can be recovered, enabling at least 50% more gels to be probed with a single batch of [^{35}S]-VU-1 CaM. This recycling process should be generally applicable since the same technique used to initially purify the expressed protein could

be used to repurify it following probing. To directly repurify [^{35}S]-VU-1 from the probe buffer, apply the probe buffer, made 1 m*M* $CaCl_2$ if required, to a 0.5-mL column of phenyl-Sepharose equilibriated with 50 m*M* Tris-HCl, pH 7.5, 1 m*M* $CaCl_2$ in a 2-mL disposable polypropylene column (Bio-Rad), wash with 10 bed vol of 50 m*M* Tris, pH 7.5, 1 m*M* $CaCl_2$, and elute with 3 bed vol of 50 m*M* Tris, pH 7.5, 2 m*M* EDTA.

15. Expression and radiolabeling of [^{35}S]-VU-1 CaM produces an easily detected, biochemically active probe for calmodulin function. Minor modifications of the labeling protocol described for [^{35}S]-VU-1 CaM would provide cost-effective quantities of other labeled proteins for use in applications either requiring or facilitated by a radiolabeled protein.

References

1. Reznikoff, W. and Gold, L., eds. (1986) *Maximizing Gene Expression.* Butterworths, Boston. 375p.
2. Butler, M. O., Harwood, C. R., and Mosely, B. E. B., eds. (1989) *Genetic Transformation and Expression.* Itercept, Andover. 574p.
3. Goedell, D. V., ed. (1990) *Methods in Enzymology, vol. 185: Gene Expression Technology.* Academic, New York. 681p.
4. Cohen, P. and Klee, C. B. (1988) *Calmodulin.* Elsevier, Amsterdam. 371p.
5. Amann, E. and Brosius, J. (1985) 'ATG vectors' for regulated high-level expression of cloned genes in Escherichia coli. *Gene* **40,** 183–190.
6. Roberts, D. M., Crea, R., Malecha, M., Alvarado-Urbina, G., Chiarello, R. H., and Watterson, D. M. (1985) Chemical synthesis and expression of a calmodulin gene designed for site-specific mutagenesis. *Biochemistry* **24,** 5090–5098.
7. Amann, E., Brosius, J., and Ptashne, M. (1983) Vectors bearing a hybrid trp-lac promoter useful for regulated expression of cloned genes in Escherichia coli. *Gene* **25,** 167–178.
8. Maniatis, T., Fritsch, E. F., and Sambrook, J. (1982) *Molecular Cloning: A Laboratory Manual.* Cold Spring Harbor Laboratory, Cold Spring Harbor, NY. 545p.
9. Slaughter, G. R. and Means, A. R. (1987) Use of the ^{125}I-labeled protein gel overlay technique to study calmodulin-binding proteins. *Meth. Enzymol.* **139,** 433–444.
10. Asselin, J., Phaneuf, S., Watterson, D. M., and Haiech, J. (1989) Metabolically ^{35}S-labeled calmodulin as a ligand for the detection of calmodulin-binding proteins. *Anal. Biochem.* **178,** 141–147.
11. Deutscher, M. P., ed. (1990) *Methods in Enzymology, vol. 182: Guide to Protein Purification.* Academic, New York. 894p.
12. Lydan, M. A. and O'Day, D. H. (1992) Calmodulin and calmodulin-binding proteins during cell fusion in *Dictyostelium discoideum*: developmental regulation by calcium ions. *Exp. Cell Res.* **205,** 134–141.
13. Kameshita, I. and Fujisawa, H. (1989) A sensitive method for detection of calmodulin-dependent protein kinase II activity in sodium dodecyl sulfate-polyacrylamide gel. *Anal. Biochem.* **183,** 139–143.

CHAPTER 37

Preparation and Ligand Screening of a λgt11 Lysogen Library

Rupert Mutzel

1. Introduction

The λgt11 system has been designed to allow expression of cDNAs either in phage plaques or in bacteria lysogenized with recombinant λgt11 phages *(1,2)*. cDNAs are cloned into a unique *Eco*RI restriction site near the 3' end of a truncated *E. coli* lactose operon *(lacP/O–lacZ)*, allowing inducible, high-level transcription of *lacZ*-cDNA-encoded fusion mRNAs that can be translated into fusion proteins if the cDNA coding strand is properly oriented with respect to *lac* transcription, and in phase with the β-galactosidase reading frame. A host strain is chosen that carries a *lacI*^{q} allele on a high-copy number plasmid (*p*MC9), assuring efficient repression of *lac* transcription in the absence of an inducer. Since λgt11 encodes a temperature-sensitive *cI* repressor (*cI*857), it can be lysogenized and stably maintained as a lysogen (i.e., as a "single copy vector") at the permissive temperature, and subsequently made to switch to a "multi copy vector" on inducing lytic growth of the prophage by a temperature shift. An amber mutation in the *S* gene (*S100*amb) renders the phage lysis-defective in a nonsuppressor host, allowing the accumulation of phage-encoded products inside the bacterial host cells. To maximize stability of *lacZ*-cDNA-encoded fusion protein, a host strain is used that carries a deletion in the gene for the *lon* protease. λgt11 can also form plaques if grown on a host that carries an amber suppressor mutation (for a recent review on the λgt11 system, *see* ref. *3*).

From: *Methods in Molecular Biology, Vol. 31: Protocols for Gene Analysis*
Edited by: A. J. Harwood Copyright ©1994 Humana Press Inc., Totowa, NJ

For immunoscreening, Young and Davis *(1)* first proposed lysogenization of the λgt11 library in a high-frequency of lysogenization (*hflA*) strain and screening for cDNA-encoded proteins accumulated in the lysogens after induction of lytic growth and induction of *lac* transcription. This approach was soon abandoned because λgt11 phage plaques were shown to produce sufficient amounts of cDNA-encoded protein for detection with antisera. In addition, phage plaques can be plated at higher densities than lysogen colonies, allowing more clones to be screened per plate.

It was later recognized that eukaryotic proteins can retain functionality when expressed in bacteria as λgt11-encoded fusion proteins *(4–6)*. As a consequence, λgt11 cDNA expression libraries have also proven useful for isolating cDNAs by screening phage plaques for the function of their encoded proteins *(7–9)*. Functional isolation of a gene is particularly useful whenever specific antibodies or oligodeoxyribonucleotide probes are not available, or heterologous antibodies or DNA probes fail to specifically recognize their counterparts from evolutionary distant organisms. Analysis of the gene and its product may then be accelerated by overexpression of the protein in a suitable heterologous organism, preparation of antibodies against the recombinant protein, and the use of cDNA and antibodies as tools for the elucidation of the structure, expression, and function of the gene.

A major drawback of functional plaque screening is, however, the low amount of cDNA-encoded protein that is synthesized in a phage plaque (about 1 fmol, *7*). Although this is sufficient for ligand screening with high-affinity ligands ($K_d < 10^{-9} M$) for which sensitive assays exist, plaque screening will fail whenever the K_d for the ligand or the sensitivity of the assay falls below a certain limit. This problem was encountered during functional screening of a *Dictyostelium discoideum* λgt11 cDNA phage library *(10)* with ^{125}I-labeled *D. discoideum* calmodulin. Whereas this approach had previously been successful for the isolation of cDNA clones for a calmodulin-binding protein from a rat brain cDNA library *(8)*, it failed for the *Dictyostelium* system. The dilemma was solved by expressing the cDNAs in colonies of lysogenic bacteria instead of phage plaques, increasing the amount of cDNA-encoded protein at least 20-fold *(11)*. In addition, it could be shown that lysogen screening allowed the detection of a cDNA-encoded cyclic nucleotide-binding protein (the regulatory subunit of *Dictyostelium* cAMP-dependent protein kinase; *7,12*) with tritiated cAMP of low specific radioactivity.

This chapter will describe how a lysogen library consisting of nearly 100% recombinant bacteria can be prepared from an existing phage library and screened with a protein or low-Mr ligand. An example for the detection of lysogens encoding a ligand-binding protein is shown in Fig. 1.

2. Materials

2.1. Lysogenization of the Library

For standard procedures on the propagation and maintainance of λ phages and *E. coli* strains, refer to ref. *13*.

1. Bacterial strains: Y1089 (Δ*lacU*169, Δ*lon*, *araD*139, *strA*, *hflA* [chr::*Tn*10], *p*MC9); Y1089 is used as the host strain for the preparation of the lysogen library. Y1088 (*supE*, *supF*, *metB*, *hsdR*$^-$ [M^+], *tonA*21, *strA*, *lacU*169, *proC*::*Tn*5, *p*MC9); λgt11 will form clear plaques on Y1088; the strain is recommended for titration and amplification of λgt11 and λ*cI*h80.
2. Phages: λgt11 cDNA library; challenging phage: λ*cI*h80.
3. L-broth/ampicillin (LB/amp): 10 g bacto tryptone, 5 g yeast extract, 5 g NaCl/L, pH 7.3; autoclave, add 100 mg/L ampicillin from a 250 mg/mL stock prior to use. LB/amp agar plates contain 1.8% agar.
4. 20% maltose (autoclave).
5. 10 m*M* $MgSO_4$ (autoclave).
6. SM: 5.8 g NaCl, 2 g $MgSO_4$, 50 mL 1*M* Tris-HCl, pH 7.5, 5 mL 2% gelatin/L.
7. Dimethyl sulfoxide (DMSO): Filter-sterilize.

2.2. Plating of the Library and Preparation of Nitrocellulose Replicas for Functional Screening

8. IPTG: 1*M* isopropyl-β-D-thiogalactopyranoside in water. This can be stored at 4° or 8°C for several weeks.
9. 20 m*M* K-phosphate, pH 7.2, 0.15*M* NaCl, 0.05% Tween 20.
10. Whatman 3MM filter paper: ca. 85 mm circles can be cut from sheets.
11. Nitrocellulose (NC) filters: (Schleicher and Schuell [Keene, NH] BA 85, 0.45 μm), 85 mm circles.
12. Bovine serum albumin (fraction V, Serva, Heidelberg, Germany).

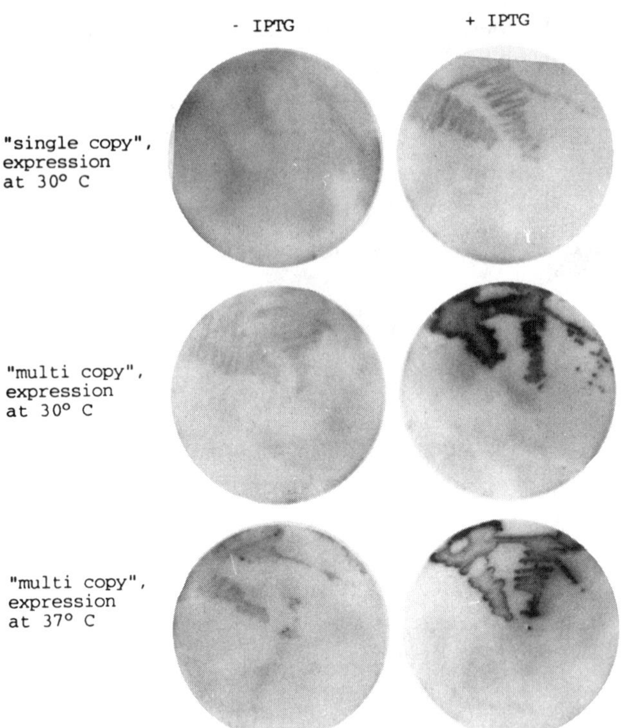

Fig. 1. The effect of λgt11 copy number, induction of *lac* transcription, and incubation temperature on the expression of a *lacZ*-cDNA-encoded calmodulin-binding protein by filter-bound λgt11 lysogens. Y1089 cells lysogenized with a recombinant λgt11 phage encoding a *Dictyostelium* calmodulin-binding protein (clone AB4.1, ref. *11*) (streaked on the upper half of each plate) or nonrecombinant λgt11 (streaked on the lower half of each plate) were probed with ^{125}I-labeled calmodulin. NC filter replica were prepared and treated as described in Sections 3.2. and 3.3.1. with the following modifications: upper two filters, incubation at 30°C throughout the procedure to show IPTG-induced expression from the integrated prophage; middle two filters, induction of lytic growth of the prophage for 15 min at 42°C, expression at 30°C; lower two filters, conditions as described in the protocol.

2.3. Detection of Recombinant Lysogens

13. TBSII buffer: 10 m*M* Tris-HCl, pH 7.5, 0.15*M* NaCl, 0.05% Tween 20—make from stocks of 1*M* Tris, 3*M* NaCl, and 10% Tween 20.2.3. Detection of Recombinant Lysogens
14. "En^{3}Hance" autoradiography enhancer spray (New England Nuclear, Boston, MA).

15. Ligands: ^{125}I-labeled calmodulin (3–10 mCi/mg, *see* ref. *14* for details); ^{3}H-labeled cAMP (30–40 Ci/mmol, Amersham, Arlington Heights, IL).
16. Autoradiography film (Kodak X-OMAT AR).

3. Methods

3.1. Lysogenization of the Library

Preparation of the library is performed in two steps: First, the λgt11 phage library is lysogenized in Y1089, and second, residual nonrecombinant cells are eliminated. There is no direct selection against nonrecombinant parental Y1089 cells. Lysogens are, however, immune against a further λ infection; therefore the library is challenged with a λ*cI* mutant that will kill virtually all of the nonlysogenic bacteria.

1. Grow Y1089 overnight at 37°C in LB/amp + 0.2% maltose.
2. Dilute the culture to A_{600} = 0.2 in 10 m*M* $MgSO_4$ (about 30-fold) and transfer 1 mL of this suspension to a sterile 18-mL tube. Add 5 × 10^8 pfu of the λgt11 phage library plus SM to 2 mL final volume, swirling gently to mix. Incubate for 25 min at room temperature (*see* Note 1).
3. To counterselect nonrecombinant Y1089 cells, add 5 × 10^8 pfu λ*cI*h80 and incubate 15 min at room temperature (*see* Note 2). Dilute the suspension with 4 mL of LB/amp. Incubate for 40 min at 30°C on a roller or shaker to allow lysis of nonrecombinant cells.
4. Harvest by centrifugation at room temperature (e.g., 10 min at 6000 rpm in a Sorvall SS34 rotor using 15 mL Corex tubes).
5. Remove the medium carefully with a Pasteur pipet; do not decant because the bacterial pellet is very small. Resuspend the bacteria in 1 mL of LB/amp and add 0.1 mL of DMSO. Transfer aliquots (20–50 μL) to sterile Eppendorf tubes and freeze at –70°C (*see* Note 3).
6. Determine the titer of the library and the percentage of nonrecombinant Y1089 cells. Dilute 10^2, 10^3, 10^4, and 10^5 fold in LB/amp medium and plate 0.1 mL of each dilution on two LB/amp plates. Incubate one at 30°C and one at 42°C overnight. Lysogens grow at 30°C only, nonrecombinant cells at both 30 and at 42°C.

3.2. Plating of the Library and Preparation of Nitrocellulose Replicas for Functional Screening (see *Note 4)*

1. Dilute the library in LB/amp medium to 10^5/mL and plate 0.05–0.1 mL on well-dried LB/amp plates (5000–10,000 colonies/9-cm plate for functional screening, *see* Note 5). Grow for 16 h at 30°C. Colonies should be very small (1–2 mm diameter).

2. For each plate, prepare 2 plastic Petri dishes, one containing 2.5 mL of LB/amp medium and one containing 2.5 mL of LB/amp medium plus 2 m*M* IPTG. In each dish, place a 85 mm Whatman 3MM filterpaper circle. The wet filters serve as a substrate for NC filter-bound bacteria during further growth and induction of the prophage and IPTG-induction of *lac* transcription.
3. Place a NC filter circle on the plate and pierce the filter and underlying agar layer with a needle to mark its orientation on the plate.
4. Remove the filter and transfer it (with the colony side up) to the plastic Petri dish containing the LB/amp soaked Whatman 3MM filter. Take care to peel the filter off very carefully to avoid "smearing" of the colonies. Let the filter-bound bacteria grow for 1–2 h at 30°C (*see* Note 6).
5. Seal the original agar plates with parafilm and store refrigerated.
6. Prewarm the plastic dish containing the 3MM filter circle soaked with 2.5 mL LB/amp medium + 2 m*M* IPTG to 42°C for 15 min and transfer the NC filter into this dish.
7. Induce lytic growth of the prophage 15 min at 42°C, then transfer to 37°C for 1 h to allow expression of the *lacZ*-cDNA fusion genes (*see* Note 7).
8. Lyse filter-bound bacteria by two cycles of freeze-thawing. Filters are frozen for 30 s on a metal surface placed in a dry ice-ethanol bath (*see* Note 8), and thawed on a piece of Whatman 3MM paper.
9. Remove all bacterial debris with a jet of assay buffer or blocking buffer. Take care to drain off the effluent rapidly.

Filters are now ready for detection of specific activity.

3.3. Detection of Recombinant Lysogens

Detection procedures using either a protein ligand (^{125}I-labeled calmodulin) or a low-Mr ligand (^{3}H-labeled cAMP) will be described as examples. The conditions used have been optimized for each of the ligands. Screening with other ligands should be performed under conditions that are specifically optimized.

3.3.1. Screening with ^{125}I-Labeled Calmodulin

1. Block the NC filters for 1 h at room temperature in TBSII buffer containing 0.1 m*M* $CaCl_2$ and 3% BSA (*see* Note 9).
2. Transfer to a solution of 1–5 µCi/mL ^{125}I-labeled calmodulin and incubate for 1 h at room temperature (*see* Note 10).
3. Wash the filters 3 times for 1 min each in approx 30 mL of ice-cold TBSII buffer containing 0.1 m*M* $CaCl_2$.
4. Blot on a piece of Kleenex paper and then let dry on Whatman 3MM paper for about 30 min at room temperature. Fix the filters with a small

piece of adhesive tape on Whatman 3MM paper; wrap in polythene film. Autoradiograph on Kodak X-OMAT AR film overnight at –70°C using intensifying screens.

3.3.2. Screening with ^{3}H-Labeled cAMP

1. Block the filters for 1 h at 4°C in 20 m*M* K-phosphate, pH 7.2, 0.15*M* NaCl, 0.05% Tween 20 (*see* Note 9).
2. Transfer to a solution of 5 μCi/mL ^{3}H-cAMP in the same buffer and incubate for 30 min at 4°C (*see* Note 10).
3. Wash 3 times for 20 s each in buffer without cAMP at 4°C. For rapid washing, the filter should be held with foreceps and agitated in the washing solution.
4. Blot on a piece of Kleenex paper and let dry for 30 min at room temperature.
5. Impregnate with "En^{3}Hance" spray (*see* Note 11).
6. Fix the filters with a small piece of adhesive tape on a Whatman 3MM filter. Autoradiograph for 48–72 h at –70°C on Kodak X-OMAT AR film using intensifying screens. Direct contact between the filter and the film is required for autoradiography, therefore do not wrap in plastic foil.

3.4. Purification of Recombinant Clones

1. If several filters are autoradiographed on one sheet of film, it is easy to exactly position the developed autoradiogram on the filters since the contours of the filters are usually visible on the autoradiogram owing to background radioactivity. Mark the positions of the needle-holes on the film with a fine permanent lab marker. Alternatively, the holes may be marked directly on the filter with radioactive ink.
2. The holes in the original agar plate are aligned with the marks on the autoradiogram. To pick positives, try first to identify "the" positive colony as precisely as possible; pick it with a fine platin wire or a wooden toothpick, and streak on LB/amp for isolated colonies. If the exact position of the positive signal is not easy to determine, divide the region (about 5 mm) around the positive colony in several sectors, scrape off the bacteria, and streak for isolated colonies.
3. Rescreen for positive single colonies. In most cases an isolated recombinant colony can be identified during the second round of screening, and after the third round clones are usually pure (*see* Note 12).

3.5. Isolation of λgt11 Phages from Recombinant Lysogens

As a rule, phage particles should be isolated from purified recombinant lysogens and used to relysogenize Y1089 in order to prove by a rescreening experiment that the detected function is indeed encoded by the λgt11 phage. Phage will also serve for λ DNA preparation by miniprep or large-scale purification procedures or direct amplification of insert DNA by PCR techniques.

1. Grow the recombinant lysogen in LB/amp medium overnight at 30°C.
2. Dilute with 2 vol of LB/amp; allow lytic growth of the phage for 1 h at 42°C.
3. Harvest bacteria by 20 s centrifugation at room temperature in an Eppendorf centrifuge. Remove the medium with a drawn-out Pasteur pipet.
4. Resuspend in 1/10 vol of SM, add 50 μL chloroform/ethanol (50:1), vortex vigorously.
5. Centrifuge 10 min in an Eppendorf centrifuge to remove bacterial debris.
6. Save phage supernatant in a fresh tube.

For preparation of high-titer phage stocks, Y1088 should be used as host.

4. Notes

1. The number of cells and phages can be scaled down if the titer of the phage library is too low. In this case, all volumes are reduced by the same factor.
2. The efficiency of lysogenization can vary from experiment to experiment between 90–30%. It is therefore important to eliminate the nonlysogenic parental cells with a λ*cI* mutant. A library consisting of only 50% λgt11 lysogens would require screening of twice the number of colonies than a library consisting of only lysogens. More than 90% of the nonrecombinant cells are usually killed by treatment of the library with λ*cI*h80. Note that other *cI* mutants are also suited for challenging the library.
3. DMSO cultures can be stored for years at –70°C without significant loss of viability. Viability drops on freezing and thawing and so the titer of the lysogen library should be determined with an aliquot that has been frozen. Thawed aliquots should not be refrozen for further screening experiments. It is possible to amplify the library by growing colonies at low density on agar plates (e.g., 25,000 colonies/14 cm plate), eluting bacteria with LB/amp medium containing 10% DMSO and freezing at –70°C. The risk of certain clones being counterselected owing to any property of their cDNA insert (size, DNA secondary structure, expression of cDNA-encoded products that could reduce bacterial viability) should be rather low compared to amplification of phages because the prophage is part of the *E. coli* genome and expression of cloned cDNAs is minimal in this single copy status (*see* Fig. 1).

4. It is also possible to perform functional screening by a stepwise enrichment technique *(11)*. Aliquots of the library corresponding to 1000 individual lysogens are diluted in LB/amp medium and grown to early logarithmic phase (A_{600} = 0.2). Lytic growth of the prophage and *lac* expression are induced by temperature shift and the addition of IPTG, respectively. Crude extracts are then prepared as described in ref. *15*. Functional tests can be performed with these extracts. Positive "sublibraries" are then further screened.
5. For screening of a lysogen library with antibodies or DNA probes a nearly confluent lawn of up to 30,000 colonies can be plated owing to the high sensitivity of these techniques. Plaque screening however is a better choice for use with antisera or DNA probes.
6. Up to three NC replica may be prepared from the same plate for screening with different probes if great care is taken to avoid smearing of the colonies when the filter is removed. For a first replica, 1 h growth is sufficient; the second and third replica should be incubated for a longer time prior to phage induction to allow regrowth of the bacteria on the filters.
7. In some cases allowing expression at 30°C rather than at 37°C could help stabilize expressed proteins. With IPTG-induced lysogens expressing a *Dictyostelium* calmodulin-binding protein there is no significant difference in signal strength on expression at 30°C or at 37°C (*see* Fig. 1).
8. A smooth metal surface is required for optimal temperature conductance. This can be a "replica block" immersed in a dry ice-ethanol bath. The most convenient and least expensive device is a thin-walled (ca. 1 mm) stainless steel dish (20 cm diameter) with a flat bottom that will float in the bath, protect the filters from ethanol spills, and cool down very rapidly. It should also be noted that frozen NC filters are very fragile and will break if bent during removal from the metal surface.
9. Several filters can be processed in the same tray if the volume of reagents is sufficiently large to assure that they can freely float. Fourteen centimeter plastic Petri dishes with 30–40 mL of solution can accommodate up to six 85 mm filters. During all incubation and washing steps the filters are constantly agitated either on a "Rocker" or on a rotary shaker at 50–60 rpm. For final washes (which have to be very short to avoid dissociation of bound ligand) treating each filter separately will reduce background.
10. Like antibody solutions, diluted working solutions of many ligands may be reused several times if appropriately stored between experiments. For example, ^{125}I-labeled calmodulin can be stored at 4–8°C in TBSII buffer containing 3% BSA and 0.2% NaN_3, and ^{3}H-labeled cAMP can be stored frozen. If solutions are reused, they should be routinely ana-

lyzed for degradation, i.e., radiolabeled protein ligands by SDS-polyacrylamide gel electrophoresis and autoradiography, low-Mr ligands by suitable analytical techniques (*see* ref. *11*).

11. The manufacturer recommends applying "three light coats" of "En^{3}Hance" spray on the surface. I obtained best results by spraying until the filter appeared just wet. The spray should be applied under a well-ventilated hood, and gloves should be worn. The product releases a very strong odor that will impregnate the autoradiography cassette, and so a cassette should be reserved for exposing "En^{3}Hance" treated filters.
12. Fusion proteins from purified lysogens can be analyzed as described *(15)*. There is also the possibility to determine specificity and affinity of the cDNA-encoded proteins either directly on colony blots, or on dot blots or protein blots prepared from crude extracts of the lysogens (cf refs. *7, 11*).

References

1. Young, R. A. and Davis, R. W. (1983) Efficient isolation of genes by using antibody probes. *Proc. Natl. Acad. Sci. USA* **80,** 1194–1198.
2. Young, R. A. and Davis, R. W. (1983) Yeast RNA polymerase II genes: isolation with antibody probes. *Science* **222,** 778–782.
3. Young, R. A. and Davis, R. W. (1991) Gene isolation with λgt11 system. *Meth. Enzymol.* **194,** 230–238.
4. Kaufman, D. L., McGinnis, J. F., Krieger, N. R., and Tobin, A. J. (1986) Brain glutamate decarboxylase cloned in λgt-11: fusion protein produces γ-aminobutyric acid. *Science* **232,** 1138–1140.
5. Bernier, L., Alvarez, F., Norgard, E. M., Raible, D. W., Mentaberry, A., Schembri, J. G., Sabatini, D. D., and Colman, D. R. (1987) Molecular cloning of a 2',3'-cyclic nucleotide 3'-phosphodiesterase: mRNAs with different 5' ends encode the same set of proteins in nervous and lymphoid tissues. *J. Neurosci.* **7,** 2703–2710.
6. Mutzel, R., Simon, M. N., Lacombe, M. L., and Véron, M. (1988) Expression and properties of the regulatory subunit of *Dictyostelium* cAMP-dependent protein kinase encoded by λgt11 cDNA clones. *Biochemistry* **27,** 481–486.
7. Lacombe, M. L., Ladant, D., Mutzel, R., and Véron, M. (1987) Gene isolation by direct in situ cAMP binding. *Gene* **58,** 29–36.
8. Sikela, J. M. and Hahn, W. E. (1987) Screening an expression library with a ligand probe: isolation and sequence of a cDNA corresponding to a brain calmodulin-binding protein. *Proc. Natl. Acad. Sci. USA* **84,** 3038–3042.
9. Singh, H., LeBowitz, J. H., Baldwin Jr., A. S., and Sharp, P. A. (1988) Molecular cloning of an enhancer binding protein: isolation by screening of an expression library with a recognition site DNA. *Cell* **52,** 415–423.
10. Lacombe, M. L., Podgorski, G. J., Franke, J., and Kessin, R. H. (1986) Molecular cloning and developmental expression of the cyclic nucleotide phosphodiesterase gene of *Dictyostelium discoideum*. *J. Biol. Chem.* **261,** 16,811–16,817.

11. Mutzel, R., Baeuerle, A., Jung, S., and Dammann, H. (1990) Prophage lambda libraries for isolating cDNA clones by functional screening. *Gene* **96,** 205–211.
12. Mutzel, R., Lacombe, M. L., Simon, M. N., de Gunzburg, J., and Véron, M. (1987) Cloning and cDNA sequence of the regulatory subunit of cAMP-dependent protein kinase from *Dictyostelium discoideum. Proc. Natl. Acad. Sci. USA* **84,** 6–10.
13. Sambrook, J., Fritsch, E. F., and Maniatis, T. (1989) *Molecular Cloning. A Laboratory Manual (Part I)*, Cold Spring Harbor Laboratory, Cold Spring Harbor, NY.
14. Winckler, T., Dammann, H., and Mutzel, R. (1991) Ca^{2+}/calmodulin-binding proteins in *Dictyostelium. Res. Microbiol.* **142,** 509–519.
15. Huynh, T. V., Young, R. A., and Davis, R. W. (1985) Constructing and screening cDNA libraries in λgt10 and λgt11, in *DNA Cloning Vol. I* (Glover, D. M., ed.), IRL, Oxford, pp. 49–78.

Index